THE MULTIVARIATE NORMAL DISTRIBUTION
Theory and Applications

THE MULTIVARIATE NORMAL DISTRIBUTION
Theory and Applications

THU PHAM-GIA
Université deMoncton, Canada

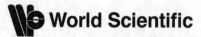
World Scientific

NEW JERSEY · LONDON · SINGAPORE · BEIJING · SHANGHAI · HONG KONG · TAIPEI · CHENNAI · TOKYO

Published by

World Scientific Publishing Co. Pte. Ltd.

5 Toh Tuck Link, Singapore 596224

USA office: 27 Warren Street, Suite 401-402, Hackensack, NJ 07601

UK office: 57 Shelton Street, Covent Garden, London WC2H 9HE

Library of Congress Control Number: 2021934048

British Library Cataloguing-in-Publication Data
A catalogue record for this book is available from the British Library.

THE MULTIVARIATE NORMAL DISTRIBUTION
Theory and Applications

ISBN 978-981-123-528-3 (hardcover)
ISBN 978-981-123-529-0 (ebook for institutions)
ISBN 978-981-123-530-6 (ebook for individuals)

For any available supplementary material, please visit
https://www.worldscientific.com/worldscibooks/10.1142/12237#t=suppl

Desk Editors: Balamurugan Rajendran/Daniele Lee

Typeset by Stallion Press
Email: enquiries@stallionpress.com

Printed in Singapore

Preface

Most disciplines have contacts with probability and statistics. In fact, one or two courses in statistics are often required in most undergraduate curricula, especially in fields leaning toward the quantitative approach. After all, statistics can be defined loosely as *a collection of methods to process data* and it is very important to learn some of these basic methods.

This brings us to discuss first *probability*, a domain that Russell (1927) has, sarcastically, commented on as "Probability is amongst the most important science, not least because no one understands it". Good, a well-known contemporary researcher, suggested that there are five sorts of probabilities. These theories are discussed in the book by Fine (1973). Although the real meaning of the probability of an event is still elusive, its mathematical basis is solid now after the work of generations of probabilists using Kolmogoroff path-breaking work. We discuss this topic in Chapter 8.

In everyday life, we all recognize non-numerical concepts of probability, expressed in terms, such as "likely", "highly likely", "at least as probable as", "chances are that", and "very unlikely". While almost all agree on probability as a numerical quantity, some believe that we need a wider range of mathematical concepts. There is no consensus as to the meaning of probability, but my belief is that all of these accounts of meaning are viable within appropriate domains of application.

Probabilistic reasoning includes both probability theory and the methodology by which we choose specific probability models and apply them. The different types of meanings that have been held for probability are roughly divided as objective and subjective. More detailed discussion can be found in Chapter 6.

Objective meanings relate probability to the following: idealized frequencies of occurrences of events in long runs of repeated, causally unlinked repetitions of a random experiment; propensities for occurrences of events that relate to frequencies through laws of large numbers; epistemic or knowledge-centered ideas of the degree of inductive support that one proposition provides to another.

Subjective meanings, degrees of personal belief about the occurrences of events, have become quite popular and find strong proponents among Bayesian statisticians. The usual mathematical choice for probability is that it takes values in the unit interval, satisfies an axiom of additivity for disjoint events or propositions, and may also have certain limiting properties (e.g. σ-additivity or monotone continuity using the Kolmogorov theory) when infinitely many events are involved.

Classical statistics, often called frequentist statistics, are frequently based on a particular distribution, called the Normal, or the Gaussian, or the Gauss–Laplace distribution. Informally, it is a bell-shaped curve precisely given by a mathematical equation, the expression of which can be conveniently manipulated and used in different topics of statistics. It also responds to our intuitive feeling that any unknown quantity, when estimated in some way, has a central value, with variations on either side, which are taken symmetrically if there is no reason to privilege one side at the expense of the other. Other bell-shaped curves exist, but their expressions are more complicated and hence are less convenient to handle.

In several variables, it is called the multinormal distribution, which is however often handled using matrices for convenience. But this convenience has a price. The clarity of the arguments can suffer, except for cases of two or three variables. In this book, we do our best to make the arguments less abstract and, hence, we start with the univariate case and move progressively toward the vector and matrix cases. We wish to present some aspects of this important distribution without being too serious on the theory. To encourage newcomers, we will not include proofs when they are too elaborate and tend to "scare away" readers. But we will indicate the main steps of certain proofs judged of importance. An important mathematical tool, like the

G- or *F*-function, is explained in two places: in Chapter 4, with respect to its applications, and in Chapter 13, with respect to its theoretical roots. The book can be considered as a compendium of results related to the normal, some at the contemporary research level, and does not propose any exercise. The main goal is to let you appreciate all the beauty and usefulness of this distribution, its numerous applications in so many different domains, and its potential to be used in other applications by researchers including yourself. Some of the topics presented are based on our own experience while working at a university, or a research center, or in the industry.

There are a few books dealing with the normal, either exclusively or having an important chapter on it. We have a deep respect for these works and, when needed, will refer to them for the appropriate information. We will try to be complementary to these published works. But we believe that it is the applied part which is missing in most of them, while the very large majority of students and researchers having contact with the normal distribution are not in mathematics or statistics but in an applied field that uses these two disciplines. The term *applied* should be understood here in a broad sense. Undoubtedly, there are also numerous applications of the normal in the field of statistics itself, theoretical or applied. A book like ours can easily be drawn into the statistical field, which will be an ungrateful task because we cannot address satisfactorily all the applications there within the few dozen pages. So, we present a broad landscape of a presence of the normal in statistics without going into details, such as into precise hypothesis testing. Similarly, we have ignored the whole important topic of regression using the normal, as it will take us several dozen pages. In short, this is not a treatise in statistics.

Basically, the normal is presented here under two complementary forms: (1) as a positive definite quadratic form in p-variables $Q(x_1, \ldots, x_p)$ and (2) as the Mahalanobis distance $\Delta_{\mathbf{X}}$ from $\mathbf{X}$ to its mean. Up to now, mathematical and statistical properties of the multinormal are derived using the first approach, while some applications in various fields are developed using the second one. Furthermore, the role that the normal in one or several dimensions plays is shown to be varied: It could be the sole distribution of data (education, sociology, management science), one among several factors under study (engineering, statistics), or one of the entries of a large random matrix (Wigner matrix). Some innovative approaches are presented, together with graphs and tables, to relate different dimensions of the normal distribution. In particular, animated graphs on our website allow one to have a better appreciation of the figure seen from different angles.

The approach used in the book is a gradual one, going from one scalar variable to a vector variable and to a matrix variable. As the subject becomes more abstract, and hence more difficult, I believe that some readers will not feel like going into later chapters. For a basic comprehension in one variable, reading can stop at Chapter 5, where several applications in one dimension are presented. Chapter 6 is a necessary read if one is interested in Bayesian statistics and the applications of the normal in this domain. Chapter 7 deals with bivariate and trivariate normal distributions, where graphical illustrations are still possible. Some new ideas are developed there, with the Computational Appendix providing numerical support to the chapter. Chapter 8 discusses the meaning of probability and its uses in decision theory and Bayesian statistics. This chapter can be skipped if one is not interested in these topics. Chapters 9 and 10 extend the results to $\mathbb{R}^p$, $p \geq 4$ and give several applications there, with, however, graphical illustrations limited to $\mathbb{R}^3$. Chapter 11 deals with Jacobians and functions of matrix arguments while Chapter 12 is useful if one wants to learn about random matrices, their transformations and their distributions, and the normal random matrix. But they need, as a prerequisite, some solid background in classical analysis and a course in mathematical probability and statistics. Finally, Chapter 13 consists of complements, i.e. information and knowledge that are not required, but very helpful to better appreciate the book.

The book can be useful to a varied clientele, with different interests and motivations. Applications are presented whenever appropriate for illustration purpose only. A number of our own published research articles figure among them, not for self-promotion purpose, but because we understand them well and will be able to discuss the results with any interested reader. Other well-known applications, such as Herrnstein and Murray's book, deserve to be mentioned and discussed.

In the introduction of his book, William Feller wrote that there are three parts in probability: formal logical content, intuitive background, and applications. They should complement and interact with each other, a view that I certainly share. *Theoretical progress opens new fields of applications, and in turn, applications lead to new problems and fruitful research.* In his book, Farrell (1985) stated on p. 3 of the introduction, "A major area of concern today should be computation", after acknowledging that weighted sums of zonal polynomials converge very slowly. By computation, he meant applied computation leading to numerical results. The situation has hardly changed during the last 40 years, and I fully agree with Farrell.

On theory and applications, one agrees that applications must have some solid theoretical basis and conversely, theory with interesting applications gets more importance. In fact, personally, at the end of my several talks, seminars, presentations at meetings, at universities or colleges, one question frequently came up: "How do you apply this result?". This result could be a new theorem that I just proved, a new method that I wish to popularize, or an approach that somebody else has established. The questioner was invariably disappointed if I could not come up with interesting and important applications. For applied maths, CT scans, or computer tomography, based on the Radon transform and 3D image reconstruction is certainly the most impressive application of integral transforms into non-invasive medical imaging. For probability and statistics, there are equally interesting and important applications to mathematics, physics and to real life. But the problem is also how to present them in layman's terms, and that's another challenge.

Finally, this book could even be titled modestly as *The Multivariate Normal Distribution: Some Theory and a Few Applications*, since we could only present some basic results in theory and applications as we try to guide you through this immense field. But we hope that your trip will be an enjoyable and useful one.

About the Author

Thu Pham-Gia is Emeritus Professor at the Université de Moncton in New Brunswick, Canada, where he chaired the Department of Mathematics and Statistics for 6 years. He obtained his BA from the University of Saigon in 1966, his MA from the University of Hawaii in 1969 and his PhD in probability from the University of Toronto in 1972. Active in research, he has published over 80 papers in refereed journals and served as a referee to a dozen reputable journals. He won the Thomas Saaty Prize (USA) in 1998, and the Canadian Excellence Award in Research (the Natural Science and Engineering Research Council, NSERC, Ottawa) in 2004. He has organized, or helped organize, several conferences in mathematics and statistics at home and abroad.

To acquire real-life experiences in statistics, he has taken leave from universities to work at the Industrial Material Research Institute (Boucherville, QC, Canada), where he served as an adviser in stochastic processes, at Pratt and Whitney Canada (Longueuil, QC, Canada), where he worked as production of powerplants cost controller, and Bell Canada (Montreal, QC, Canada), where he was manager of statistics service. He was also an on-campus consultant in statistics for several years.

He has directed several development projects overseas, in statistics and management sciences, under the sponsorship of the Canadian

International Development Agency (CIDA), with a development grant close to 1 million CAD to the National Economics University, Hanoi, VN, AUPELP-UREF and FICU and worked with several universities and institutions abroad.

He is a senior member of both the Institute of Electrical and Electronics Engineers (IEEE, USA) and the Institute of Industrial Engineers (IIE, USA), and Fellow of the Institute of Statisticians (IS, UK) and the Royal Statistical Society (RSS, UK).

Since his retirement in 2011, he spends his winters at Bradenton, Florida, USA, where he divides his time between statistics, fishing, stock trading, and line dancing.

Acknowledgments

I wish to thank several members of the Applied Multivariate Statistical Analysis Research Group (AMSARG), at the Université de Moncton, Canada, and Ho Chi Minh University of Sciences, Vietnam, for their helpful cooperation, especially in setting up very instructive computer programs to illustrate various applications of the Normal. Dr Dinh Ngoc Thanh is the main coordinator of the project, which runs smoothly thanks to his dedication and experience. Phong Nguyen used his talents to produce beautiful informative graphics in three dimensions. Phong Duong, of Ton Duc Thang University, read many chapters of the book associated with matrix transformations and zonal polynomials and made helpful comments. His painstaking work of converting my barely readable manuscript from Word format into Latex is very much appreciated. Thanks a bundle, guys and girls!

Foremost, thanks to my *better half*, Jeannette LeBlanc, for being extremely patient while I spent hours at the keyboard at our winter home in Bradenton, Florida. She also volunteered to do those housework and yardwork that traditionally fall on a man's shoulders and managed to obtain better results, much to my annoyance, sometimes. As a final touch, Jeannette also helped me greatly with editing the author query form of the complete book. Many of my fellow residents at the *Bradenton Tropical Palms* mobile park in Florida also gave me moral support, when I was confined to my reading desk instead of going out enjoying the sun and the heat. I am thinking about Lloyd, with whom I would rather go fishing. In particular, several ladies in the park have been very helpful in proofreading. Jane Galbraith, for example, read Chapters 5, 8 and 10 entirely and put forth several constructive criticisms. Janice McMillan and Beth Bechler reviewed two particular long chapters. Nathaly and Celso Guiang helped out in many ways. Dr. Ahcene Brahmi, Director of our departmental computer laboratory, has helped in setting up several animated programs on our web page to support the book. Some other colleagues at the Université de Moncton, such as Vartan, Claude, Salah-Eddine, and others that I might have forgotten (sorry friends!), have provided support and encouragement. A spacious office Université de Moncton with internet connection, reserved to me for three full years by Professor Claude Gauthier, also helped me much in my work. I say thanks to them *all*. Thanks to Balamurugan Rajendran and Daniele Lee Zhen Rong, from World Scientific, for being very helpful. Your technical staff makes wonders on our technical documents, and this book would not be the same without them.

My family, Lan, David, Vinh, Janelle, Guy, and Joel have provided emotional support for which I am very appreciative.

Finally, there are quotations, valid and truthful too, that most authors know about:

(1) *In writing this book, my merit and luck are to be able to stand on shoulders of giants.*
(2) *You copy from one book, that's outright plagiarism. If you copy from a dozen books, that's scholarship, and if you copy from one hundred books, that's outstanding knowledge.*

Abbreviations, Notations, and Symbols

rv	random variable
wrt	with respect to
chf	characteristic function of rv X: $\quad \varphi_X(t) = \mathbb{E}(\exp(itX))$
Frt	Fourier transform
$H^{m\ n}_{q\ p}(.)$	Fox function with four parameters
$G^{m\ n}_{q\ p}(.)$	Meijer function with four parameters
Δ^2_X	Mahalanobis square distance of $\mathbf{X}$ to the mean $\boldsymbol{\mu}$
Lpt	Laplace transform
Mgf	moment-generating function of X: $M_X(t)$
Mlt	Mellin transform
$\mathbb{R}^p$	p-dimensional Euclidean space ($\mathbb{R}^1 = \mathbb{R}$)
$\mathbf{a}, \mathbf{a}^t$	column and row vectors, respectively
$\mathbf{a} = (a_1, \ldots, a_p)^t$	column vector with components $a_1, \ldots, a_p$
$1_p = (1, \ldots, 1)^t$	column vector consisting of p ones
$A(p \times q)$	matrix with p rows and q columns
$[a_{ij}]$	matrix with elements a_{ij}'s
$\mathbf{I}_p$	unit matrix of order p

$\mathbf{A}^t$	transposes of matrix $\mathbf{A}$
$\mathrm{diag}(\theta_1, \ldots, \theta_p)$	diagonal matrix with diagonal elements $\theta_1, \ldots, \theta_p$
$\mathrm{diag}(\mathbf{A}_1, \ldots, \mathbf{A}_k)$	block-diagonal matrix with elements $\mathbf{A}_1, \ldots, \mathbf{A}_k$
$|\mathbf{A}|$	determinant of matrix $\mathbf{A}$
$\mathrm{tr}(\mathbf{A})$	trace of square matrix $\mathbf{A}$
$\mathbf{A} \otimes \mathbf{B}$	Kronecker product of matrices $\mathbf{A}$ and $\mathbf{B}$
$\mathbf{0}$	zero matrix consisting of 0's
$\mathbf{A} > \mathbf{B}$	the matrix $\mathbf{A} - \mathbf{B}$ is positive definite
$\mathbf{A} \geq \mathbf{B}$	the matrix $\mathbf{A} - \mathbf{B}$ is positive semi-definite
dof	degrees of freedom
iid	independent and identically distributed
$X, Y, \ldots$	random variable (rv)
$\mathbf{X}, \mathbf{Y}, \ldots$	random vector (r.vec.)
$A, B, C, \ldots$	constant matrix
$\mathbf{X}, \mathbf{Y}, \mathbf{S}, \ldots$	random matrix
$\mathrm{VEC}(\mathbf{X})$	np-column vector $\begin{pmatrix} \mathbf{X}_{(1)} \\ \mathbf{X}_{(2)} \\ \cdots \\ \mathbf{X}_{(p)} \end{pmatrix}$ obtained from the $(n \times p)$ matrix $\mathbf{X} = (\mathbf{X}_{(1)}, \ldots, \mathbf{X}_{(p)})$ by empiling column vectors $\mathbf{X}_{(k)} = \begin{pmatrix} x_{k,1} \\ x_{k,1} \\ \cdots \\ x_{k,n} \end{pmatrix}$
$\mathbb{E}(X), \mathrm{Var}(X)$	expectation and variance of random variable X
$\mathbb{E}(\mathbf{X}), \mathrm{Var}(\mathbf{X})$	expectation and covariance matrix of random vector $\mathbf{X}$
$\mathbb{E}(\mathbf{X}), \mathrm{Var}(\mathbf{X})$	expectation and covariance matrix of random matrix $\mathbf{X}$
$\mathrm{Cov}(\mathbf{X}, \mathbf{Y})$	covariance matrix of random matrices $\mathbf{X}$ and $\mathbf{Y}$
cdf	cumulative distribution function: $F_X(t) = \mathbb{P}(X \leq t)$

$N(\mu, \sigma^2)$	the univariate normal distribution with mean μ and variance σ^2, $((N_1(\mu, \sigma^2) = N(\mu, \sigma^2)))$
$\phi(x)$	pdf of the standard normal distribution $N(0, 1)$
$\Phi(x)$	cdf of the standard normal distribution $N(0, 1)$
$\varphi(t)$	chf of the standard normal $N(0, 1)$
$N_p(\boldsymbol{\mu}, \boldsymbol{\Sigma})$	the normal distribution in p dimensions, with mean vector $\boldsymbol{\mu}$ and covariance matrix $\boldsymbol{\Sigma}$
$\mathrm{Gam}(\sigma, \lambda)$	the gamma distribution with scale and shape parameters σ and λ
$\mathrm{Beta}(p, q)$	the beta function
$\beta_p^I(a, b)$ and $\beta_p^{II}(a, b)$	the beta central distributions of type I and type II in p variables
$t(n)$	the t-distribution with n dof
χ_p	the chi distribution with p dof
χ_p^2	the chi-square distribution with p dof
$\chi_{p,\tau}^2$	the non-central chi-square distribution with p dof and non-centrality parameter τ
$F_{m,n}$	the F-distribution with (m, n) dof
$F_{m,n;\tau}$	the non-central F-distribution with (m, n) dof and non-centrality parameter τ
$\Lambda(m, n, p)$	the distribution of Wilks' lambda criterion of dimension p with (m, n) dof
$W_p(\boldsymbol{\Sigma}, n)$	the *Wishart distribution* with n dof freedom and covariance matrix Σ in p variables
LR	likelihood ratio, likelihood ratio test
ML, MLE	maximum likelihood, maximum likelihood estimator
MANOVA	multivariate analysis of variance
$\|\mathbf{x}\|$	the Euclidean norm of the vector $\mathbf{x} = (x_1, \dots, x_p)^t$: $\sqrt{\sum_{i=1}^p x_i^2}$

$\|f(x)\|_1$ the L_1-norm of real function f on $\mathbb{R}^p$:
$$\int_{\mathbb{R}^p} |f(x)|dx$$

$_2F_1(.)$ Gauss hypergeometric function

$_1F_1(.)$ Kummer confluent hypergeometric function of the first kind
$$_1F_1(\alpha, \beta; x) = \sum_{n=0}^{\infty} \frac{(\alpha,n)}{(\beta,n)} \frac{x^n}{n!}, \text{ with}$$
$$(\alpha, n) = \frac{\Gamma(\alpha+n)}{\Gamma(\alpha)}.$$

$\psi(.)$ Kummer confluent hypergeometric function of the second kind

Critical value: We will use the following meaning for χ_α^2 and for any critical value C_α:

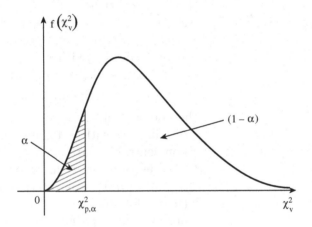

Percentiles of the chi-square distribution with p dof: By $\chi_{p,\alpha}^2$, we mean the $100 \times \alpha\%$ percentile of $Y \sim \chi_p^2$. Hence, $\mathbb{P}(Y \leq \chi_{p,\alpha}^2) = \alpha$ and $\mathbb{P}(Y > \chi_{p,\alpha}^2) = 1 - \alpha$. For example, $\mathbb{P}(Y \leq \chi_{2,0.20}^2 = 3.219) = \alpha = 0.20$, and $\mathbb{P}(Y > 3.219) = 1 - \alpha = 0.80$.

Hence, in this book, α is the probability between the origin and the numerical value of $\chi_{2,\alpha}^2$. $\chi_{p,\alpha}^2$ is also called the $(1 - \alpha) \times 100\%$ upper percentile of χ_p^2.

WARNING: It is no doubt that a volume full of mathematical symbols like this book will still contain legions of errors, typos, misspellings, even formula or logic errors, in its first edition, in spite of the efforts made by the author, and his collaborators, to remove them. We wish to apologize for any of these errors and ask the reader to bring them to our attention, so that the next editions will be better.

Any comments/suggestions/criticisms/questions can be made by writing to us at the website: https://www.facebook.com/thu. phamgia, tompham100@gmail.com.

Contents

List of Figures

List of Tables

Chapter 1

Review

1.1 Introduction

In this chapter, we review briefly the properties of some basic functions that we will encounter later in the book. A first course in probability/statistics is desired, as background knowledge (e.g. the first four chapters of Hogg and Craig (1995)).

1.2 The Power and Factorial Functions

In the current conversation, it is sometimes said that the cost of an item "has increased exponentially" to mean a very large increase.

There are other increasing functions: the power function $f(x) = x^N$, for example, increases fast for $x > 1$ and $N > 1$.

But the factorial function $5! = 1 \times 2 \times 3 \times 4 \times 5 = 120$ increases faster than the power function. We have

$$69! = 1.71122 \times 10^{98},$$

which is the maximum value for most simple hand-held calculators since 70! would give an overflow. The approximative number of digits of a factorial, $d(x)$, can be obtained from a convenient formula:

$$d(x) = \text{int}(\log_{10} x) + 1,$$

where int is the integral part. For example, the number of ways a set of cards can be shuffled is 52! and there are $d(52!) = 68$ digits in 52!.

Since the value of $n!$ can become very large, Stirling's formula can be used to approximate it: $n! \sim \sqrt{2\pi n}(n/e)^n$. More precisely, we have

$$\sqrt{2\pi n}\left(\frac{n}{e}\right)^n \leq n! \leq \sqrt{2\pi n}\left(\frac{n}{e}\right)^n e^{\frac{1}{12n}}. \qquad (1.1)$$

The laws of exponents, which apply to positive real numbers a:

$$a^x a^y = a^{x+y}, \ (ab)^x = a^x b^x, \ (a^x)^y = a^{xy}, \ a^{-x} = \frac{1}{a^x}$$

also apply to e and we have

$$(e^x)^y = e^{xy}, \ \frac{e^x}{e^y} = e^{x-y}.$$

Applications:

(1) **A very large number of possibilities:** In total, there are

$$53{,}644{,}737{,}765{,}488{,}792{,}839{,}237{,}440{,}000$$

or $\left(5.36 \times 10^{28} = \frac{52!}{(13!)^4}\right)$ different deals possible in the card game of Bridge (Deal in bridge: One particular allocation of 52 cards to the four players including the bidding, the play of the cards and the scoring based on those cards).

The immenseness of this number can be understood by answering this question: "How large an area would you need to spread all Bridge deals if each deal would occupy only one square millimeter?".

The answer is: "An area more than one hundred million times the surface area of the earth" (Wikipedia).

(2) **Factorials are used in permutations and combinations:** There are

$$\binom{52}{13} = \frac{52!}{13!(52-13)!} = 635{,}013{,}559{,}600$$

different hands in bridge (13 cards from a deck of 52 cards) and

$$\binom{52}{5} = \frac{52!}{5!(52-5)!} = 2{,}598{,}960$$

different hands in poker (5 cards from a deck of 52 cards).

1.3 The Double Factorial

The double factorial symbol is sometimes encountered: The product of all the integers from 1 up to some non-negative integer n that have the same parity (odd or even) as n is called the double factorial, or semi-factorial, of n and is denoted by $n!!$. A consequence of this definition is that $0!! = 1$ as an empty product. Therefore, for even n, the double factorial is

$$n!! = \prod_{k=1}^{n/2} (2k) = n \times (n-2) \times (n-4) \times \cdots \times 4 \times 2,$$

and for odd n, it is

$$n!! = \prod_{k=1}^{(n+1)/2} (2k-1) = n \times (n-2) \times (n-4) \times \cdots \times 3 \times 1.$$

For example, $9!! = 9 \times 7 \times 5 \times 3 \times 1 = 945$.

1.4 The Exponential Function

The exponential function is thought to be the most useful function in mathematics. It is defined as a series, where the general term is the ratio of the two functions just considered above:

$$\exp(x) = \sum_{n=0}^{\infty} \frac{x^n}{n!} = 1 + \frac{x}{1!} + \frac{x^2}{2!} + \frac{x^3}{3!} + \cdots \tag{1.2}$$

The general term is $x^k/k!$, where $k!$ is much larger than x^k and hence provides convergence to the series for all values of x. Usually, $\exp(70) = 2.5154 \times 10^{30}$ gives an overflow value for hand-held calculators. Also, we have

$$\exp(a) \times \exp(b) = \exp(a+b), \quad \forall a, b \in \mathbb{R}.$$

Hence, $\exp(x) = e^x$, with

$$e = \sum_{n=0}^{\infty} \frac{1}{n!},$$

is called Euler's constant or Euler's number. It is an irrational number, with approximate value $2.718281828\ldots$ and precise value unknown.

Remarks:

(1) Euler number e can be shown to be a transcendental number, i.e. it is the solution of no algebraic equation. Also, we have approximations to its value, $e \approx 87/32$, or $e \approx 878/323$, and, with respect to the other transcendental number π, we have $e^\pi = 23.140\ldots > \pi^e = 22.459\ldots$.

(2) When $p = n/m$, with n, m positive integers,

$$[\exp(p)]^m = \exp(mp) = e^{mp}. \tag{1.3}$$

The limit relation

$$\lim_{n\to\infty} \left(1 + \frac{x}{n}\right)^n = \exp(x) \tag{1.4}$$

serves to establish several results, in particular the Central Limit Theorem. Also, $\lim_{x\to\infty} (x^n e^{-x}) = 0, \forall n$. Frequently, for $a > 0$, we use $a^x = \exp(x \ln a)$.

(3) Several integral transforms (Fourier, Laplace, moment-generating functions, etc.) use the exponential function in their definitions to have the function to be transformed benefit from all properties of the exponential (see Chapter 13).

1.5   Applications

We give some simple applications in the following sections.

1.5.1   *Use in finance*

If a capital of 1\$ earns interest at an annual rate of $x\%$ compounded monthly, then the interest earned each month is $x/12$, so each month the total value is multiplied by $(1 + x/12)$, and the value at the end of the year is $(1 + x/12)^{12}$. If interest is compounded daily, this becomes $(1 + x/365)^{365}$. Letting the number of time intervals per year grow without bound leads to the limit definition (1.4) of the exponential function, usually called continuously compounded rate, and gives the highest interest that a capital can make. Hence, a

Table 1.1 Growth of 10,000$ capital, placed at 5%.

10,000$	1 year	3 years	5 years
Annually	10,500	11,576	20,789
Quarterly	10,509	11,602	21,072
Monthly	10,511	11,614	21,137
Daily	10,512	11,616	21,158
Hourly	10,512	11,618	21,169
Continuously	10,512	11,619	21,170

capital of 10,000$ placed at 5% interest computed continuously would become $10,000 \times \exp(0.05) = 10,512.71$$ after 1 year and $10,000 \times \exp(0.05 \times 3) = 11,618.34$$ after 3 years (see Table 1.1).

1.5.2 *The exponential function* $\exp(-ax)$

If $f(x)$ is a positive real-valued function with property $f(x + y) = f(x) \times f(y), x, y > 0$, and bounded in any finite interval, then either $f(x)$ vanishes identically or $\exists \lambda$ such that $f(x) = \exp(-\lambda x), x > 0$.

1.6 The Gamma Function

Using the power function, together with the exponential, we define

$$\Gamma(z) = \int_0^\infty t^{z-1} \exp(-t) dt,$$

for all $z \in \mathbb{R}^+$, or $z \in \mathbb{C}, Re(z) > 0$. It is continuous on $\mathbb{R}^+$, but has discontinuities at zero and at all negative integers $-k$, and it has no zero (see Fig. 1.1). On this graph, we can see that 0 and $\{-n\}$ are discontinuous points, with $\Gamma(x)$ alternately negative and positive on successive segments

$$\{[-j, -j - 1]\}_{j=0}^\infty$$

and $\Gamma(x) \to \infty$ for $x \to 0+$, and $\Gamma(x) \to -\infty$ for $x \to 0-$.

We have $\Gamma(z + 1) = z\Gamma(z)$, hence $\Gamma(n + 1) = n!$, and the gamma function can be regarded as an extension of the factorial, with

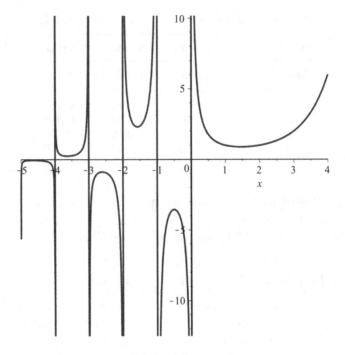

Fig. 1.1   Graph of function Gamma.

$\Gamma(2) = \Gamma(1) = 1$. Also, $\log(\Gamma(x))$ is convex on $(0, \infty)$ and

$$\Gamma\left(\frac{x}{p} + \frac{y}{q}\right) \le \Gamma(x)^{1/p}\Gamma(y)^{1/q}.$$

We have

$$\Gamma(z)\Gamma(1-z) = \frac{\pi}{\sin \pi z},$$

$$\Gamma(1/2) = \sqrt{\pi},$$

$$\Gamma\left(n+\frac{1}{2}\right) = \frac{1.3.5...(2n-1)}{2^n}\sqrt{\pi}, \quad n = 1, 2, \ldots,$$

$$\Gamma(x) = \frac{2^{x-1}}{\sqrt{\pi}}\Gamma\left(\frac{x}{2}\right)\Gamma\left(\frac{x+1}{2}\right).$$

Stirling formula, as previously encountered, gives

$$\lim_{x \to \infty} \frac{\Gamma(x+1)}{\sqrt{2\pi x}(x/e)^x} = 1.$$

Rudin (1978) can be consulted for other properties of the Gamma as well as for various questions in mathematical analysis.

1.7 The Beta Function $B(x, y)$

The beta function, $B(x, y)$, $x, y > 0$, is defined by the integral

$$B(x, y) = \int_0^1 t^{x-1}(1 - t)^{y-1} dt = \frac{\Gamma(x)\Gamma(y)}{\Gamma(x + y)} = B(y, x).$$

Also,

$$B(x + 1, y) = \frac{x}{x + y} B(x, y).$$

Applications:

(1) **How can one generalize the function exp?**
By replacing $n!$ by the gamma function, where x can take any positive value, and we have the Mittag-Leffler function:

$$E_a(x) = \sum_{n=0}^{\infty} \frac{x^n}{\Gamma(an + 1)}$$

$$= 1 + \frac{x}{\Gamma(a + 1)} + \frac{x^2}{\Gamma(2a + 1)} + \frac{x^3}{\Gamma(3a + 1)} + \cdots \quad (1.5)$$

This function has important applications in fractional analysis, where we do not limit ourselves to $f'(x)$, $f''(x)$, ..., but also consider $f^{(0.56)}(x)$, $f^{(3.25)}(x)$, etc.

(2) **Statistical distributions:** They are often obtained very simply by limiting a mathematical expression to a domain where it is positive and by computing the integral over that fixed domain. We then divide that mathematical expression by the integral. This is the case of the exponential distribution, and the Gamma distribution, as seen in the following section.

1.8 The Exponential Distribution

It has as density, limited to $[0, \infty]$, $f(x) = \lambda \exp(-\lambda x)$, $\lambda > 0$. Its cumulative distribution function (cdf) is $F(x) = 1 - \exp(-\lambda x)$, $x \geq 0$.

It enjoys the memoryless property, i.e.

$$\mathbb{P}(X > x + y | X > x) = \mathbb{P}(X > y),$$

whereby the past $(X > x)$ has been forgotten. It is related to the Poisson process, where it is the distribution of the inter-arrival time.

1.9   The Gamma Distribution

1.9.1   *In one parameter α*

It has as density, limited to $[0, \infty]$,

$$f(x) = \frac{x^{\alpha-1} \exp(-x)}{\Gamma(\alpha)}, \quad \alpha > 0. \tag{1.6}$$

1.9.2   *In two parameters α and β*

We have the shape parameter α and the scale parameter β,

$$f(x) = \frac{1}{\beta^{\alpha} \Gamma(\alpha)} x^{\alpha-1} \exp\left(-\frac{x}{\beta}\right), \quad \alpha, \beta > 0. \tag{1.7}$$

Computations show that $\mathbb{E}[X] = \alpha\beta$, $Var[X] = \alpha\beta^2$.

Two special cases:

(1) When $\alpha = 1$, we have the *exponential distribution* with density

$$f(y) = \frac{1}{\beta} \exp\left(-\frac{y}{\beta}\right), \quad \beta, y > 0,$$

with mean β and variance β^2, as seen in Section 1.8. Setting $\beta = 1/\lambda$, where λ is the rate, we have another form of the density

$$f(y) = \lambda \exp(-\lambda y), \quad \lambda, y > 0,$$

with mean $1/\lambda$ and variance $1/\lambda^2$.

(2) When the shape parameter is a positive integer n, we call the distribution an **Erlang distribution** with shape n and rate λ.

Taking $\alpha = n$, $\beta = 1/\lambda$ in (1.7), we have its density as

$$f(x) = \frac{\lambda^n}{(n-1)!} x^{n-1} \exp(-\lambda x),$$

with $\lambda, (n-1) > 0$. It is interesting to note that $X \sim Erl(n, \lambda)$ can be written as a sum of n independent identically distributed exponential variables, and vice versa, i.e.

$$X \sim Erl(n, \lambda) \Leftrightarrow X = \sum_{j=1}^{n} X_j, \text{with } X_j \overset{iid}{\sim} \exp(\lambda).$$

The gamma can hence be seen as a generalization of the Erlang distribution, where the integer n is replaced by a positive constant.

Note: The *hypoexponential* and *hyperexponential* are two other continuous distributions that can be derived from independent exponential variables

$$\{X_j \sim \exp(\lambda_j),\, j = 1, \ldots, n\}.$$

Their densities are more complex to derive (see Ross (1997)).

1.10 The Beta Distribution of the First Kind, $X \sim \beta_1^I(\alpha, \gamma)$

It is defined on the interval $[0, 1]$. Its density is

$$f(x) = \frac{x^{\alpha-1}(1-x)^{\gamma-1}}{B(\alpha, \gamma)}, \quad \alpha, \gamma > 0.$$

It is among the most used distributions, under its scalar forms and matrix forms, and it will be discussed further in later chapters.

Similarly, $Y \sim \beta_1^{II}(\alpha, \gamma), \alpha, \gamma > 0$, is a beta variable of the second kind if its density is $f(y) = \frac{y^{\alpha-1}}{B(\alpha,\gamma)(1+y)^{\alpha+\gamma}}, y \geq 0$.

1.11 Conclusion

Readers have learned the above facts in previous courses in mathematics, and this short chapter just serves to review them.

Feller (1957) contributed the first major work on applied probability and gave a wealth of information.

Bibliography

Feller, W. (1957). *An Introduction to Probability Theory and Its Applications*, Vol. 1 (John Wiley and Sons, New York).

Hogg, R. and Craig, A. T. (1995). *Introduction to Mathematical Statistics* (Prentice-Hall, New York).

Ross, S. (1997). *Introduction to Probability Models* (Academic Press, Toronto).

Rudin, W. (1978). *Principles of Mathematical Analysis* (Wiley, New York).

Chapter 2

The Univariate Normal Distribution

2.1 Introduction

In this chapter, we present the normal distribution in one dimension, first under its most basic form, then under its general form, and study different ways that will bring the second form to the first one and vice versa. Basic properties of the general univariate normal are presented and we will discuss about different operations related to the normal, including checking the normality of the distribution from which we have a sample and performing data transformation so that we can reach normality.

2.2 How Could the Normal Distribution be Derived?

Some textbooks start the study of the standard normal by giving the formula on its density, without much comment on its meaning, preventing the reader from having a deeper feeling for the subject. Let us start by giving a completely different motivation to the expression of the density by relating it to some well-known principles in physics.

We are searching for a distribution which is vaguely concentrated around the origin and symmetric about that point. We follow Newton's principle that the attraction between two points is proportional to their masses and inversely proportional to the square of their distance.

The standard normal on $(-\infty, +\infty)$**:** Here, the masses at zero and z are unities, and the attraction is proportional to the inverse of the exponential of half the square distance from 0 to z, i.e.

$$df(x) = c \cdot \frac{dz}{\exp\left(\frac{z^2}{2}\right)}, \quad -\infty < z < \infty.$$

With the value of the normalizing constant being $c = \left(\sqrt{2\pi}\right)^{-1}$, this leads to the density

$$f(z) = \frac{1}{\sqrt{2\pi}} \exp\left(-\frac{z^2}{2}\right), \quad -\infty < z < \infty. \tag{2.1}$$

This density is the normalized attraction of point z to the origin under the above hypotheses. It is taken as the definition of the density of the standard normal, denoted by $Z \sim N_1(0,1)$ or $Z \sim N(0,1)$ if there is no confusion possible. We will use (2.1) as the density of $Z \sim N(0,1)$ from now on.

Notes:

(1) Tradition wants that we take half of the square distance so that the variance is 1. But if we take the whole square distance, as Gauss did, we will have a more convenient formula:

$$f(z) = \frac{1}{\sqrt{\pi}} \exp(-z^2), \quad -\infty < z < \infty,$$

with $\sigma^2 = 1/2$ as variance. Several authors have expressed similar points of view.

(2) The standard normal was not really derived as we suggest above, but came through a long evolution process via the work of Gauss and theory of errors in physics (see Chapter 5).

2.3   The General Case

In the general case, our density could be built through four steps:

(1) On $\mathbb{R}$, we fix a center α and a variation β, and a relative change about α is recorded as the relative distance $(x - \alpha)/\beta$, which could be positive or negative.

(2) We follow Newton's principle as before, or can use directly the fact that the density is proportional to the inverse of the exponential of half of this square distance, i.e. proportional to

$$\exp\left(-\frac{1}{2}\frac{(x-\alpha)^2}{\beta^2}\right).$$

(3) The normalizing constant to make the area unity has been found to be $\left(\beta\sqrt{2\pi}\right)^{-1}$.

(4) The density of the general normal is then as follows:

$$f(x) = \frac{1}{\beta\sqrt{2\pi}}\exp\left(-\frac{(x-\alpha)^2}{2\beta^2}\right), \quad x \in (-\infty, \infty).$$

We write $X \sim N_1(\alpha, \beta^2)$. By computing the mean and the variance, we find

$$\mathbb{E}(X) = \frac{1}{\sqrt{2\pi}}\int_{-\infty}^{\infty}(\alpha + \beta u)\exp\left(\frac{-u^2}{2}\right)du = \alpha$$

and

$$\text{Var}(X) = \mathbb{E}[(X - \mathbb{E}[X])^2] = \beta^2$$

by integration by parts. Hence, the density of a general normal variable is expressed, very logically, as a function with its mean and standard deviation as parameters.

$$f(x) = \frac{1}{\sigma\sqrt{2\pi}}\exp\left(-\frac{(x-\mu)^2}{2\sigma^2}\right), \quad -\infty < x, \mu < \infty, \sigma \in \mathbb{R}^+. \quad (2.2)$$

2.4 Alternative Formulations

Some authors use the term *precision* instead of variance and define it as $\tau^2 = \sigma^{-2}$ or $\tau = \sigma^{-1}$, and we have

$$f(x) = \frac{\tau}{\sqrt{2\pi}}\exp\left(-\frac{\tau^2(x-\mu)^2}{2}\right), \quad -\infty < x, \mu < \infty, \tau \in \mathbb{R}^+.$$

This parameterization has its advantages and disadvantages. One advantage of using this parameterization is in the study of conditional distributions in the multivariate normal case. Alternately, we can define $\tau = \sigma^{-2}$, i.e. the precision is now the inverse of the variance, and have

$$f(x) = \sqrt{\frac{\tau}{2\pi}} \exp\left(-\frac{\tau(x-\mu)^2}{2}\right), \quad -\infty < x,\ \mu < \infty,\ \tau \in \mathbb{R}^+.$$

The question about which normal distribution should be called the "standard" one is also answered differently by various authors.

Stigler (1982) insists the standard normal should have variance $\sigma^2 = 1/(2\pi)$, resulting in a very convenient form for the density

$$f(z) = \exp(-\pi z^2), \quad -\infty < z < \infty,$$

which is a much simpler and easier-to-remember formula, with sd 0.40 and quartiles ± 0.25 approximately. Also, the pdf now has unit height at zero and simple approximate formulas for the quantiles of the distribution.

2.5   Properties of the Standard Normal

(a) When $\mu = 0$, $\sigma = 1$, we have the standard normal $Z \sim N_1(0,1)$, with density denoted $\phi(.)$, already mentioned.
(b) We first have

$$\int_{-\infty}^{\infty} \exp\left(\frac{-u^2}{2}\right) du = \sqrt{2\pi}.$$

It is to be noted that the function $\exp(-u^2)$ does not have a primitive, and computation related to this function has to be done numerically. Hence, the cumulative values of the standard normal, denoted $\Phi(.)$, have to be tabulated for convenient use.

$$\Phi(z) = \int_{-\infty}^{z} \phi(t)dt,$$

with

$$\phi(t) = \frac{1}{\sqrt{2\pi}} \exp\left(-\frac{t^2}{2}\right), \quad -\infty < t < \infty,$$

being the density.

(c) It is possible to define the normal, univariate, and multivariate, as Breiman (1969, p. 106) did, as the sum of small independent components and by making some basic assumptions.

2.6 Graphics

For the standard normal density with equation (2.1), we have a smooth curve, with the shape of a bell, with the characteristics shown in Figs. 2.1 and 2.2.

2.6.1 *Density*

We have a curve in $\mathbb{R}^2$, which is in the form of a symmetric bell-shaped curve.

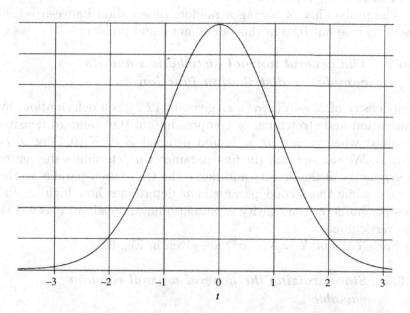

Fig. 2.1 Density $\phi(z)$ of the standard normal $Z \sim N_1(0,1)$.

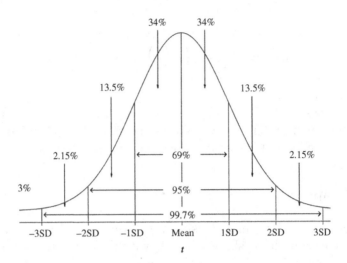

Fig. 2.2   Probabilities between evenly spaced points of the Z-axis (Mean $= 0$, SD $= 1$) or X-axis (Mean $= \mu$ and SD $= \sigma$).

The density of Z is a curve symmetric wrt to the vertical axis $z = 0$, having its summit at $z = 0.4$, approximately, and $z = 0$ as left and right asymptotes.

The probability of having a random observation between -1.26 and 1.26 is about 0.80 in this one-dimensional problem.

2.6.2   *The general normal distribution and its cumulative distribution function*

The density of $X \sim N(\mu, \sigma^2)$, as given by (2.2), is a deformation, by translation and stretching, or compressing, of the standard density, obtained when $\mu = 0$, $\sigma = 1$, and denoted $Z \sim N_1(0, 1)$ or $Z \sim N(0, 1)$. We can see that the first parameter μ determines the center of symmetry of the density and hence the translation parallel to the x axis, while the second parameter σ determines how high on the axis of symmetry the density is, and also how spread out it is wrt to the vertical axis.

Some cdf's of $X \sim N(\mu, \sigma^2)$ are given in Fig. 2.3.

2.6.3   *Standardizing the general normal random variable*

It is possible to relate all normal variables $X \sim N(\mu, \sigma^2)$ to the standard normal. $Z = (X - \mu)/\sigma$ has mean zero and unit variance,

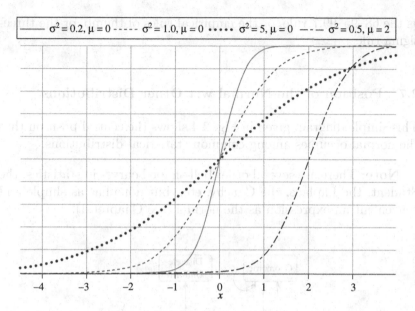

| $\sigma^2 = 0.2, \mu = 0$ | $\sigma^2 = 1.0, \mu = 0$ | $\sigma^2 = 5, \mu = 0$ | $\sigma^2 = 0.5, \mu = 2$ |

Fig. 2.3 Cumulative Distribution Functions $\Phi(x)$ of $N(\mu, \sigma^2)$ for different values of μ and σ^2.

i.e. $Z \sim N_1(0,1)$, with density $\phi(z)$ and cdf $\Phi(z)$. Conversely, with the standard normal random variable Z, we can always construct another normal random variable with specific mean μ and variance σ^2 by a linear transformation $X = \sigma Z + \mu$.

The two pdf's and cdf's are related by

$$F_X(x) = \Phi\left(\frac{x-\mu}{\sigma}\right), \quad f_X(x) = \frac{1}{\sigma}\phi\left(\frac{x-\mu}{\sigma}\right). \qquad (2.3)$$

The pdf ϕ and cdf Φ of the standard normal can hence be used for the general normal as well.

Hence, to compute $\mathbb{P}(X \leq a)$ for $X \sim N(\mu, \sigma^2)$, we merely have to write as follows:

$$\mathbb{P}(X \leq a) = \mathbb{P}\left(\frac{X-\mu}{\sigma} \leq \frac{a-\mu}{\sigma} = \theta\right) = \mathbb{P}(Z \leq \theta).$$

Tabulated values of this probability, in function of θ, for $-4 \leq \theta \leq 4$, exist in most textbooks. About 68% of the values drawn from a normal distribution are within one standard deviation σ from the mean; about 95% of the values lie within two standard deviations; about 99.7% are within three standard deviations. This fact is known as the 68–95–99.7 rule, or the empirical rules of thumb, or the three-sigma rule.

2.7   Position of the Normal wrt Other Distributions

This simple diagram given in Fig. 2.4 shows the central position that the normal occupies among common statistical distributions.

Note: There are several other bell-shaped curves in statistics: the Student, the Laplace, the Cauchy, etc., but none has as simple and convenient an expression as the normal (see Chapter 4).

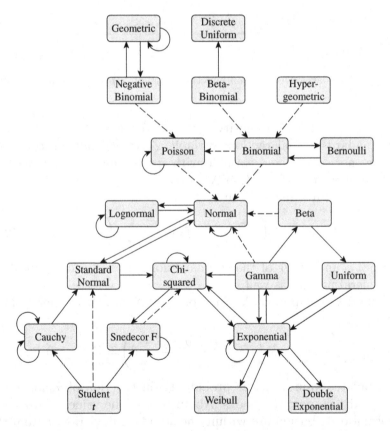

Fig. 2.4   The univariate normal's position among common distributions.

2.8 Main Properties of the General Normal Distribution, $X \sim N(\mu, \sigma^2)$

We have the following:

(a) **Mode, median, mean:** μ.

(b) **Coefficients of skewness:** 0; Kurtosis: 3.

(c) **Mean deviation:** $\sqrt{\frac{2\sigma^2}{\pi}} = 0.797885\sigma$.

(d) **Entropy:** $\text{Ent}(X) = -\int_{-\infty}^{\infty} f(x) \ln[f(x)] dx = \frac{1}{2}(1 + \ln(2\sigma^2 \pi))$.
Of all distributions with specified μ and σ^2, the normal $N(\mu, \sigma^2)$ has maximum entropy.

(e) **Cumulative distribution function:** The cumulative distribution function (cdf) gives the probability that X falls in the interval $(-\infty, x]$. The cdf of the general normal distribution can be computed as

$$\mathbb{P}(X \le x) = F(x; \mu, \sigma^2) = \frac{1}{\sigma\sqrt{2\pi}} \int_{-\infty}^{x} e^{-(t-\mu)^2/(2\sigma^2)} dt$$

$$= \Phi\left(\frac{x-\mu}{\sigma} = \theta\right) = \frac{1}{\sqrt{2\pi}} \int_{-\infty}^{\theta} e^{-t^2/2} dt$$

$$= \frac{1}{2}\left[1 + \text{erf}\left(\frac{x-\mu}{\sigma\sqrt{2}}\right)\right], \quad x \in \mathbb{R}.$$

(f) **Gauss error function** erf$(.)$ is defined by

$$\text{erf}(x) = \frac{1}{\sqrt{\pi}} \int_{-x}^{x} e^{-t^2} dt = \frac{2}{\sqrt{\pi}} \int_{0}^{x} e^{-t^2} dt.$$

(g) The integral cannot be expressed in terms of elementary functions. We have $\Phi(-x) = 1 - \Phi(x)$.

(h) The antiderivative of $\Phi(x)$ is $\int \Phi(x) dx = x\Phi(x) + \phi(x) + c$.

(i) We have numerical probabilities for x outside some symmetric intervals containing μ:

$$\mathbb{P}(|x - \mu| \ge 1.96\sigma) = 0.05,$$
$$\mathbb{P}(|x - \mu| \ge 2.33\sigma) = 0.02,$$

$$\mathbb{P}(|x - \mu| \geq 2.58\sigma) = 0.01,$$
$$\mathbb{P}(|x - \mu| \geq 3.00\sigma) = 0.00230.$$

The last inequality is frequently used. For example, in the Six Sigma Quality Program (see Chapter 5): If conforming items occupy three standard deviations on either side of the mean, non-conforming items only occupy about a quarter of 1%, or 0.23%, or, in other terms, roughly 2.3 out of 1,000 items or 2,300 non-conforming items out of one million.

2.9   Other Properties of $X \sim N(\mu, \sigma^2)$

(1) The family of normal distributions is closed under linear transformations, i.e. if X is normally distributed with mean μ and variance σ^2, then a linear transform $aX + b$ (for some real numbers a and b) is also normally distributed: $aX + b \sim N_1(a\mu + b, a^2\sigma^2)$. Also, if X_1 and X_2 are two independent normal random variables, with means μ_1 and μ_2 and standard deviations σ_1 and σ_2, then their linear combination will also be normally distributed:

$$aX_1 + bX_2 \sim N_1(a\mu_1 + b\mu_2, a^2\sigma_1^2 + b^2\sigma_2^2). \qquad (2.4)$$

(2) The converse of (1) is also true: If X_1 and X_2 are independent and their sum $X_1 + X_2$ is distributed normally, then both X_1 and X_2 must be normal. This is known as Cramer's Decomposition Theorem. The interpretation of this property is that a normal distribution is only divisible by other normal distributions. Another application of this property is in connection with the Central Limit Theorem: Although the CLT asserts that the distribution of a sum of arbitrary non-normal iid random variables is approximately normal, Cramer's Theorem shows that it can never become exactly normal.

(3) If X and Y are jointly binormal and uncorrelated, then they are independent. The requirement that X and Y should be jointly normal is essential since for non-jointly normal random variables, uncorrelatedness does not imply independence (see Chapter 7).

(4) If X and Y are independent $N_1(\mu, \sigma^2)$ random variables, then $X + Y$ and $X - Y$ are also independent and identically distributed.

This property uniquely characterizes the normal distribution, as can be seen from Bernstein's Theorem: If X and Y are independent and such that $X + Y$ and $X - Y$ are also independent, then both X and Y must necessarily have normal distributions.

(5) Normal distribution is infinitely divisible: For $X \sim N_1(\mu, \sigma^2)$, a normally distributed X with mean μ and finite variance σ^2, we can find n independent random variables $\{X_1, \ldots, X_n\}$, each distributed normally with mean μ/n and variance σ^2/n such that $X_1 + X_2 + \cdots + X_n \sim N_1(\mu, \sigma^2)$.

(6) The Kullback–Leibler divergence between two normal distributions $X_1 \sim N_1(\mu_1, \sigma_1^2)$ and $X_2 \sim N_1(\mu_2, \sigma_2^2)$ is given by

$$D_{KL}(X_1 \| X_2) = \frac{(\mu_1 - \mu_2)^2}{2\sigma_2^2} + \frac{1}{2}\left(\frac{\sigma_1^2}{\sigma_2^2} - 1 - \ln\frac{\sigma_1^2}{\sigma_2^2}\right). \tag{2.5}$$

(7) Every normal distribution is the result of exponentiating a quadratic function (just as an exponential distribution results from exponentiating a linear function):

Let $g(x) = \exp(f(x))$, where $f(x) = ax^2 + bx + c$, with $a < 0$. The conditions under which $\exp(f(x))$ is the pdf of a normal are as follows:

$$a = -\frac{1}{2\sigma^2}, \quad b = \frac{\mu}{\sigma^2}, \quad c = -\frac{\mu^2}{2\sigma^2}.$$

This yields the classic "bell curve" shape, provided that $a < 0$ so that the quadratic function is concave.

One can adjust a to control the "width" of the bell, then adjust b to move the central peak of the bell along the x-axis, and finally, one must choose c so that

$$\int_{-\infty}^{\infty} f(x)dx = 1$$

(which is only possible when $a < 0$).

(8) In Chapter 4, we will see some distributions derived from the univariate normal, which are important by their own right.

Note: For several additional properties, see Patel and Read (1996).

2.10   Test on Normality

As we see, the normal hypothesis for the population is almost made every time that some elaborate probabilistic method is used. The normal model usually comes with parametric statistics. Non-parametric statistics do not make this supposition, but the tools available become much more limited.

Hence, there is a real need to know whether the data we have gathered did come from a normal population, or by using a milder conclusion, that we cannot reject the hypothesis that it comes from a normal population. If it is so, then we can proceed to the next step.

Naturally, the use of a computer is required to carry out all the computations. Most statistical software have these routines. Graphical procedures include the $Q - Q$ plot, the probability paper sketch, and could be helpful too. We review some of these procedures. Checking on multivariate normality will be discussed in Chapter 9.

Let $\{X_1, X_2, \ldots, X_n\}$ be a sample of observations of size n. We wish to test whether this sample comes from a normal population. In other words, we wish to consider the following:

H_0: The population from which the sample is collected is normal.
H_a: The population is not normal.

At the α significance level, we would be satisfied that H_0 is NOT rejected (which does not mean, however, that we have proven that the sample comes from a normal population). The following statistical tests are frequently used:

(1) Chi-square goodness of fit,
(2) Kolmogorov–Smirnov,
(3) Shapiro–Wilk,
(4) Anderson–Darling.

Information on these tests can be found in the statistical literature (see, e.g. Ahsanullah *et al.* (2014)) and the most common statistical software (SAS, Minitab, SPSS, MATLAB) are available for their application. Usually, they give the same conclusion.

2.11 Quantile Function

The inverse of the standard normal cdf, called the quantile function or probit function, is expressed in terms of the inverse error function:

$$\Phi^{-1}(p) \equiv z_p = \sqrt{2}\,\text{erf}^{-1}(2p-1), \quad p \in (0,1). \tag{2.6}$$

Quantiles of the standard normal distribution are usually denoted as z_p. The quantile z_p is a value such that $Z \sim N_1(0,1)$ has the probability of exactly p of falling inside the $(-\infty, z_p]$ interval. Since $z_{0.95} = 1.645$, a standard normal variable is greater than 1.645 in 5% of the cases.

Quantiles are used in hypothesis testing, construction of confidence intervals, and $Q - Q$ plots.

(a) **Quantile function:** For a normal random variable with mean μ and variance σ^2, the quantile function is

$$F^{-1}(p; \mu, \sigma^2) = \mu + \sigma \Phi^{-1}(p) = \mu + \sigma\sqrt{2}\,\text{erf}^{-1}(2p-1), \quad p \in (0,1).$$

(b) $Q - Q$ **plot (Q is for quantiles):** The general objective is to plot the quantiles of the ordered dataset against those of the standard normal. This should give a straight line if data come from a normal population. The procedure is then as follows:

(1) Order the dataset from smallest to largest, obtaining $\{x_{(1)}, x_{(2)}, \ldots, x_{(n)}\}$.
(2) Associate to each number its probability value after correction for continuity. They are, respectively,

$$\left\{ \frac{1 - \frac{1}{2}}{n}, \frac{2 - \frac{1}{2}}{n}, \ldots, \frac{n - \frac{1}{2}}{n} \right\}.$$

The last probability is close to 1.
(3) Find, from a standard normal table, the quantiles $\{q_{(1)}, q_{(2)}, \ldots, q_{(n)}\}$ having these probabilities.
(4) Plot the pairs

$$\left\{ \left(q_{(1)}, x_{(1)}\right), \left(q_{(2)}, x_{(2)}\right), \ldots, \left(q_{(n)}, x_{(n)}\right) \right\}$$

and examine how close to a straight line they are.

We can then use the correlation analysis coefficient r as a tool to study the problem, where

$$r = \frac{\sum_{j=1}^{n} (x_j - \bar{x})(y_j - \bar{y})}{\sqrt{\sum_{j=1}^{n} (x_j - \bar{x})^2}\sqrt{\sum_{j=1}^{n} (y_j - \bar{y})^2}},$$

with $x_j = x_{(j)}$, $y_j = q_{(j)}$.

2.12   Transformation of Data to Reach Normality

When plotting a dataset, it could be obvious that no attempt should be made to make it look more normal. A bath tube shape, or J-shape, or a multimodal shape are clear examples of non-normality. Even bell-shaped frequency curves with thick tails should not be accepted as normal.

It is well known that the following transformations can make data closer to normal, so that it can pass the normality test. However, as in changing variables in general, the challenge will be to give a sensible meaning to the new variables. They are as follows:

(a) Square root, and powers, for count data:

$$x \to \sqrt{x}, \quad \text{and} \quad x \to x^\lambda, \ \lambda \in \mathbb{R}.$$

Several values of λ can be used to plot the corresponding set $\{x^\lambda\}$ to find which one is closest to normality.

(b) Transformation logit for proportions p:

$$p \to \frac{1}{2}\log\left(\frac{p}{1-p}\right).$$

(c) Fisher's transformation for the correlation coefficient r:

$$r \to \frac{1}{2}\log\left(\frac{1+r}{1-r}\right).$$

Other *ad hoc* transformations can also prove to be very helpful.

(d) A very useful transformation has been suggested by Box and Cox as a modified family of power transformations:

$$f_\lambda(x) = \begin{cases} \dfrac{x^\lambda - 1}{\lambda}, & \lambda \neq 0, \\ \ln x, & \lambda = 0. \end{cases}$$

For a given dataset, the choice of λ is the value that maximizes the function

$$g(\lambda) = (\lambda - 1) \sum_{i=1}^{n} \ln x_i - \frac{n}{2} \ln \left[\frac{1}{n} \sum_{j=1}^{n} (f_\lambda(x_j) - \xi)^2 \right],$$

$$\xi = \sum_{j=1}^{n} f_\lambda(x_j)/n.$$

The procedure is to be carried out by computer.

More details can be found in Patel and Read (1996).

2.13 Characterization

In the previous section, the normal distribution was defined by specifying its probability density function. However, there are other ways to characterize a probability distribution as normal. They include the following: the cumulative distribution function, the moments, the cumulants, the characteristic function, and the moment-generating function.

2.13.1 *Probability density function (pdf)* $\phi(x)$

Other properties of the density

(a) Function $\phi(x)$ is unimodal and symmetric around the point $x = \mu$, which is at the same time the mode, median, and mean of the distribution.
(b) Inflection points of the curve occur one standard deviation away from the mean on either side (i.e. at $x = \mu - \sigma$ and $x = \mu + \sigma$).

(c) The function $\phi(x)$ is log concave.

(d) The function is supersmooth of order 2, implying that it is infinitely differentiable.

(e) It is closely related to Hermite polynomials, which it generates. Setting

$$\phi(x) = \frac{1}{\sqrt{2\pi}} \exp\left(-\frac{x^2}{2}\right) \quad \text{and} \quad D = \frac{d}{dx},$$

we have

$$D\phi(x) = -x\phi(x),$$
$$D^2\phi(x) = (x^2 - 1)\phi(x),$$
$$D^3\phi(x) = (3x - x^3)\phi(x).$$

We define the Hermite polynomials:

$$H_r(x) = -\frac{D^r\phi(x)}{\phi(x)},$$

with the coefficient of x^r being unity. We obtain

$$\begin{cases}
H_0 = 1, \\
H_1 = x, \\
H_2 = x^2 - 1, \\
\quad \cdots \\
H_6 = x^6 - 15x^4 + 45x^2 - 15, \\
H_7 = x^7 - 21x^5 + 105x^3 - 105x.
\end{cases}$$

Among the numerous properties of these polynomials, we note the following:

- $\frac{d}{dx} H_r(x) = r H_{r-1}(x), r \geq 1.$
- They satisfy the differential equation

$$\frac{d^2}{dx^2} H_r(x) - x \frac{d}{dx} H_r(x) + r H_r(x) = 0, \quad r \geq 2.$$

- They are orthogonal, i.e.

$$\int_{-\infty}^{\infty} H_m(x)H_n(x)\phi(x)dx = \begin{cases} 0, & m \neq n, \\ n!, & m = n. \end{cases}$$

2.13.2 Characteristic function $\varphi_X(t)$ and moment-generating function $M_X(t)$

(1) When finite moments do not exist, the moment-generating function (mgf) might not exist (see (2)). But the characteristic function (chf), based on the Fourier transform of the density, always does and can determine a probability distribution function uniquely by the use of an inversion formula in complex analysis.

The characteristic function $\varphi_X(t)$ of an rv X is defined as the expected value of $\exp(itX)$, where i is the imaginary unit, and $t \in \mathbb{R}$ is the argument of the characteristic function, i.e. $\varphi_X(t) = \mathbb{E}[\exp(itX)]$. Thus, the characteristic function is the Fourier transform of the density $\phi(x)$ (see Chapter 13), but with a positive sign. For $X \sim N_1(\mu, \sigma^2)$, the characteristic function is as follows:

$$\varphi_X(t; \mu, \sigma^2) = \int_{-\infty}^{\infty} \exp(itx)\frac{1}{\sqrt{2\pi\sigma^2}} \exp\left(-\frac{1}{2\sigma^2}(x-\mu)^2\right) dx$$

$$= \exp(i\mu t) \cdot \exp\left(-\frac{1}{2}\sigma^2 t^2\right). \tag{2.7}$$

This form involves only the exponential function, with the imaginary part affecting only the mean. This fact makes the chf of X a very powerful tool in several manipulations of the normal rv X. The characteristic function can be analytically extended to the entire complex plane.

(2) In statistics, we rather stay in the real domain and consider the moment-generating function (mgf) of X, which is defined as the expected value of $\exp(tX)$, or $M_X(t) = \mathbb{E}[\exp(tX)]$. This is the Laplace transform but with a positive sign (see Chapter 13). For a normal distribution, the moment-generating function exists

and is equal to

$$M_X(t; \mu, \sigma^2) = \mathbb{E}[\exp(tX)] = \exp(\mu t + \frac{1}{2}\sigma^2 t^2). \qquad (2.8)$$

2.13.3 *The cumulant generating function*

The cumulant generating function (cgf), also called *the second characteristic function*, is the logarithm of the moment-generating function:

$$\mathrm{cgf}(t; \mu, \sigma^2) = \psi(t) = \ln M_X(t; \mu, \sigma^2) = \mu t + \frac{1}{2}\sigma^2 t^2. \qquad (2.9)$$

Since this is a quadratic polynomial in t, only the first two cumulants are non-zero.

Note: Some authors define the second characteristic function as the logarithm of the characteristic function.

2.13.4 *Moments of order k*

We know that if the moment of order k of X exists, then

$$\frac{d^k}{dt^k}\varphi_X(t) = i^k \int\limits_{-\infty}^{\infty} x^k e^{itx} dF(x),$$

and hence, at the limit, $\varphi_X^{(k)}(0) = i^k m_k$, where m_k is the moment of order k.

The normal distribution has moments of all orders, i.e. for $X \sim N_1(\mu, \sigma^2)$, the expectation $\mathbb{E}[|X|^p]$ exists and is finite for all p such that $Re(p) > -1$. Usually, we are interested only in moments of integral orders $p = 1, 2, 3, \ldots$.

2.13.4.1 *Central moments*

Central moments are the moments of X around its mean μ. Thus, a central moment of order p is the expected value of $(X - \mu)^p$. Using standardization of normal random variables, this expectation will be equal to $\sigma^p \mathbb{E}[Z^p]$, where Z is the standard normal.

$$\mathbb{E}[(X - \mu)^p] = \begin{cases} 0 & \text{if } p \text{ is odd,} \\ \sigma^p (p-1)!! & \text{if } p \text{ is even.} \end{cases}$$

Here, $n!!$ denotes the double factorial (see Chapter 1), i.e. the product of even or odd numbers from n to 1.

2.13.4.2 *Central absolute moments*

Central absolute moments are the moments of $|X - \mu|$. They coincide with regular central moments for all even orders, but are zero for all odd p's.

$$\mathbb{E}[|X - \mu|^p] = \sigma^p \frac{2^{\frac{p}{2}} \Gamma\left(\frac{p+1}{2}\right)}{\sqrt{\pi}}. \tag{2.10}$$

The last formula is true for any non-integer p.

2.13.4.3 *Raw moments and absolute moments*

Raw moments and absolute moments are the moments of X and $|X|$, respectively. The formulas for these moments are much more complicated and are given in terms of Kummer confluent hypergeometric functions.

$$\mathbb{E}(X^p) = \begin{cases} \sigma^p 2^{p/2} \dfrac{\Gamma\left(\frac{p+1}{2}\right)}{\sqrt{\pi}} \cdot {}_1F_1\left(-\dfrac{p}{2}, \dfrac{1}{2}; -\dfrac{1}{2}(\mu/\sigma)^2\right), & p \text{ even,} \\[4mm] \mu \sigma^{p-1} 2^{(p+1)/2} \dfrac{\Gamma\left(\frac{p}{2} + 1\right)}{\sqrt{\pi}} \\[2mm] \qquad \cdot {}_1F_1\left(\dfrac{1-p}{2}, \dfrac{3}{2}; -\dfrac{1}{2}(\mu/\sigma)^2\right), & p \text{ odd,} \end{cases}$$

$$\tag{2.11}$$

with ${}_1F_1(\alpha, \beta; x) = \sum_{n=0}^{\infty} \frac{(\alpha, n)}{(\beta, n)} \frac{x^n}{n!}$, being the Kummer confluent hypergeometric function of the first kind, where $(\alpha, n) = \frac{\Gamma(\alpha+n)}{\Gamma(\alpha)}$.

$$\mathbb{E}[|X|^p] = \sigma^p 2^{\frac{p}{2}} \frac{\Gamma\left(\frac{1+p}{2}\right)}{\sqrt{\pi}} \cdot {}_1F_1\left(-\frac{p}{2}, \frac{1}{2}, -\frac{1}{2}(\mu/\sigma)^2\right). \tag{2.12}$$

These expressions remain valid even if p is not an integer. The proofs of these results are laborious.

The first two cumulants are equal to μ and σ^2, respectively, whereas all higher-order cumulants are zero. We have Table 2.1.

Conversely, if the random variable X has moments of all orders, its chf is

$$\varphi_X(t) = \mathbb{E}(\exp(itX)) = \int \exp(itx)dF(x)$$

$$= 1 + itm_1 + \frac{(it)^2}{2}m_2 + \cdots + \frac{(it)^k}{k!}m_k + \cdots$$

by using

$$\exp(itX) = 1 + itX + \frac{(it)^2}{2!}X^2 + \cdots + \frac{(it)^k}{k!}X^k + \cdots$$

and the fact that the series is absolutely convergent.

The chf determines the distribution of the rv X, as well as the cumulative distribution function does.

Using $\varphi_X(t)$ and complex analysis tools can be very effective and elegant to derive several results related to the normal. Similarly, we can use the mgf to stay in the real domain. The following is an example: Let $X_1, X_2, \ldots, X_n$ be n independent normal variables $X_i \sim N(\mu_i, \sigma_i^2)$. The second chf of X_i (or cumulants generation function) is

$$\psi_j(t) = t\mu_j + \frac{t^2\sigma_j^2}{2}. \tag{2.13}$$

Table 2.1   Moments and cumulants of the general normal.

Order	Raw moment	Central moment	Cumulant
1	μ	0	μ
2	$\mu^2 + \sigma^2$	σ^2	σ^2
3	$\mu^3 + 3\mu\sigma^3$	0	0
4	$\mu^4 + 6\mu^2\sigma^2 + 3\sigma^4$	$3\sigma^4$	0
5	$\mu^5 + 10\mu^3\sigma^2 + 15\mu\sigma^4$	0	0

Hence, the sum $Y = \sum_{j=1}^{n} X_j$ has as second chf the corresponding sum

$$\sum_{j=1}^{n} \psi_j(t) = tM + \frac{\sigma^2 t^2}{2},$$

with $M = \sum_{j=1}^{n} \mu_j$, $\sigma^2 = \sum_{j=1}^{n} \sigma_j^2$, and Y is also a normal variable.

2.14 Exponential Family of Distributions

This is a family of distributions that includes most of the common distributions in statistics. It has the general form:

$$f(x) = \exp\{A(\theta)B(x) + C(x) + D(\theta)\},$$

with A, B, C and D being arbitrary functions, continuous or discrete, and θ is the parameter. This exponential family of distributions is called *Natural Exponential family* if $B(x) = x$.

Taking

$$A(\theta) = \theta, \quad C(x) = -\frac{x^2 + \log(2\pi)}{2}, \quad D(\theta) = -\frac{\theta^2}{2}, \quad (2.14)$$

we can see that $N_1(\theta, 1)$ is a member of the Natural Exponential Family.

2.15 Sampling from the Normal

When a random sample $\{X_1, \ldots, X_n\}$ is taken from the $X \sim N(\mu, \sigma^2)$, two statistics are usually considered: the sample mean

$$\bar{X} = n^{-1} \sum_{j=1}^{n} X_j$$

and the sample variance

$$s^2 = \frac{1}{(n-1)} \sum_{j=1}^{n} (X_j - \bar{X})^2 = \frac{1}{(n-1)} \left[\sum_{j=1}^{n} X_j^2 - \frac{1}{n} \left(\sum_{j=1}^{n} X_j \right)^2 \right].$$

They are maximum likelihood estimates (MLE) of μ and σ^2, respectively, and also their unbiased estimators. What is particular is that they are independent of each other, which is also a feature that characterizes the normal. We have $\bar{X} \sim N(\mu, \sigma^2/n)$ while $(n-1)\,S^2 \sim \chi^2_{(n-1)}$. These results can be obtained by using Helmert's approach, as shown in the following.

Helmert's transformation: The orthogonal transformation presented in the following is especially useful for samples from a normal population, for the Jacobian is unity and the density function remains of the same form. Let us define a square orthonormal matrix $C = [c_{ij}]$ as a matrix such that $C^t C = C C^t = I_p$ and the sum of the squares in each column is 1. Then we also have $|C| = 1$, or $|C| = -1$, and

$$(u = Cx) \Rightarrow \sum_{j=1}^{n} u_j^2 = \sum_{j=1}^{n} x_j^2.$$

Furthermore, if $\{X_1, \ldots, X_n\}$ are iid $N(0,1)$, then $V = CX$ has its components V_i iid $N(0,1)$.

Now, we consider a $(n \times n)$ matrix C with the following form:

$$C = \begin{bmatrix} 1/\sqrt{n} & 1/\sqrt{n} & \cdots & 1/\sqrt{n} & \cdots & 1/\sqrt{n} \\ 1/\sqrt{2} & -1/\sqrt{2} & \cdots & 0 & \cdots & 0 \\ \cdots & \cdots & \cdots & \cdots & \cdots & \cdots \\ 1/\sqrt{j(j-1)} & 1/\sqrt{j(j-1)} & \cdots & (1-j)/\sqrt{j(j-1)} & \cdots & 0 \\ \cdots & \cdots & \cdots & \cdots & \cdots & \cdots \\ 1/\sqrt{n(n-1)} & 1/\sqrt{n(n-1)} & \cdots & 1/\sqrt{n(n-1)} & \cdots & (1-n)/\sqrt{n(n-1)} \end{bmatrix}.$$

$$(2.15)$$

Let

$$X = \begin{pmatrix} X_1 \\ \vdots \\ X_n \end{pmatrix}.$$

We have

$$V = \begin{pmatrix} V_1 \\ \vdots \\ V_n \end{pmatrix} = C \begin{pmatrix} X_1 \\ \vdots \\ X_n \end{pmatrix}.$$

C is orthonormal and

$$V_1 = \frac{1}{\sqrt{n}}X_1 + \cdots + \frac{1}{\sqrt{n}}X_n = \left(\sum_{j=1}^{n} X_j/\sqrt{n}\right) = \sqrt{n}\bar{X}_n = U.$$

Also,

$$\sum_{j=2}^{n} V_j^2 = \sum_{j=1}^{n} V_j^2 - V_1^2 = \sum_{j=1}^{n} X_j^2 - \frac{1}{n}\left(\sum_{j=1}^{n} X_j\right)^2 = (n-1)\,s_n^2 = W.$$

By the above results on orthonormal transformations, U and W are independent, $U \sim N_1(0,1)$, and so are $V_2, \ldots, V_n$. So, $W \sim \chi_{n-1}^2$, a chi-square with $(n-1)$ dof, with density

$$f(w) = \frac{1}{2^{\frac{1}{2}(n-1)}\Gamma\left(\frac{n-1}{2}\right)}e^{-\frac{1}{2}w}w^{\frac{1}{2}(n-3)},$$

as shown in Section 4.9.

This method of proof was used by Helmert in 1876. This interesting and important result also characterizes the normal distribution, i.e. if the mean and variance of a random sample from a population are independent, the population must be normal.

2.16 Conclusion

In this chapter, we have presented some properties of the univariate normal distribution without going into advanced topics. Properties of theoretical interest are, in general, not included here. As we can see, these properties are varied and can lend themselves to operations and applications of different kinds, which we will see in Chapters 4 and 5.

Bibliography

Ahsanullah, M., Golam Kibria, B.M. and Shakil, M. (2014). *Normal and Student's t-Distributions and Their Applications* (Atlantis Press, Amsterdam).

Breiman, L. (1969). *Probability and Stochastic Processes: With a View Toward Applications* (Houghton-Mifflin, New York).

Patel, J. and Read, C. (1996). *Handbook of the Normal Distribution* (Marcel Dekker, New York).

Stigler, S.M. (1982). A modest proposal: A new standard for the normal, *The American Statistician*, **36**(2), 137–138.

Chapter 3

Limit Theorems

3.1 Introduction

In this chapter about *limit theorems*, we treat a major topic on the normal and explain why the normal is so important in statistics. It is no doubt that without the results presented here the normal would be just another distribution, with no particular superior influence in large sample theory and statistical applications. In general, limit theorems are under the topics of *laws of large numbers* (LLN) and *Central Limit Theorem* (CLT), with multiple variations and generalizations. We will note that, in LLN, there is no mention of the nature of the limit distribution, while in CLT, the normal distribution plays an important role. Also, LLN has important applications in the Monte Carlo simulation, whereas CLT is the source of many approximations of a discrete, or continuous, statistical distribution by the normal.

Although there are several kinds of convergence for a sequence of random variables, we will deal mainly with two kinds: *convergence in distribution, or in law*, and *convergence in probability*. Other types of convergence include the following: *almost sure convergence*, or *convergence with probability one*, and *convergence in mean*. The reason there are so many types of convergence is that a random variable is also a function, and a sequence of rv's can eventually take a constant value, such as convergence in probability, or approach an unchanging probability distribution, such as convergence in distribution.

Two applications of the normal distribution as approximations are presented. We conclude the chapter with some notions of spectral limits.

3.2   Convergence in Distribution

Definition 3.1. Let $\{X_n\}_{n=1}^{\infty}$ be a sequence of rv's with cdf $F_n(x)$ and chf $\varphi_n(t)$. We say that $\{X_n\}_{n=1}^{\infty}$ converges in distribution toward the rv X, with cdf $F(x)$, if

$$\lim_{n \to \infty} F_n(x) = F(x),$$

except perhaps at points of discontinuity of $F(x)$. We write $X_n \overset{d}{\to} X$.

It is to be noted that the limit has to be a cdf too, i.e. $F(x)$ is an increasing function, with $F(-\infty) = 0$ and $F(\infty) = 1$. Moreover, this definition does not say anything on the proximity between X_n and X, only between $F_n(x)$ and $F(x)$. The study of the related sequence of chf $\{\varphi_n(t)\}_{n=1}^{\infty}$ is very helpful since we have a correspondence between the convergences of $F_n(x)$ and of $\varphi_n(t)$.

We have Levy's theorem as follows.

Theorem 3.1. *If $\{X_n\}_{n=1}^{\infty}$ converges in distribution toward X, then the chf $\{\varphi_n(t)\}$ converges uniformly in all finite interval of t. Conversely, if $\{\varphi_n(t)\}_{n=1}^{\infty}$ converges toward a limit $\varphi(t)$ with uniform convergence in a neighborhood of the origin, then $\{X_n\}_{n=1}^{\infty}$ converges in distribution toward X.*

Several well-known researchers have tried to modify the condition on the function $\varphi(t)$ to improve and generalize this important result. We report here two of these interventions:

(a) **Glivenko:** The limit function $\varphi(t)$ is a characteristic function.
(b) **Cramer:** The limit function $\varphi(t)$ is continuous at zero.

3.3   Convergence in Probability (Stochastic or Weak Convergence)

In this type of convergence, we are interested in the probability that the sequence approaches a fixed limit value.

Definition 3.2. $\{X_n\}_{n=1}^{\infty}$ converges in probability toward a for $n \to \infty$ if

$$\forall \eta \in (0,1) \text{ and } \varepsilon > 0, \ \exists N \in \mathbb{N} \text{ s.t. for } n \geq N,$$
$$\mathbb{P}\left(|X_n - a| < \varepsilon\right) \geq 1 - \eta.$$

We write $X_n \overset{\mathrm{P}}{\to} a$ for $n \to \infty$.

The limit distribution can be seen as completely concentrated at the point a. However, it does not involve the normal distribution and neither do the weak and strong laws in the following sections.

3.4 Laws of Large Numbers

In the probabilistic/statistical literature, there are two kinds of *laws of large numbers*, both related to the convergence of the relative frequency toward the expected value. They differ by the way this convergence occurs: *convergence in probability* for the weak law and *convergence almost surely* for the strong law. But for applications in this book, this difference matters very little.

Let $\{X_i\}_{i=1}^{\infty}$ be sequence of iid rv's with finite expected value μ. The following two laws are considered fundamental to all empirical sciences, and these justify the intuitive foundation of probability as frequency. Tchebycheff's inequality provides a very simple version of the two laws of large numbers,

$$\mathbb{P}\left(|\bar{X} - \mu| \leq \varepsilon\right) > 1 - \frac{\sigma^2}{n\varepsilon^2},$$

for a sample of size n in a population of finite mean μ and variance σ^2, since it states that the probability that the sample mean falls within a small interval about the population mean can be made as near to one as desired.

3.4.1 *Weak law of large numbers (Bernoulli–Khintchine)*

The weak law of large numbers (WLLN) says that the relative frequency of an event in n independent experiences converges in probability toward the probability of that event. Generally, the sample average converges in probability toward the expected value.

We note

$$\bar{X}_n \xrightarrow{p} \mu \text{ for } n \to \infty, \text{ or } \forall \varepsilon > 0, \ \lim_{n\to\infty} \mathbb{P}\left(\left|\bar{X}_n - \mu\right| > \varepsilon\right) = 0.$$

Proof. Let the event A have probability p. In n repetitions of the experience, let Z_n be the absolute frequency of A, or the number of times A has arrived, while $z_n = Z_n/n$ is the relative frequency. Let $Y_n = z_n - p$, it has 0 as mean and pq/n as variance. Using Tchebycheff inequality,

$$p = \mathbb{P}\left(|Y_n| < \varepsilon\right) > 1 - \frac{\sigma_n^2}{\varepsilon^2} = 1 - \eta, \text{ with } \eta = \frac{pq}{n\varepsilon^2}. \qquad \square$$

Similarly, we can prove that the mean of a sample converges in probability toward $\mathbb{E}(X)$ or the frequency function for a sample of size n converges in probability toward $F(x)$. No condition on the variance is required.

3.4.2   *Strong law of large numbers*

The strong law of large numbers (SLLN) states that the sample average converges "almost surely" or "strongly" toward the expected value,

$$\bar{X}_n \xrightarrow{as} \mu \text{ when } n \to \infty \text{ or } \mathbb{P}\left(\lim_{n\to\infty} \frac{S_n}{n} = \mu\right) = 1.$$

If the variables are not identical, then each must have a finite second moment s.t.

$$\sum_{k=1}^{\infty} \frac{\text{Var}(X_k)}{k^2} < \infty.$$

This law was discovered by Emile Borel in 1909, which is an improvement of the Bernoulli–Khintchine weak law.

Remark 3.1.

(1) There are instances when the weak LLN holds but not the strong LLN, but the converse is not true.

(2) The difference between these two laws, the weak and the strong, is rather difficult to explain in layman's terms: The weak law says that the absolute difference will be ultimately small, but can take some larger value infrequently. The strong law says that the probability that such an event can happen is very small.

An example is that it is logically possible that a coin shows head every time it is tossed, but such an event almost never happens.

It is, however, quite surprising to read the remarks made by two leading probabilists of the last century, as quoted by Chung (1974, p. 233): *Van der Waerden*: The strong law of large numbers scarcely plays a role in mathematical statistics; *William Feller*: The weak law of large numbers is of very limited interest and should be replaced by the more precise and more useful strong law of large numbers.

As one suspects, simulation using the Monte Carlo methods is based on this strong law. Extremely accurate results obtained in simulation confirm the validity of this law, which seems to be underestimated sometimes.

3.4.3 *SLLN for simulation*

The following version of the SLLN is directed toward the Monte Carlo simulation.

Let $\{X_j\}$, $j = 1, 2, \ldots$ be a sequence of iid rv's with values in $\mathbb{R}$, s.t. $\mathbb{E}(X_j) < \infty$. For $N \geq 1$, let $S_N = N^{-1} \sum_{j=1}^{N} X_j$ denote the empirical mean of $\{X_j\}$, $j = 1, 2, \ldots, N$. Then the SLLN applies and we have $\lim_{N \to \infty} S_N = \mathbb{E}(X_j)$ almost surely.

Hence, let θ be a quantity that must be approximated numerically. We suppose that there exists a family of iid rv's $\{X_1, \ldots, X_N\}$ which can be computer-simulated and a function f s.t. $\mathbb{E}[f(X_j)] = \theta$. Then, except on an event of zero probability, θ can be simulated by using the Monte Carlo approach as follows:

(1) Draw a sample $\{X_1(\omega), \ldots, X_N(\omega)\}$.

(2) Compute the empirical mean: $S_N(\omega) = \sum_{j=1}^{N} f(X_j(\omega))/N$.

(3) Approximate θ by this mean.

This approximation is good if N is large enough, but the SLLN does not give precision on the convergence.

3.4.4   *A famous application of the LLN*

A famous application of the LLN is the Buffon needle experiment, performed to have an approximate value of π. More concretely, a two-dimensional space is divided into vertical strips of 1 cm width, and we throw a needle of 1 cm at random and call it a success S if it intersects one of the vertical lines. With an appropriate system of coordinates, it can be proven that the probability for success is $2/\pi$.

Back in 1850, Wolf from Zurich performed 5,000 throws, setting $Y_j = 1$ if the throw is a success, and obtained $\hat{\pi} = 3.1596$, a surprisingly good result in view of the primitive conditions of his experience. This experience has been repeated under more sophisticated conditions lately and results reported in the statistical literature. Estimates for π get much closer to $3.14159\ldots$.

Whereas there is no mention of any distribution in either of these large numbers laws, we have the normal distribution explicitly mentioned in the following limit theorem.

3.5   Central Limit Theorem

The earliest result is the de Moivre–Laplace approximation published in 1802.

3.5.1   *De Moivre–Laplace theorem*

Let X be a binomial rv, i.e. $X \sim \mathrm{Bin}\,(n,p)$. Then for any numbers c, d,

$$\lim_{n\to\infty} \mathbb{P}\left(c \le \frac{X - np}{\sqrt{npq}} \le d \right) = \frac{1}{\sqrt{2\pi}} \int_c^d e^{-x^2/2} dx.$$

As early as 1783, Laplace recognized the need for tabulation of this integral, which cannot be expressed in closed form, and the first tables were published by the French physicist, Kramp, in 1799.

The central limit theorem states that under certain (fairly general) conditions, the sum of a large number of random variables will have an approximately normal distribution. For example, we have the following.

3.5.2   Lindenberg–Levy central limit theorem

If $\{X_1, \ldots, X_n\}$ is a sequence of iid random variables, each having finite mean μ and variance σ^2, then

$$\sqrt{n}\left(\frac{1}{n}\sum_{i=1}^{n} X_i - \mu\right) \xrightarrow{d} N\left(0, \sigma^2\right).$$

The general idea is as follows: As n gets larger, the distribution of the difference between the sample average $\bar{X}_n$ and its limit μ when multiplied by the factor $\sqrt{n}$ (i.e. $\sqrt{n}\left(\bar{X}_n - \mu\right)$) approximates the normal distribution with mean 0 and variance σ^2. For large enough n, the distribution of $\bar{X}_n$ is close to the normal distribution with mean μ and variance σ^2/n.

The usefulness of the theorem is that the distribution of $\sqrt{n}\left(\bar{X}_n - \mu\right)$ approaches normality regardless of the shape of the distribution of the individual X_i's. It also implies that probabilistic and statistical methods that work for normal distributions can be applied to problems that rely on other types of distributions, provided that the samples used are large enough. It is often stated that a sample size of 30 or more can be considered large enough for any case. But if the parent population displays some symmetry, the required sample size can be much smaller.

For a theorem of such fundamental importance to applied probability and statistics, the central limit theorem has a remarkably simple proof, using characteristic functions. It is similar to the proof of the (weak) law of large numbers. The moment-generating function can also be used.

The theorem will hold even if the summands x_i are not iid, although some constraints on the degree of dependence and the growth rate of moments have to be imposed. The following section provides an example.

3.5.3   Liapounoff theorem

If $\{X_i\}_{i=1}^{\infty}$ is a sequence of independent rv's with finite means $\{\mu_i\}_{i=1}^{\infty}$ and finite variances, $\{\sigma_i^2\}_{i=1}^{\infty}$ all different. Suppose that all absolute deviations of order 3, $\delta_i^3 = \mathbb{E}\left(|X_i - \mu_i|^3\right)$, $i = 1, \ldots, \infty$, exist. Suppose that $E_n/\Sigma_n \to 0$ for $n \to \infty$, where $E_n^3 = \sum_{j=1}^n \delta_j^3$, $\Sigma_n^2 = \sum_{j=1}^n \sigma_j^2$. Then S_n converges in distribution toward a normal.

3.5.4 *Limit theorems for densities*

Gnedenko Theorem (1968): Let $\{X_i\}_{i=1}^\infty$ be a sequence of iid absolutely continuous rv's with common finite mean μ and finite variance σ^2. Let the standardized sum, $\left(\sum_{i=1}^n X_i - n\mu\right)/\sqrt{n}\sigma$, $n \geq n_0$ have pdf $f_n(x)$. Then $f_n(x) \to \varphi(x)$ uniformly in $|x| < \infty$ as $n \to \infty$ iff $\exists m$ s.t. $f_m(x)$ is bounded.

Renyi, Gnedenko, Kolmogoroff, and Feller have succeeded in replacing the condition of boundedness of $f_n(x)$ by other conditions.

There are several results in the literature on limit theorems in different contexts. Some of them are presented in Chapter 6 of Patel and Read (1996).

3.6   Other Types of Convergence

3.6.1   *Almost sure convergence*

$\{X_n\}$ converges almost surely, or strongly, or with probability 1, toward X if

$$\mathbb{P}\left(\lim_{n\to\infty} X_n = X\right) = 1,$$

i.e. events for which X_n does not converge to X have probability zero. We then write $X_n \overset{a.s.}{\to} X$. In the probability space $(\Omega, \mathfrak{F}, \mathbb{P})$, this means that

$$\mathbb{P}\left(\omega \in \Omega : \lim_{n\to\infty} X_n(\omega) = X(\omega)\right) = 1.$$

This type of convergence is already used in SLLN (Section 3.4).

If $\lim_{n\to\infty} X_n(\omega) = X(\omega)$ for all $\omega \in \Omega$, we have *sure convergence*. However, this notion is very seldom used since it is different from the notion of *almost sure convergence* only on a set of probability zero.

3.6.2 *Convergence in mean*

We say that $\{X_n\}$ converges in the rth mean toward X if

$$\lim_{n\to\infty} \mathbb{E}\left(|X_n - X|^r\right) = 0.$$

For $r = 1$, we have convergence in mean, and for $r = 2$, convergence in mean square which plays an important role in stochastic processes and stochastic calculus. We have the following inclusions.

3.6.3 *Relations between types of convergence*

(Convergence in rth mean) $\to$ (Convergence in sth mean, $r \geq s$)

$\downarrow$

(Convergence almost surely) $\to$ (Convergence in probability)

$\downarrow$

(Convergence in distribution).

The above theorems give the justification why the normal is used when the sample size is large, usually larger than 30 for very asymmetric parent distributions, and why the normal model is appropriate when in the presence of a large number of unknown variables.

3.7 Approximations

The importance of the central limit theorem cannot be overemphasized.

Another practical consequence of the central limit theorem is that certain distributions can be approximated by the normal distribution by using the proximity of a distribution to its limit. If $\{X_i\}_{i=1}^{\infty}$ have same finite mean and finite variance, $U_n = (S_n - n\mu)/(\sigma\sqrt{n})$ converges in distribution to $N(0,1)$. Equivalently,

$$Z_n = \left(\sqrt{n}\frac{\bar{X}_n - \mu}{\sigma}\right) \xrightarrow{d} N(0,1).$$

For example, by matching the mean and variance of the distribution with the approximating normal, we have the following:

- The binomial distribution, Bin(n,p), is approximately normal $N(np, np(1-p))$ for large n and p not too close to zero or one.
- The Poisson(λ) distribution is approximately normal $N(\lambda, \lambda)$ for large values of λ.

Since these two distributions are discrete, there is some correction for continuity to be done when using the normal.

- The chi-square distribution χ_k^2 is approximately normal $N(k, 2k)$ for large k.
- The Student's t-distribution with ν dof, $t(\nu)$, is approximately normal $N(0,1)$ when ν is large.

Whether these approximations are sufficiently accurate depends on the purpose for which they are needed and the rate of convergence. It is typically the case that such approximations are less accurate in the tails of the distribution. A general upper bound for the approximation error in the CLT is given by the Berry–Esseen theorem (see Section 3.9).

3.8 Central Limit Theorem for a Bivariate Distribution

The CLT that we have seen earlier can be generalized to any dimension. To avoid notation complexities, let's treat the bivariate case.

Let $\{X_i, Y_i\}_{i=1}^{\infty}$ be iid random vectors, with $\mathbb{E}(X_i) = \mathbb{E}(Y_i) = 0$, $i = 1, 2, \ldots$, and same variances and correlation coefficient σ_1^2, σ_2^2, ρ. From the univariate case, we know that the cdf of

$$\left(\frac{X_1 + \cdots + X_n}{\sigma_1/\sqrt{n}} \right) \xrightarrow{d} \Phi(.)$$

and similarly for

$$\left(\frac{Y_1 + \cdots + Y_n}{\sigma_2/\sqrt{n}} \right).$$

Then in $\mathbb{R}^2$, we have

$$\left(\frac{X_1 + \cdots + X_n}{\sigma_1/\sqrt{n}}, \frac{Y_1 + \cdots + Y_n}{\sigma_2/\sqrt{n}} \right) \xrightarrow{d} W \sim N_2(0, 0; 1, 1; \rho),$$

where $N_2(0,0;1,1;\rho)$ is the standard bivariate normal distribution (see Thomasian (1969)).

Again, there are generalizations of this theorem, with less conditions imposed on the variances and the coefficient of correlation. For example, we have the following.

3.9 Limit Theorems and Distributions in the Multivariate Case

Similar to the univariate case, we have limit theorems for a sample from a multivariate distribution. In particular, Varadarajan (1958) proved this existence result.

Theorem 3.2. *Let F_n be the distribution function of the random vector $\mathbf{X}_n = (X_{1,n}, \ldots, X_{p,n})$ and $F_{\lambda n}$ be the distribution function of the linear function $\sum_{j=1}^{p} \lambda_j X_{jn}$. A necessary and sufficient condition that F_n converges to a p-variate df is that $F_{\lambda n}$ converges to a limit for each vector $\lambda = (\lambda_1, \ldots, \lambda_p)$.*

3.10 The Berry–Esseen Theorems

These theorems give the errors between the limit and the approximating sum, based on very simple hypotheses. They are as follows.

Theorem 3.3 (Case of identically distributed rv's). *Let $\{X_j\}_{j=1}^{\infty}$ be iid random variables with zero mean, same finite variance σ^2, and same finite third absolute moment ρ. Let $Y_n = \sum_{j=1}^{n} X_j/n$, $\Phi(.)$ be the cdf of $Z \sim N_1(0,1)$ and F_n be the cdf of $Y_n \sqrt{n}/\sigma$. Then we have*

$$|F_n(x) - \Phi(x)| \leq \frac{C\rho}{\sigma^3 \sqrt{n}}.$$

The constant C has been determined as less than 0.4748 in the latest officially recognized result (2012).

Theorem 3.4 (Case of non-identically distributed rv's). *Let* $\{X_j\}_{j=1}^{\infty}$ *be random variables with zero mean, finite variances* σ_j^2, *and finite third absolute moment* $\mathbb{E}(|X_j|^3) = \rho_j$. *Then, setting*

$$S_n = \frac{\sum_{j=1}^{n} X_j}{\sqrt{\sum_{j=1}^{n} \sigma_j^2}}, \quad \Theta_1 = \underset{1 \leq j \leq n}{Max} \left\{ \frac{\rho_j}{\sigma_j^2} \right\} \bigg/ \sqrt{\sum_{j=1}^{n} \sigma_j^2},$$

$$\Theta_2 = \left(\sum_{j=1}^{n} \rho_j \right) \bigg/ \left(\sum_{j=1}^{n} \sigma_j^2 \right)^{3/2},$$

we have two inequalities: Berry (1941)

$$|F_n(x) - \Phi(x)| \leq K_1 \Theta_1,$$

and, similarly, Esseen (1942)

$$|F_n(x) - \Phi(x)| \leq K_2 \Theta_2,$$

with $0.5600 \geq K_2 \geq \frac{\sqrt{10}+3}{6\sqrt{2\pi}} = 0.409 \ldots.$

Note: We have $\Theta_2 \leq \Theta_1$, and when X_j are identical, we have

$$\Theta_1 = \Theta_2 = \frac{\rho}{\sqrt{n}\sigma_1^3},$$

with $\mathbb{E}(|X_j|^3) = \rho$.

3.11   Limiting Spectral Distribution

In the last 60 years, there have been some important new concepts in convergence, motivated by the fact that in statistics, there could now be more variables than observations. Also, in theoretical physics, random matrices can converge in ways different from the classical ones (see also Random Matrix Theory in Chapter 13).

So, whereas in the classical multivariate analysis, we are interested in the convergence of the sample matrix of covariance toward the population covariance matrix, here we are concerned with another mode of convergence. Let $\mathbf{X}$ be a random vector, normal or not, in $\mathbb{R}^p$

and $\{\mathbf{X}_i\}_{i=1}^n$, a random sample from $\mathbf{X}$, with $\mathbf{X}_j = (X_{1j}, \ldots, X_{pj})^t \sim N_p(\boldsymbol{\mu}, \boldsymbol{\Sigma})$, X_{ij} being iid with mean zero and variance σ^2. We define the sample covariance matrix

$$\mathbf{S}^* = \frac{1}{n} \sum_{j=1}^n \mathbf{X}_j \mathbf{X}_j^t,$$

whereas the usual sample covariance matrix is

$$\mathbf{S} = \frac{1}{n} \sum_{j=1}^n (\mathbf{X}_j - \bar{\mathbf{X}})(\mathbf{X}_j - \bar{\mathbf{X}})^t.$$

$\mathbf{S}^*$ is $(p \times p)$ symmetric, with eigenvalues $\ell_1 \geq \cdots \geq \ell_p$. We consider the empirical distribution of these roots:

$$F_n(x) = \frac{\#(\ell_i : \ell_i \leq x)}{p},$$

where $\#(.)$ denotes the number of elements of the set. If $F_n(x)$ converges to F as $p \to \infty$, we call F the limiting spectral density (LSD). We have the following result.

Theorem 3.5. *Let F_n be the spectral distribution of the sample covariance matrix $\mathbf{S}^*$. Let $n \to \infty$ s.t. $p/n \to c$. Then $F_n \overset{a.s.}{\to} F$ under the following conditions on the limit c:*

(a) *If $0 < c < 1$, we have*

$$f(x) = \frac{dF(x)}{dx} = \begin{cases} \dfrac{\sqrt{(x-a)(b-x)}}{(2\pi)\, c\sigma^2 x}, & a < x < b, \\ 0, & \text{otherwise,} \end{cases}$$

  with $a = \sigma^2(1 - \sqrt{c})^2$, $b = \sigma^2(1 + \sqrt{c})^2$.
(b) *If $1 < c < \infty$, we have*

$$f(x) = \frac{dF(x)}{dx} = \begin{cases} \dfrac{\sqrt{(x-a)(b-x)}}{(2\pi)\, c\sigma^2 x}, & a < x < b, \\ 1 - \dfrac{1}{c}, & x = 0. \\ 0, & \text{otherwise} \end{cases}$$

The above results were obtained by Marcenko and Pastur (1967).

Concerning the law of large numbers and the central limit theorem, a different aspect is given by the following results on the asymptotic distribution of $\left(\mathrm{tr}S^*, \mathrm{tr}\left\{(S^*)^2\right\}\right)$ under a high-dimension framework.

Theorem 3.6. *Let S^* be the above sample covariance matrix, with $N = n + 1$ being the sample size, taken from $N_p(\boldsymbol{\mu}, \boldsymbol{\Sigma})$. Let us consider the 2-vector $\left(\mathrm{tr}S^*, \mathrm{tr}\left\{(S^*)^2\right\}\right)$ and make the following simplified assumptions:*

$$p/n \to c \in (0, \infty),$$

$$k = 1, 2, 3, 4 \to \frac{\mathrm{tr}\boldsymbol{\Sigma}^k}{p} = O(1).$$

Also, for the eigenvalues of $\boldsymbol{\Sigma}$, $\lambda_1 \geq \cdots \geq \lambda_p$, let

$$\alpha = \sum_{j=1}^{p} \lambda_j/p, \quad \delta^2 = \sum_{j=1}^{p} (\lambda_j - \alpha)^2/p,$$

i.e. their mean and variance, then

(a) ***Spectral law of large numbers:***

$$\mathrm{tr}(S^*)/p \xrightarrow{p} \alpha, \quad \text{and} \quad \mathrm{tr}\left\{(S^*)^2\right\}/p \xrightarrow{p} (1 + c)\alpha + \delta^2.$$

(b) ***Spectral central limit theorem:*** *Under some mild conditions, we have, for the 2-vector,*

$$\mathbf{V} = \begin{pmatrix} \dfrac{\mathrm{tr}S^*}{p} - \alpha \\[2mm] \dfrac{\mathrm{tr}\left\{(S^*)^2\right\}}{p} - \dfrac{n + p + 1}{n}\alpha^2 \end{pmatrix},$$

its convergence in distribution toward the 2-normal vector:

$$n\mathbf{V} \xrightarrow{d} N_2\left(\begin{pmatrix} 0 \\ 0 \end{pmatrix}, \begin{bmatrix} \gamma_{11} & \gamma_{12} \\ \gamma_{21} & \gamma_{22} \end{bmatrix}\right),$$

with $\gamma_{11} = \dfrac{2\alpha^2}{c}$, $\gamma_{12} = \gamma_{21} = 4\alpha^3\left(1 + c^{-1}\right)$, $\gamma_{22} = 4\alpha^4 \left(2c^{-1} + 5 + 2c\right)$.

For more details, see Fujikoshi *et al.* (2010).

We conclude this chapter with two applications which are simple but useful.

3.12 Applications

3.12.1 *Die rolling*

A balanced die is rolled and we have a success when the face shown is • (or one). We want the frequency of success to be real close to $1/6$, say differing from $1/6$ by not more than 0.01. We know that the more times we roll the die, the closer to $1/6$ is the probability of success.

(1) *How many times do we have to roll the die to be at least 95% confident of the result?*

By using the *Bernoulli law of large numbers*. We have

$$\mathbb{P}\left(|f_n - \pi| \leq \varepsilon\right) \geq 1 - \eta, \quad f_n = \frac{S}{n} \Rightarrow n \geq \frac{\pi(1-\pi)}{\varepsilon^2 \eta}.$$

Here, $\eta = 0.05$, $\varepsilon = 0.01$, and hence,

$$n \geq \frac{1}{6}\frac{5}{6}\frac{1}{(0.01)^2(0.05)} \approx 28,000.$$

So, we have to throw the die that high a number of independent rolls, if we use *Chebychev inequality*, which applies to any distribution.

(2) *Same question, now using the binomial distribution approximated by the normal.*

We have, again,

$$\mathbb{P}\left(\left|\frac{S}{n} - \frac{1}{6}\right| \leq \frac{1}{100}\right) \geq 0.95 \Rightarrow n \geq?.$$

Using normal approximation,

$$\frac{S}{n} \approx N\left(\mu = \frac{1}{6}, \sigma^2 = \frac{1}{6}\frac{5}{6}\Big/ n\right),$$

we have the inequality

$$\Phi\left(0.06\sqrt{\frac{n}{5}}\right) \geq 0.975,$$

and using normal tables, we obtain $0.06\sqrt{n/5} \geq 1.96$ or $n \geq 5335$, a much lower number.

3.12.2   *The binomial distribution and multiple choice tests*

Most teachers have had the chance (or malchance?) of teaching large classes of several hundred students in a huge amphitheater. Although supported by a few graduate assistants, when exam times come, they, undoubtedly, find that marking and grading papers require some herculean work.

So, multiple choice tests, marked by computer, seem to be a solution. But how can we plan the exams, and the answers, so that students who just guess the answers and mark them on their answer sheets, have a very low probability of passing the course? Undoubtedly, producing a test that is suitable requires careful planning.

We consider, for example, a test with 20 independent questions, where for each question students have to choose one of the 10 answers given. We suppose then that each question answered correctly is worth 5 points and students pass the test only if they have at least 55 points. The probability that a "guessing student" answers rightly one question is $1/10$. Let S be the rv giving the number of right answers. We have S as a binomial variable,

$$S \sim \text{Bin}(20, 0.10), \quad n = 20, \ p = 0.01.$$

The probability that he/she passes the test is then

$$\mathbb{P}(S \geq 11) = \sum_{k=11}^{20} C(20, k)(0.10)^k (0.90)^{20-k} = 7.088606332 \times 10^{-7}.$$

This is probably too low, and we wish to choose n and p so that

$$\mathbb{P}(\text{passing test by guessing}) \leq 0.10.$$

Using the normal approximation, we have $S \cong N(np, npq)$ if np, $nq > 5$.

If we have $n = 50$ questions with 10 answers each ($p = 0.10$), each good answer is worth 2 points and the passing test will have at least 28 questions with good answers. Here, we can use normal approximation $S \cong N(np, npq) = N(5, 4.5)$ since $np = 5$ and $nq = 45$. We

Table 3.1   Probability of passing by guessing with 20 questions, with variable p.

$n = 20$			
$p = 1/10$	$7.088606332 \times 10^{-7}$	$p = 1/5$	0.0005634136977
$p = 1/8$	0.000006568233415	$p = 1/4$	0.003942141664
$p = 1/7$	0.00002416107793	$p = 1/3$	0.03763657109
$p = 1/6$	0.0001050193565	$p = 1/2$	0.4119014740

have now

$$\mathbb{P}\left(S \geq 28\right) \cong \mathbb{P}\left(S \geq 27.5\right) = \mathbb{P}\left(Z \geq \frac{27.5 - 5}{\sqrt{4.5}} = 10.6\right) = 0.00,$$

which is acceptable, but this scenario requires making up 50 questions with 10 answers each, which is a lot of work.

Suppose that we decide to have 20 questions. Then Table 3.1 provides the related probabilities. *Hence, 20 questions with 3 answers each seems to be a good compromise solution.* We can also see that going from two to three answers lowers the probability of passing the course by guessing by more than tenfold.

If we set $n = 40$, $p = 1/8$ (8 answers instead of 10), we need at least 22 good answers and a similar probability of $\mathbb{P}\left(S \geq 22\right)$ can be computed by normal approximation.

By varying n and p, but keeping the passing grade for at least 55%, we can find a good combination using Table 3.2:

$$\mathbb{P}\left(S \geq 0.55n\right) = \sum_{k=\lceil 0.55n \rceil}^{n} \frac{n! p^k (1-p)^{n-k}}{(n-k)! k!},^{[1]}$$

where $\lceil 0.55n \rceil$ denotes the smallest integer greater than or equal to $0.55n$.

---

[1]The reader is invited to use this formula to find the best combination (number of questions and number of answers given in each question).

Table 3.2   Probability of passing by guessing, with variables (n, r).

n	$p = 0.1$	$p = 0.2$	$p = 0.25$	$p = 0.5$
$n = 10$	0.0001469026000	0.006369382400	0.01972770691	0.3769531250
$n = 15$	0.000002846482453	0.0007849853911	0.004193014464	0.3036193848
$n = 20$	0	0.0005634136974	0.003942141665	0.4119014740
$n = 25$	0	0.00007629717024	0.0009158314411	0.3450189827
$n = 30$	0	0.00001047205480	0.0002156938259	0.2923323559
$n = 35$	0	0.000001451784689	0.00005131686442	0.2497799166
$n = 40$	0	0.000001059503016	0.00004863915837	0.3179140013
$n = 45$	0	0	0.00001189686478	0.2757421651
$n = 50$	0	0	0.000002916826240	0.2399438310
$n = 60$	0	0	0	0.2594790016
$n = 70$	0	0	0	0.2014815376
$n = 80$	0	0	0	0.2170211278
$n = 90$	0	0	0	0.1714165597
$n = 100$	0	0	0	0.1841008086

3.13   Conclusion

The notions presented in this chapter have had a deep influence on classical statistics, modern statistical simulation and applied probability models of all types. Large sample theory has found applications in other domains as well, and simulation has become a powerful tool whenever analytic methods cannot be applied. Gambling theory and the planification of games of chance for profit rely on the SLLN, for their development, and have proven to be very successful (see also Chapter 8). New concepts of spectral convergence and limits are beginning to make headway into classical probability and statistics.

Bibliography

Chung, K.L. (1974). *Elementary Probability Theory with Stochastic Processes* (Springer-Verlag, New York).

Fujikoshi, Y., Ulyanov, V.V. and Shimizu, R. (2010). *Multivariate Statistics, High-Dimensional and Large-Sample Approximations* (Wiley, New York).

Marcenko, V.A. and Pastur, L.A. (1967). Distribution of eigenvalues in certain sets of random matrices, *Matematicheskii Sbornik*, **114**(4), 507–536.

Patel, J. and Read, C. (1996). *Handbook of the Normal Distribution* (Marcel Dekker, New York).

Thomasian, A. (1969). *The Structure of Probability Theory with Applications* (McGraw-Hill, New York).

Varadarajan, V.S. (1958). A useful convergence theorem, *Sankhya*, **20**, 221–222.

Chapter 4

Distributions Derived from the Normal

4.1 Introduction

Several well-known distributions are obtained from the normal by unitary or binary operations. Some of these operations are simple, but others are much more elaborate. Some derived distributions have important applications in probability/statistics. The general beta distribution, for example, is one such case. Defined on a finite interval, with only two parameters for definition, it is very convenient to use. Numerous questions and requests for precision, coming from researchers in applied fields using our AMSARG research results on the operations on these betas, bear witness to the usefulness of this family in one variable. A complete volume on the univariate beta is available (see Gupta and Naradajah (2004)). Extensions to several variables, and to matrix variates, start to have applications too, and they are presented in further chapters of this book. In this chapter, we start with the log-normal distribution and give its main properties. We give the densities of rv's obtained from the univariate normal by unitary operations and then by binary operations. More advanced functions, such as Meijer G-functions and Fox H-functions, are presented and their uses in the derivation of the densities of products of more than two normal rv's, and ratio of two normal variables, are presented. We have to deal with matrix variate distributions here and will use matrix notations and operations whenever convenient presentation requires them.

4.2   Overview of Approaches Used to Obtain Derived Distributions

Methods used by authors to derive a new density include the following:

(1) **Multivariate calculus approach using the Jacobian of the transformation:** This direct approach is often the most effective.

(2) **Characteristic function or moment-generating function:** Using the rich properties of these functions, we work in the complex domain before going back to densities (for the chf). Identification with well-known expressions is usually made.

(3) **Mellin transforms:** This method applies only to products and ratios, i.e. $Y = X_1 X_2$, $W = X_1/X_2$. We consider the $(p-1)$-th moment of X_1 and X_2 and use the identity for independent rv's: $\mathbb{E}\left(Y^{p-1}\right) = \mathbb{E}\left(X_1^{p-1} X_2^{p-1}\right) = \mathbb{E}\left(X_1^{p-1}\right) \mathbb{E}\left(X_2^{p-1}\right)$. By using inverse Mellin transform (see Chapter 13), we have $\varphi(y) = \frac{1}{2\pi i} \int_\Omega y^{-p} \mathbb{E}\left(Y^{p-1}\right) dy$. The theorem of residues is usually applied at this last step to derive the expression of $\varphi(y)$ in terms of common functions, hence avoiding the use of unreliable numerical computation. For the product of two positive rv's, $W = X.Y$, we can use the Mellin convolution of the two densities:

$$f(w) = \int\limits_0^\infty \frac{1}{y} f_1\left(\frac{w}{y}\right) f_2(y)\, dy = \int\limits_0^\infty \frac{1}{x} f_2\left(\frac{w}{x}\right) f_1(x)\, dx.$$

Similarly, for the quotient $T = X/Y$, we have

$$g(t) = \int\limits_0^\infty \frac{1}{(1/y)} f_1\left(\frac{t}{1/y}\right) f_2(y)\, dy = \int\limits_0^\infty y f_1(yt) f_2(y)\, dy.$$

(4) **Fourier and Laplace transforms:** They apply rather to linear combinations of rv's. Other integral transform methods can be effective too.

(5) **G- and H-functions:** Based on Mellin–Barnes integral and Mellin transforms, these functions provide powerful tools to derive densities for products, ratios, and powers of two, or several, positive rv's, and also for their cumulative distribution

functions (cdf). We note that this is an extension of method (3) since these two functions are generalizations of hypergeometric functions (see Pham-Gia and Dinh (2016)).

(6) Differential forms and exterior algebras (usually with method (1)).

(7) Other approaches (cumulative distribution function, limit theorems) or a combination of these methods.

(8) **Monte Carlo and quasi-Monte Carlo simulations:** This is often the last resort, but very useful to deal with untractable mathematical expressions.

4.3 Probability Laws with Density Expressions Close to the Normal

The following laws have bell-shaped densities and/or density expressions close to the normal:

(a) The Cauchy distribution
The standard Cauchy has density,

$$f(x) = \frac{1}{\pi(1+x^2)}, \quad -\infty < x < \infty, \tag{4.1}$$

while the generalized Cauchy, $X \sim \mathrm{CAU}(\alpha, \mu)$, has density,

$$f(x) = \frac{\alpha}{\pi\left(\alpha^2 + (x-\mu)^2\right)}, \quad -\infty < x, \mu < \infty, \quad \alpha > 0. \tag{4.2}$$

Its mean and variance do not exist.

(b) Inverse Gaussian (or Wald's distribution)
For $X \sim \mathrm{IGD}(\mu, \lambda)$, $x > 0$, μ, $\lambda > 0$,

$$f(x; \mu, \lambda) = \sqrt{\frac{\lambda}{2\pi x^3}} \exp\left(-\frac{\lambda(x-\mu)^2}{2\mu^2 x}\right),$$

where λ is the shape parameter and μ the location parameter. Its mean and variance are μ and μ^3/λ, respectively.

(c) The half-normal distribution $W \sim \mathrm{HNL}(0, \sigma^2)$
We start with a normal variable with mean 0 and consider the positive part of its density. Two times that part gives the density on

$(0, \infty)$ of the HNL $(0, \sigma^2)$. Hence, we have

$$f(x) = \frac{2}{\sigma\sqrt{2\pi}} \exp\left(-\frac{x^2}{2\sigma^2}\right), \quad x \geq 0,$$

and

$$F(x) = \text{erf}\left(\frac{\theta x}{\sqrt{\pi}}\right), \theta^2 = \frac{\pi}{2\sigma^2}.$$

Its mean is $\mathbb{E}(X) = \sigma\sqrt{\frac{2}{\pi}}$ and its variance $\frac{\pi-2}{\pi}\sigma^2$.

(d) The folded normal distribution

For $Y \sim N(\mu, \sigma^2)$, the distribution of $X = |Y|$ is called the folded normal, $X \sim \text{FNL}(\mu, \sigma^2)$, since its density is obtained by folding the negative side on the positive side. If $\mu = 0$, we have the half-normal studied above.

It can be seen that if $\mu > 0$ is large, folding will have little effect on the normal density, but if it is small or negative, this effect will be important. We have

$$f(x) = \frac{1}{\sigma\sqrt{2\pi}}\left\{\exp\left(-\frac{(x-\mu)^2}{2\sigma^2}\right) + \exp\left(-\frac{(x+\mu)^2}{2\sigma^2}\right)\right\}, \quad x \geq 0.$$

(e) The Laplace distribution

A random variable has a Laplace distribution, $X \sim \text{LAP}(\mu, a), a > 0$ distribution if its pdf on $(-\infty; \infty)$ is

$$f(x) = \frac{1}{2a}\exp\left(-\frac{|x-\mu|}{a}\right).$$

Its mean, median, and mode are all μ and its variance $2a^2$. Its mgf is

$$M(t) = \frac{\exp(\mu t)}{1 - a^2 t^2}, \quad |t| < \frac{1}{a}. \tag{4.3}$$

The Laplace has fatter tails than the normal. The difference between the two densities is in the absolute value expression, which replaces the square one, in the exponent.

(f) The Student *t*-distribution

Perhaps the distribution which is closest to the normal, for the bell shape of its density as well as for its limit when the degree of freedom increases, is the Student *t*-distribution.

The standard Student with $n \geq 1$ dof, $t \sim \text{SST}(n)$, has density

$$f_n(t) = \frac{1}{\sqrt{n}\text{Beta}(n/2, 1/2)} \left(1 + \frac{t^2}{n}\right)^{-\frac{n+1}{2}}, \quad -\infty < t < \infty.$$

(4.4)

The general Student, $t \sim \text{GST}(\nu, \mu, \sigma)$, $\nu, \sigma > 0$, has density

$$g(t) = \frac{1}{\sigma\sqrt{\nu}\text{Beta}(\nu/2, 1/2)} \left(1 + \frac{1}{\nu}\left(\frac{t-\mu}{\sigma}\right)^2\right)^{-\frac{\nu+1}{2}},$$

$$-\infty < t < \infty.$$

(4.5)

We have $\mathbb{E}(T) = \mu$ for $\nu > 1$ and $\text{Var}(T) = \frac{\nu}{\nu-2}\sigma^2$ for $\nu > 2$.

For $\nu = 1$, it reduces to the Cauchy distribution while for $\nu \to \infty$, it converges to the normal distribution. For $\mu = 0$, $\sigma = 1$, it reduces to the standard t. Its density has fatter tails than the standard normal.

4.4 Distributions Obtained by Performing Operations on the Normal Variable

4.4.1 *Operations on a single normal random variable*

These operations are as follows: exponentiation, inverse, square power, absolute value, truncation, etc. We just give the main information on these distributions. Details can be found in Ahsanullah *et al.* (2014).

4.4.1.1 *Log-normal distribution*

(a) Origin and applications

X is a direct transform of the normal Y, i.e. $X = \exp(Y)$.

A positive rv is log normal, $X \sim \text{LN}(.)$, if its logarithm $Y = \ln(X)$ is normal (hence it is the exponential of a normal). A variable might be modeled as log normal if it can be thought of as the product of many independent random variables each of which is positive, just like the normal is the sum of many rv's.

There are numerous applications of this distribution found in the health sciences, such as the *length* of *inert* appendages (hair, claws, nails, and teeth), physiological measurements, such as blood pressure and adult measures (length, height, skin area, and weight).

The ongoing COVID-19 outbreak in March 2020 shows the applicability of the log normal. Olsson (2020) found that data for new confirmed infections for Hubei, Hubei outside Wuhan, China excluding Hubei, as well as Zhejiang and Fujian provinces, all follow a log-normal distribution that can be used to make a rough estimate of the date of the last new confirmed cases in the respective areas. In engineering, the log-normal distribution is used to analyze extreme values of monthly and annual maximum values of daily rainfall and river discharge volumes.

(1) In reliability analysis, the log-normal distribution is often used to model times to repair a maintainable system. It has been proposed that coefficients of friction and wear may be treated as having a log-normal distribution. We present in Chapter 5 the stress–strength reliability model, with both *stress* and *strength* being log-normally distributed.
(2) In finance, in particular in the Black–Scholes model, changes in the logarithm of exchange rates, price indices, and stock market indices are assumed normal.
(3) The logarithm of city-size distribution is normally distributed. There is also evidence of log normality in the firm size distribution of Gibrat's law.
(4) In wireless communication, the local-mean power expressed in logarithmic values has a normal distribution. If $\log_a (Y)$ is normally distributed, then so is $\log_b (Y)$ for any two positive numbers $a, b \neq 1$.

(b) Density and cumulative distribution function of $X \sim LN(\mu, \sigma^2)$

$$f_X \left(x; \mu, \sigma^2\right) = \frac{1}{x\sigma\sqrt{2\pi}} \exp\left(-\frac{(\ln x - \mu)^2}{2\sigma^2}\right), \quad x > 0,$$

$$F_X \left(x; \mu, \sigma^2\right) = \frac{1}{2}\operatorname{erfc}\left[-\frac{\ln x - \mu}{\sigma\sqrt{2}}\right] = \Phi\left(\frac{\ln x - \mu}{\sigma}\right),$$

where erfc is the complementary error function and Φ is the standard normal cdf.

(c) Mean and variance

Writing $X = \exp(\mu + \sigma Z)$, where μ and σ are the mean and standard deviation, of the variable's natural logarithm, with Z a standard normal variable, we have

$$\mathbb{E}[X] = \exp\left(\mu + \frac{1}{2}\sigma^2\right),$$

$$\mathrm{Var}[X] = \left[\exp(\sigma^2) - 1\right]\exp(2\mu + \sigma^2),$$

$$\mathrm{s.d.}[X] = \sqrt{\mathrm{Var}[X]} = \exp\left(\mu + \frac{1}{2}\sigma^2\right)\sqrt{\exp(\sigma^2) - 1},$$

$$\mathrm{Mode}[X] = \exp(\mu - \sigma^2),$$

$$\mathrm{Med}[X] = \exp(\mu).$$

Equivalently, parameters μ and σ can be obtained if the expected value and variance are known:

$$\mu = \ln(\mathbb{E}[X]) - \frac{1}{2}\ln\left(1 + \frac{\mathrm{Var}[X]}{(\mathbb{E}[X])^2}\right),$$

$$\sigma^2 = \ln\left(1 + \frac{\mathrm{Var}[X]}{(\mathbb{E}[X])^2}\right).$$

(d) The characteristic function

The characteristic function $\mathbb{E}[\exp(itX)]$ has a number of representations. The integral itself does not converge for complex t with $\mathrm{Im}(t) \leq 0$. Its Taylor formal series

$$\varphi_X(t) = \sum_{n=0}^{\infty} \frac{(it)^n}{n!}\exp\left(n\mu + \frac{n^2\sigma^2}{2}\right)$$

diverges.

The moment-generating function for the log-normal distribution does not exist on $\mathbb{R}$, but only exists on the half-interval $(-\infty, 0]$. A log-normal distribution is not uniquely determined by its moments $\mathbb{E}(X^k)$, $k > 1$. In fact, there is a whole family of distributions with the same moments as a log-normal distribution.

(e) Maximum likelihood estimation of parameters

Hence, using the formulas for the normal distribution, we have

$$\hat{\mu} = \frac{\sum_k \ln x_k}{n}, \quad \hat{\sigma}^2 = \frac{\sum_k (\ln x_k - \hat{\mu})^2}{n}.$$

(f) Generating log-normally distributed random variates

Given $Z \sim \mathrm{LN}(0,1)$, a random variate Z drawn from this normal distribution, then the variate $X = \exp(\mu + \sigma Z)$ is such that $X \sim \mathrm{LN}(\mu, \sigma^2)$.

(g) Related distributions

(1) If $X_j \sim \mathrm{LN}\left(\mu_j, \sigma_j^2\right)$, $j = 1, \ldots, n$ are independent log-normally distributed variables, and $Y = \prod_{j=1}^{n} X_j$, then Y is also distributed log-normally, $Y \sim \mathrm{LN}(\sum_{j=1}^{n} \mu_j, \sum_{j=1}^{n} \sigma_j^2)$.

(2) If $X \sim \mathrm{LN}\left(\mu, \sigma^2\right)$, then $aX \sim \mathrm{LN}\left(\mu + \ln a, \sigma^2\right)$, for $a > 0$, $\frac{1}{X} \sim \mathrm{LN}\left(-\mu, \sigma^2\right)$ and $X^a \sim \mathrm{LN}\left(a\mu, a^2\sigma^2\right), a \neq 0$.

4.4.1.2 *Square of the standard normal: the chi-square distribution*

If $Z \sim N(0,1)$, its square Z^2 has the chi-square distribution with 1 dof, or $Z^2 \sim \chi_1^2$.

For $X \sim N\left(\mu, \sigma^2\right)$ the square of X/σ has the non-central chi-square distribution with 1 dof and non-centrality parameter $\theta = \mu^2/\sigma^2$: $X^2/\sigma^2 \sim \chi_1^2(\theta)$.

We have the sum of n independent chi-squares χ_1^2 as the variable χ_n^2 with n dof and, consequently for independent chi-squares, we have $\chi_{k_1}^2 + \chi_{k_2}^2 = \chi_{k_1+k_2}^2$. More details on the chi-square is given in Section 4.9.

4.4.1.3 *Truncated normal distributions*

A density can be truncated on the left, on the right, or on both sides. The truncated variable is then defined on the remaining interval by using the same expression of the density, but with a new normalizing constant.

For the normal which is defined on $(-\infty, \infty)$, truncation permits us to focus our attention on a smaller interval. Details on truncated normal can be obtained in Ahsanullah *et al.* (2014), where other lesser used distributions are also presented (skew-normal, generalized normal, etc.).

4.4.2 *Operations on two (or more) independent normal variables*

(1) If $Z_1, Z_2 \overset{\text{iid}}{\sim} N(0,1)$, (with iid meaning independent and identically distributed), then $Z_1 \pm Z_2 \sim N(0,2)$.

(2) $\frac{Z_1}{Z_2} \sim \text{CAU}(1,0)$ (standard Cauchy).

(3) $X_1 \sim N(\sigma_1^2, 0)$, $X_2 \sim N(0, \sigma_2^2) \to \frac{X_1}{X_2} \sim \text{CAU}(\sigma_1/\sigma_2, 0)$ (general Cauchy (see Section 4.3)).

(4) Their Euclidean norm $\sqrt{X_1^2 + X_2^2}$ has the Rayleigh distribution, also known as the *chi distribution* with 2 dof (see Section 4.9).

(5) If $Z_1, Z_2, \ldots, Z_n \overset{\text{iid}}{\sim} N(0,1)$, then $\sum_{i=1}^n Z_i^2 \sim \chi_n^2$.

(6) If $X_1, X_2, \ldots, X_n \overset{\text{iid}}{\sim} N(\mu, \sigma^2)$, then $\bar{X} = \sum_{i=1}^n X_i/n$, and S^2 are independent, which can be demonstrated using the Helmert transformation (see Chapter 2). Their ratio will give the Student's t-distribution with $n-1$ dof, SST $(n-1)$:

$$t = \frac{\bar{X} - \mu}{S/\sqrt{n}} = \frac{\frac{1}{n}(X_1 + \cdots + X_n) - \mu}{\sqrt{\frac{1}{n(n-1)}\left[(X_1 - \bar{X})^2 + \cdots + (X_n - \bar{X})^2\right]}}$$

$$\sim t_{n-1}.$$

(7) If $X_1, \ldots, X_n; Y_1, \ldots, Y_m \overset{\text{iid}}{\sim} N(0,1)$, then the ratio of their normalized sums of squares will have the F-distribution with (n, m) dof:

$$F = \frac{(X_1^2 + X_2^2 + \cdots + X_n^2)/n}{(Y_1^2 + Y_2^2 + \cdots + Y_m^2)/m} \sim F_{n,m}.$$

(8) Let X_1, X_2 be two independent normal variables $X_i \sim N_1(\mu_i, \sigma_i^2)$, $i = 1, 2$. We have $X_1 \pm X_2 \sim N(\mu_1 \pm \mu_2, \sigma_1^2 + \sigma_2^2)$ and $AX_1 + BX_2 \sim N(A\mu_1 + B\mu_2, A^2\sigma_1^2 + B^2\sigma_2^2)$, with A and B being real constants.

(9) Convolution product to determine the density of a sum of two independent random variables:

(a) The formal approach uses the convolution (Faltung) of two densities, well known in mathematical analysis. The sum $Y = X_1 + X_2$ of two independent rv's means that a value of X_1 can be added to any value of X_2, according to the two probability distributions, resulting in a third rv Y with density

$$f(y) = \int_{-\infty}^{\infty} f_1(y - t) f_2(t) \, dt = \int_{-\infty}^{\infty} f_1(t) f_2(t - y) \, dt.$$

Similarly, for the difference $W = X_1 - X_2$, we have

$$f(w) = \int_{-\infty}^{\infty} f_1(w + t) f_2(t) \, dt = \int_{-\infty}^{\infty} f_1(t) f_2(w + t) \, dt.$$

The integration interval $(-\infty, \infty)$ is usually replaced by the range of the integration variable.

(b) For a function $y = g(x_1, x_2)$, let the *joint* distribution to be determined from $h(x_1, x_2) = f_1(x_1) f_2(x_2)$ for independent variables, with disjoint domains of definition for x_1 and x_2. We compute the joint density $\varphi(x_1, y)$ of (X_1, Y), using the formula of change of variables, $(X_1, X_2) \to (X_1, Y)$, based on Jacobian $|J|$ and by determining appropriate domains giving 1-1 transformations from $Dom(X_2)$ to $Dom(Y)$. We integrate wrt x_1 to obtain the marginal density of Y, defined in a domain to be determined.

(10) Maximum of two rv's: Occasionally, we have to deal with the rv obtained by taking the maximum, or minimum, value of two rv's. We can prove that, for X_1, X_2 independent,

$$Y = Max(X_1, X_2) \to f(y) = F_{X_1}(y) f_{X_2}(y)$$
$$+ f_{X_1}(y) F_{X_2}(y),$$

which uses the cdf of the normal variables.

(11) For the minimum, $Y = min(X_1, X_2)$, we have

$$f(y) = f_{X_1}(y) + f_{X_2}(y) - F_{X_1}(y) f_{X_2}(y) - f_{X_1}(y) F_{X_2}(y).$$

(12) For $Y = X_1^2 + X_2^2$, we have

$$f(y) = \int_{-\sqrt{y}}^{\sqrt{y}} \frac{1}{2\sqrt{y-z^2}} \left\{ f_{X_1,X_2}\left(\sqrt{y-z^2}, z\right) \right.$$

$$\left. + f_{X_1,X_2}\left(-\sqrt{y-z^2}, z\right) \right\} dz.$$

In the case of $N_1\left(0, \sigma^2\right)$ for both variables, we have $Y \sim \exp\left(2\sigma^2\right)$.

(13) For the positive square root of the above variable, $Y = \sqrt{X_1^2 + X_2^2}$, we have

$$f(y) = \int_{-y}^{y} \frac{y}{2\sqrt{y^2-z^2}} \left\{ f_{X_1,X_2}\left(\sqrt{y^2-z^2}, z\right) \right.$$

$$\left. + f_{X_1,X_2}\left(-\sqrt{y^2-z^2}, z\right) \right\} dz,$$

which, in the case of both $N(0,1)$, gives the Rayleigh distribution.

(14) Hence, for a complex rv, $W = X_1 + iX_2$, with independent $X_i \sim N\left(0, \sigma^2\right)$, $i = 1, 2$, the magnitude has a Rayleigh distribution, while the phase θ, defined by $\theta = \tan^{-1}\left(X_1/X_2\right)$, has a uniform distribution on $[-\pi/2, \pi/2]$, and Y and θ are independent (see Chapter 13).

There are some results on densities of $X+Y$, $X-Y$, XY, X/Y, when $X \sim N\left(\mu, \sigma^2\right)$ and Y has a distribution from another family. For example, when that family is the Student with ν dof, we have the following: The density of $Z = X + Y$ with $X \sim N\left(\mu, \sigma^2\right)$, $Y \sim t(\nu)$, with ν arbitrary, has been derived by Forchini (2008), extending the work of Nason (2006). It is as follows:

$$f(z) = \frac{\exp\left(-\frac{(z-\mu)^2}{2\sigma^2}\right) \Gamma\left(\frac{\nu+1}{2}\right)}{\sigma\sqrt{2\pi}\,\Gamma\left(\frac{\nu}{2}\right)\left(\frac{2\sigma^2}{\nu}\right)^{\nu/2}}$$

$$\times \sum_{j=0}^{\infty} \frac{1}{j!} \left[\frac{(z-\mu)^2}{2\sigma^2}\right]^j \psi\left(\frac{\nu+1}{2}, \frac{\nu}{2} + 1 - j, \frac{\nu}{2\sigma^2}\right),$$

where $\psi(.)$ is Tricomi hypergeometric function of the second kind. Recall that, for Kummer function of the first kind, we have

$$_1F_1(a,b;z) = \frac{\Gamma(b)}{\Gamma(b-a)\,\Gamma(a)} \int_0^1 e^{zt} t^{a-1}(1-t)^{b\ a-1} dt,$$

and for Kummer function of the second kind,

$$\psi(a,b;z) = \frac{1}{\Gamma(a)} \int_0^\infty e^{-zt} t^{a-1}(1+t)^{b-a-1} dt.$$

4.5   Meijer *G*-Functions and Fox *H*-Functions

When considering the product of more than two positive rv's we have to use more advanced mathematical tools. One such tool is the Meijer *G*-function. Fox *H*-function, which includes *G*-functions as a special case, will allow even more operations, such as computing the density of a power of an rv and its cdf. But they only concern densities with positive support (positive rv's). However, using the expression of the exponential function,

$$e^{-z} = G_0^{1\ \ 0}_{1}[z|\,0], \quad z \in \mathbb{R},$$

we can express the density of $X \sim N(\mu, \sigma^2)$ as a Meijer *G*-function distribution, i.e.

$$f(z) = \frac{1}{\sigma\sqrt{2\pi}} G_0^{1\ \ 0}_{1}[z^2|\,0], \quad z \in \mathbb{R}, \quad z^2 = \frac{(x-\mu)^2}{2\sigma^2} \geq 0.$$

H-functions and *G*-functions can be considered as generalizations of hypergeometric functions $_pF_q(.)$ (see Pham-Gia and Dinh (2016), where it is shown that they are the result of a long development process, starting with the Mellin–Barnes integral replacing the hypergeometric series). They give representations to a wide variety of univariate densities of positive rv's, and products and ratios of these densities, as well as their powers. *H*-functions can be expressed in terms of elementary functions by using the residue theorem in complex analysis, together with Jordan's lemma, and numerically computed. This is not an easy task, and fortunately, Maple and Mathematica now compute the G-function. A brief presentation of these functions is available in Chapter 13.

To use this approach, densities of positive X_1 and X_2 have to be expressed as G- or H-function densities. Products and ratios of these variables will have their densities expressed as G- or H-functions, with parameters to be determined under specific rules (see Pham-Gia (2008)). However, for sums and differences, only an approximate approach can be used. Springer (1979) has a chapter on this topic, together with a Fortran program. For example, as presented in Pham-Gia (2008), the matrix beta variate distribution of the first kind $\mathbf{B}$, with $\mathbf{B} \sim \beta_p^I(\alpha, \beta)$, has density, expressed as matrices:

$$f(\mathbf{B}) = \frac{|\mathbf{B}|^{\alpha - \frac{p+1}{2}} |\mathbf{I}_p - \mathbf{B}|^{\beta - \frac{p+1}{2}}}{\text{Beta}_p(\alpha, \beta)}, \quad 0 < \mathbf{B} < \mathbf{I}_p,$$

where $\text{Beta}_p(\alpha, \beta) = \frac{\Gamma_p(\alpha)\Gamma_p(\beta)}{\Gamma_p(\alpha+\beta)}$ is the p-dim beta function, with $\Gamma_p(a) = \pi^{p(p-1)/4} \prod_{j=1}^p \Gamma\left(a - \frac{j-1}{2}\right)$. Its determinant has a density that can be identified with the product of p independent univariate beta densities of the first kind on $[0, 1]$, i.e. $|\mathbf{B}| \sim \prod_{j=1}^p Y_j$, with $Y_j \sim \beta_1^I\left(\alpha - \frac{j-1}{2}, \beta\right)$. We have, on the other hand, the density of the product of n independent univariate betas, $U = \prod_{j=1}^n X_j$, $X_j \sim \beta_1^I(\alpha_j, \beta_j)$, as

$$g(u) = \prod_{j=1}^n \frac{\Gamma(\alpha_j + \beta_j)}{\Gamma(\alpha_j)} G_{n\,n}^{n\,0}\left[u \left| \begin{array}{c} \alpha_1 + \beta_1 - 1, \ldots, \alpha_n + \beta_n - 1 \\ \alpha_1 - 1, \ldots, \alpha_n - 1 \end{array}\right.\right],$$

$$0 < u < 1,$$

which then gives the density of $|\mathbf{B}|$.

Note: Mellin transforms and Meijer G-functions are two complementary tools to deal with products and ratios of positive random variables. For the density of a product of several central normal variates, it can be expressed as a G-function, but not when the variables become non-central (i.e. with non-zero means). But in the latter case, Mellin transform method can be used.

4.6   Product and Ratio of Two, and More, Normal Variates

4.6.1   Product

(a) Algebra of random variables

For products and ratios, there are two complications. First, the algebraic rules known with ordinary numbers do not apply to the algebra of random variables. For example, for product $C = AB$, the corresponding ratio is $D = CA^{-1}$, but it does not necessarily mean that the distributions of D and B are the same. A product is not necessarily the inverse of a ratio of random variables. A counterexample follows: Let $X_i, i = 1, 2$ be two beta variables on $[0, 1]$. Let $X_3 = X_1.X_2$ be their product. Then X_3 is also defined on $[0, 1]$ too. Let $Y = X_3/X_1$. Then Y is defined on $[0, \infty]$ and is not X_2. This fact comes from the non-uniqueness of the inverse, like in matrices.

Indeed, a peculiar effect is seen for the Cauchy distribution: The product and the ratio of two independent Cauchy distributions (with the same scale parameter and the location parameter set to zero) will give the same distribution.

To derive the densities of products and ratios of normal variables, direct Mellin transform methods applied to components f^+ and f^- of the density are often effective. In particular cases, we can use Meijer G-functions and work with closed-form results.

(b) Product of two independent normal variables

For the product, closed-form formulas are available in the case of two independent normal variables with zero means, $X_i \sim N\left(0, \sigma_i^2\right), i = 1, 2$. Let $Y = X_1 X_2$. We have

$$f\left(y\right) = \frac{1}{\pi\sigma_1\sigma_2} K_0\left(\frac{y}{\sigma_1\sigma_2}\right), \tag{4.6}$$

where K_0 is the modified Bessel function of second kind.

The proof uses the chf of Y, which can be found to be

$$\varphi_Y\left(t\right) = \frac{1}{\sqrt{1 + t^2(\sigma_1\sigma_2)^2}}.$$

Inverting it, we have

$$f(y) = \frac{1}{\pi} \int_0^\infty \frac{\cos ty}{\sqrt{1 + t^2(\sigma_1\sigma_2)^2}} dt = \frac{1}{\pi\sigma_1\sigma_2} K_0\left(\frac{y}{\sigma_1\sigma_2}\right).$$

Hence, for the product of two standard normal variables, we have $\sigma_1 = \sigma_2 = 1$ and $f(z) = \frac{K_0(z)}{\pi}$.

(c) Density of the product of n independent central normal variables

Let $X_i \overset{iid}{\sim} N(0, \sigma_i^2)$, $H = \prod_{i=1}^n X_i$. We then have

$$f(h) = \frac{1}{(2\pi)^{n/2}\sigma} G_{0\ n}^{n\ 0}\left(\frac{h^2}{2^n\sigma^2}\,\bigg|\,0,\dots,0\right), \quad \sigma = \prod_{j=1}^n \sigma_j.$$

Although Maple can numerically compute and graph this function, it is interesting to see its expression, in terms of elementary functions, that is obtained by applying the Residue theorem to its poles. We have

$$f(y) = \begin{cases} h^-(y), & -\infty < y \le 0, \\ h^+(y), & 0 < y < \infty. \end{cases}$$

Because of symmetry $h^-(y) = h^+(y)$ (see Springer (1979, p. 125)), with $h^+(y)$ having a complex expression containing $U(s) = [(s+j)\Gamma(s)]^n$ and its derivatives $U^{(K)}(s) = \frac{d^K}{ds^K}U(s)$.

Note: If the variables have the same variance $\sigma_1 = \dots = \sigma_n = a$, we have

$$f(h) = \frac{1}{(2\pi)^{n/2}a^n} G_{0\ n}^{n\ 0}\left(\frac{y^2}{2^na^{2n}}\,\bigg|\,0,\dots,0\right).$$

For $a = 1$, we have all standard normal variables, and $Y = \prod_{j=1}^n Z_j$, with the density of Y being

$$f(y) = \frac{1}{(2\pi)^{n/2}} G_{0\ n}^{n\ 0}\left(\frac{y^2}{2^n}\,\bigg|\,0,\dots,0\right).$$

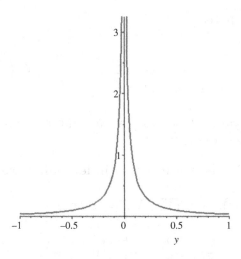

Fig. 4.1   Density of the product of six independent standard normal variables.

Figure 4.1, using Maple on

$$f(y) = \frac{1}{(2\pi)^3} G^{6\ 0}_{0\ 6}\left(\frac{y^2}{2^6}\bigg|\, 0, \ldots, 0\right),$$

shows the density of the product of six independent standard normal variables, which is exactly the same as Fig. 4.5.1 in Springer (1979, p. 129), drawn, using his very long and complex mathematical formula (4.5.44), which uses only common functions, obtained by residue theory.

However, there is no closed-form formula, when the means are different from zero. This fact is in agreement with Stein's approach in approximating the central normal with zero means, the generalization of which is made by Gaunt (2016).

For the case where the n normal variables have the same mean different from zero, i.e. $\mu_1 = \cdots = \mu_n = \mu \neq 0$, the formula might be as follows:

$$f(y) = \frac{1}{(2\pi)^{n/2}\sigma} G^{n\ 0}_{0\ n}\left(\frac{(y-\mu)^2}{2^n\sigma^2}\bigg|\, 0, \ldots, 0\right), \quad \sigma = \prod_{j=1}^{n}\sigma_j.$$

(d) Expression of the density of the product of several independent general normal variates

The product of several (and a fortiori of two) independent general normal distributions, with means different from zero, is the most complicated case to deal with, mostly because the variables can be defined on negative values and our integral transform methods do not apply to these cases. It does not have a closed-form solution and Springer's approach (see Springer (1979)) still seems to be the best one because a change of scale can be accommodated by the Mellin transform while a change of origin cannot. For the non-central normal case ($\mu_i \neq 0$), we have the density of the product of n non-central normal variables given by Springer (1979, p. 136).

Springer (1979) has two related approaches, but both use integral transforms that have to be performed on a computer. First, we have to split each density $f_i(x_i), i = 1, 2$ into f_i^+ and f_i^-, $i = 1, 2$, and have $f_i^- = f_i$ on $(-\infty, 0)$ and $f_i^- = 0$ on $(0, \infty)$ and similarly, $f_i^+ = 0$ on $(-\infty, 0)$ and $f_i^+ = f_i$ on $(0, \infty)$. We have $f_i(x_i) = f_i^-(x_i) + f_i^+(x_i), -\infty < x_i < \infty$. This technique was used by Epstein (1948) for the first time.

Let $Y = X_1 X_2$. We have $h(y) = \int_{-\infty}^{\infty} \frac{1}{x_1} f_2\left(\frac{y}{x_1}\right) f_1(x_1)\, dx_1$, which can now be written as follows:

$$h^-(-y) = \int_0^\infty \frac{1}{x_1} f_2^-\left(-\frac{y}{x_1}\right) f_1^+(x_1)\, dx_1$$

$$+ \int_0^\infty \frac{1}{x_1} f_2^+\left(\frac{y}{-x_1}\right) f_1^-(-x_1)\, dx_1$$

for $0 \leq y < \infty$ and

$$h^+(y) = \int_0^\infty \frac{1}{x_1} f_2^-\left(-\frac{y}{x_1}\right) f_1^-(-x_1)\, dx_1$$

$$+ \int_0^\infty \frac{1}{x_1} f_2^+\left(\frac{y}{x_1}\right) f_1^+(x_1)\, dx_1$$

for $0 \leq y < \infty$.

We can summarize the rest of the approach by saying that the remainder of the method concerns with taking the Mellin transforms and inverting them using the residue theorem to evaluate these integrals. Since Mellin transforms are defined on $(0, \infty)$, we have to redefine all functions so that they too are defined on $(0, \infty)$.

To carry out the rest of the operation, we have to find the order of each pole, which could be simple or multiple. Particular care should be given if parameters differ by integers. We now have to find the limits of expressions involving derivatives of those orders and apply the Leibniz rule on derivatives of products. The number of poles to be considered for a preset accuracy is another factor. Usually, poles on the left-hand plane (LHP) give values of the function for $0 \leq z \leq 1$, while those on the right-hand plane (RHP) give values for $1 \leq z < \infty$. Naturally, Jordan's lemma on the convergence to zero of the integral along the curve of the Bromwich path has to be satisfied, but this is usually so. Springer (1979) presented his method on several pages, involving elaborate formulas, in Section 4.5 of his book. It is not convenient to reproduce it here.

4.6.2   *Ratio of univariate normal variables*

For the ratio of two independent normal variables, Mellin transform methods can be applied, like for the product above. But Pham-Gia *et al.* (2006) have proposed a method that covers all cases, based on special functions, here Hermite functions and Kummer hypergeometric functions. It is no surprise that Hermite functions are present here in view of the derivation of the Hermite polynomials from derivatives of the normal density, as we have seen in Chapter 2.

Theorem 4.1. *Let* $X \sim N\left(\mu_X, \sigma_X^2\right)$ *and* $Y \sim N\left(\mu_Y, \sigma_Y^2\right)$ *be independent normal variates. Then* $W = X/Y$ *has density, for* $-\infty < w < \infty$,

$$f\left(w; \mu_X, \mu_Y, \sigma_X, \sigma_Y\right) = \frac{K_1}{\sigma_X^2 + w^2 \sigma_Y^2} \cdot {}_1F_1\left(1; 1/2; \theta_1(w)\right), \qquad (4.7)$$

where

$$\theta_1(w) = \frac{1}{2\sigma_X^2 \sigma_Y^2} \frac{\left(\mu_Y \sigma_X^2 + \sigma_Y^2 \mu_X w\right)^2}{\left(\sigma_X^2 + \sigma_Y^2 w^2\right)} \geq 0$$

and

$$K_1 = \frac{\sigma_X \sigma_Y}{\pi} \exp\left[-\frac{1}{2}\left\{\left(\frac{\mu_X}{\sigma_X}\right)^2 + \left(\frac{\mu_Y}{\sigma_Y}\right)^2\right\}\right].$$

Proof. Since the normal variates are defined on $(-\infty, \infty)$, it is essential to decompose the densities $f_i(x)$ into f_i^+ and f_i^-, like we did with the product.

We then use the Hermite function

$$H_{-2}(z) = \frac{z}{2}\int_0^\infty \frac{te^{-t^2}}{(t^2 + z^2)^{3/2}}dt = \int_0^\infty te^{-(t^2 + 2tz)}dt.$$

The proof then uses the family

$$PQF(\gamma, \eta, \varepsilon) = \left\{f(t; \gamma, \eta, \varepsilon) = C(\gamma, \eta, \varepsilon)\, t^\gamma \exp\left[-(\eta t + \varepsilon t^2)\right]\right\}$$

called the *power-quadratic exponential family of distributions* (see Pham-Gia and Turkkan (2005)), which contains the normal density when $\gamma = 0$.

The relation between Hermite function and Kummer confluent hypergeometric function is

$$H_{-2}(z) + H_{-2}(-z) = {}_1F_1\left(1, 1/2; z^2\right), \forall z,$$

where

$$ {}_1F_1(\alpha, \gamma; z) = \sum_{k=0}^\infty \frac{(\alpha, k)}{(\gamma, k)}\frac{z^k}{k!}, \gamma \neq 0, -1, -2, \ldots$$

with

$$(\alpha, k) = \alpha(\alpha + 1)\cdots(\alpha + k - 1) = \Gamma(\alpha + k)/\Gamma(\alpha).$$

Using these relations and other results in Pham-Gia and Turkkan (2005), we obtain the results given in the above theorem. □

Theorem 4.2. *Let* (X, Y) *be a binormal couple, i.e.* $(X, Y) \sim$ $N_2(\mu_X, \mu_Y; \sigma_X, \sigma_Y; \rho)$, *with density*

$$f(x, y) = \frac{1}{2\pi\sigma_X\sigma_Y\sqrt{1-\rho^2}} \times \exp\left[-\frac{1}{2(1-\rho^2)}\left\{\frac{(x-\mu_X)^2}{\sigma_X^2}\right.\right.$$
$$\left.\left. -2\rho\frac{(x-\mu_X)(y-\mu_Y)}{\sigma_X\sigma_Y} + \frac{(y-\mu_Y)^2}{\sigma_Y^2}\right\}\right].$$

Then the density of $W = X/Y$ *is*

$$f(w; \mu_X, \mu_Y; \sigma_X, \sigma_Y; \rho) = K_2 \frac{2(1-\rho^2)\sigma_X^2, \sigma_Y^2}{\sigma_Y^2 w^2 - 2\rho\sigma_X\sigma_Y w + \sigma_X^2}$$
$$\cdot {}_1F_1(1; 1/2; \theta_2(w)), \tag{4.8}$$

where

$$\theta_2(w) = \frac{\left[-\sigma_Y^2\mu_X w + \rho\mu_X\mu_Y(\mu_Y w + \mu_X) - \mu_Y\sigma_X^2\right]^2}{2\sigma_X^2\sigma_Y^2(1-\rho^2)\left(\sigma_Y^2 w^2 - 2\rho\sigma_X\sigma_Y w + \sigma_X^2\right)} \geq 0$$

and

$$K_2 = \frac{1}{2\pi\sigma_X\sigma_Y\sqrt{1-\rho^2}} \cdot \exp\left[-\frac{\sigma_Y^2\mu_X^2 - 2\rho\sigma_X\sigma_Y\mu_X\mu_Y + \sigma_X^2\mu_Y^2}{2(1-\rho^2)\sigma_X^2\sigma_Y^2}\right].$$

Proof. The proof is very similar to the proof of Theorem 4.1.   □

It is to be noted that Theorem 4.2 reduces to Theorem 4.1 when $\rho = 0$, i.e. when X and Y become independent. Moreover, when $\mu_X = \mu_Y = 0$, $\sigma_X = \sigma_Y = 1$, the standard binormal density, the formula reduces to $f(w; \rho) = \frac{\sqrt{1-\rho^2}}{\pi(1-2\rho w+w^2)}$, $-\infty < w < \infty$, a result also obtained by other authors.

Remark: This proof uses the Hermite function. But in Pham-Gia *et al.* (2006), another proof, not based on the Hermite function, is also given. Also, it is interesting to note that the density of W is either unimodal or bimodal, as shown by Marsaglia (1965).

Figure 4.2 shows the density of $W = X/Y$ and $T = X/Y$ for $(X, Y) \sim N_2\left(1.5, 1.0; (1.3416)^2, (1.98)^2; 0.65\right)$, while Fig. 4.3 shows the same densities when $\rho = 0$.

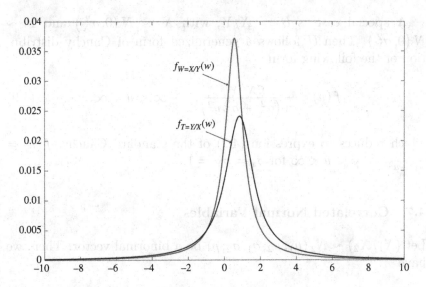

Fig. 4.2   Density of $W = X/Y$ and $T = Y/X$, with X and Y dependent $(\rho = 0.65)$.

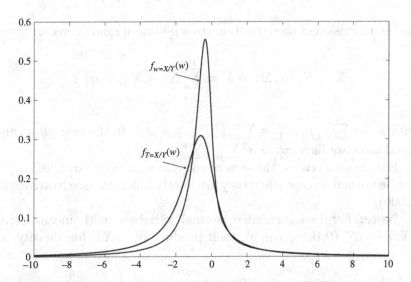

Fig. 4.3   Density of $W = X/Y$ and $T = Y/X$, with X and Y independent.

A special case is $U = X/Y$, with $X \sim N\left(0, \sigma_X^2\right)$ and $Y \sim N\left(0, \sigma_Y^2\right)$. Then U follows a generalized form of Cauchy distribution of the following form:

$$f(u) = \frac{\sigma_X \sigma_Y}{\pi \left(\sigma_X^2 + \sigma_Y^2 u^2\right)}, \quad -\infty < u < \infty, \tag{4.9}$$

which reduces to expression (4.2) of the standard Cauchy, $f(u) = \frac{1}{\pi(1+u^2)}$, $-\infty < u < \infty$ for $\sigma_x = \sigma_Y = 1$.

4.7   Correlated Normal Variables

Let $(X_1, X_2) \sim N_2\left(\mu_1, \mu_2; \sigma_1, \sigma_2; \rho\right)$ be a binormal vector. Then, we have

$$\begin{aligned} Y = X_1 + X_2 &\sim N_1\left(\mu_1 + \mu_2, \sigma_1^2 + \sigma_2^2 + 2\rho\sigma_1\sigma_2\right), \\ W = X_1 - X_2 &\sim N_1\left(\mu_1 - \mu_2, \sigma_1^2 + \sigma_2^2 - 2\rho\sigma_1\sigma_2\right), \end{aligned} \tag{4.10}$$

and an extension of these relations to a p-normal random vector gives

$$\mathbf{X} \sim N_p\left(\boldsymbol{\mu}, \boldsymbol{\Sigma}\right) \rightarrow Y = \sum_{j=1}^{p} X_j \sim N\left(\mu_Y, \sigma_Y^2\right),$$

with $\mu_Y = \sum_{i=1}^{p} \mu_i$, $\sigma_Y^2 = \sum_{j=1}^{p} \sum_{i=1}^{p} \rho_{ij} \sigma_i \sigma_j$. In the case all σ_j are equal to σ, we have $\sigma_Y^2 = \sigma^2 \sum_{j=1}^{p} \sum_{i=1}^{p} \rho_{ij}$.

But, in some cases, the sum of correlated normal variables might not be normal, except when they are jointly binormal (see Novosyolov (2006)).

Note: For two correlated normal variables, with means zero, $(X, Y) \sim N_2\left(0, 0; \sigma_X, \sigma_Y; \rho\right)$, their product $W = XY$ has density

$$h(w) = \frac{1}{\pi \sigma_X \sigma_Y \sqrt{1 - \rho^2}} \exp \frac{\rho w}{\sigma_X \sigma_Y \left(1 - \rho^2\right)} K_0\left(\frac{w}{\sigma_X \sigma_Y (1 - \rho^2)}\right), \tag{4.11}$$

which reduces to (4.6) for $\rho = 0$.

When both means are zero and variances are unity, their product reduces to

$$h\left(y\right) = \frac{1}{\pi\sqrt{1-\rho^2}} \exp\frac{\rho y}{1-\rho^2} K_0\left(\frac{y}{1-\rho^2}\right), \quad -\infty < y < \infty,$$

with K_0 being the modified Bessel of the second kind of order zero.

The density of their ratio T is then

$$g\left(t\right) = \frac{\sqrt{1-\rho^2}}{\pi} \frac{1}{t^2 - 2\rho t + 1}, \quad -\infty < t < \infty.$$

4.8 Mixture Distributions

Let $X_i \sim N\left(\mu_i, \sigma_i^2\right), i = 1, 2$, be two independent normal variables. We have the following:

(a) **Mixture of the discrete type:** It is of the form $\alpha f_1\left(x\right) + \left(1-\alpha\right) f_2\left(x\right)$. This is not the density of $\alpha X_1 + \left(1-\alpha\right) X_2$, which is obtained by convolution of the density of αX_1 with the density of $\left(1-\alpha\right) X_2$. Here, α is called mixture weight, which is affected by the density of $f_1\left(x\right)$. In the general case, we have $h\left(x\right) = \sum_{i=1}^{n} \alpha_i f_i\left(x\right)$ with $\sum_{i=1}^{n} \alpha_i = 1, 0 \leq \alpha_i$.

(b) **The general mixture:** This refers to the compound distribution by which we have the density of X depending on a parameter which, itself, has a distribution. The compound distribution of X is hence its distribution with an average value of the parameter.

$$f\left(x \mid \alpha\right) \text{ and } g\left(\alpha\right) \rightarrow f\left(x\right) = \int_{-\infty}^{\infty} f\left(x \mid \alpha\right) g\left(\alpha\right) d\alpha.$$

The same approach is used in Bayesian statistics to derive the predictive distribution of a variable, which is often observable (see Chapter 6).

(c) **Mixtures of normal distributions:** The simplest mixture is the finite mixture of two normal with density, $f_m\left(x\right)$, given by

$$f_m\left(x\right) = \left(1-\pi\right) f_1\left(x \mid \mu_1, \sigma_1\right) + \pi f_2\left(x \mid \mu_2, \sigma_2\right),$$

where $f_i, i = 1, 2$, are normal densities with parameters μ_i and σ_i, and π is the mixing proportion, as first proposed by Pearson in 1894 as an alternative to the normal. Pearson suggested fitting by the method of moments. Parameter estimation for the model is by no means simple, and the problem continues to attract attention.

If $X \sim N(\mu, \sigma^2)$ with $\mu \sim N(\lambda, \omega^2)$, then $X \sim N\left(\lambda, \sigma^2 + \omega^2\right)$. This result is used in Bayesian analysis to provide a conjugate prior distribution (see Chapter 6 on Bayesian statistics). Another result is of interest: Let X be $N\left(\mu, \sigma^2\right)$ with a mixing distribution for its precision $\tau = \sigma^{-2}$, which is $\gamma(p, \theta)$. Then

$$
\begin{aligned}
f_m\left(x\right) &= \int_0^\infty \frac{1}{\sigma(2\pi)^{\frac{1}{2}}} \exp\left[-\frac{(x-\mu)^2}{2\sigma^2}\right] \\
&\quad \times \frac{1}{\Gamma\left(p\right)} \sigma^{-2(p-1)} \theta^p e^{-\theta/\sigma^2} d\left(\sigma^{-2}\right) \\
&= \frac{\Gamma\left(p+\frac{1}{2}\right)}{\Gamma\left(p\right)(2\pi)^{\frac{1}{2}}} \frac{\theta^p}{\left[\theta + \frac{1}{2}(x-\mu)^2\right]^{p+\frac{1}{2}}}.
\end{aligned}
$$

Making the substitution $\frac{1}{2}(x-\mu)^2 = \theta t^2/p$, we obtain Student's t-distribution.

Shape of the mixture: It is possible to find the number of modes of a finite mixture of two normal densities, basing oneself on the distance between the two means and the magnitudes of the two standard deviations. It is harder to find the modes of a finite combination of several normal densities (see Stuart and Ord (1987)).

4.9   Distributions Derived From Normal Sampling

(a) The chi-square distribution

We have encountered this distribution before. It will play an important role in subsequent chapters. The chi-square with 1 dof is just the square of the standard normal, i.e.

$$
Y = Z^2 \Rightarrow Y \sim \chi_1^2, \quad \text{where } Z \sim N_1(0, 1).
$$

The chi-square with k dof χ_k^2 is the sum of k independent χ_1^2. Its density is

$$f(x) = \frac{1}{2^{k/2}\Gamma(k/2)} x^{\frac{k}{2}-1} \exp\left(-\frac{x}{2}\right), \quad x > 0,$$

and it is a special case of the Gamma distribution $\Gamma(a, \lambda)$ in two parameters

$$f_{\Gamma(a,\lambda)}(x) = \frac{\lambda^{-a}}{\Gamma(a)} x^{a-1} e^{-x/\lambda}, \quad x > 0,$$

(see Eq. (1.7)), when $\lambda = 2, a = n/2$.

For the multivariate normal vector $\mathbf{Y} \sim N_p(\mathbf{0}, \mathbf{I}_p)$, we have $Y^t Y \sim \chi_p^2$.

The non-central chi-square, with non-central parameter θ, denoted $\chi_p^2(\theta)$, is treated in Chapter 13.

In the general case of $\mathbf{Y} \sim N_p(\boldsymbol{\mu}, \boldsymbol{\Sigma})$, we have $(\mathbf{Y} - \boldsymbol{\mu})^t \boldsymbol{\Sigma}^{-1}(\mathbf{Y} - \boldsymbol{\mu}) \sim \chi_p^2$, with $\boldsymbol{\Sigma}$ positive definite. This is the same expression that enters the density of $\mathbf{Y}$, but in that density, $\mathbf{Y}$ is a vector of real variables, while here it is a random vector with normal density. Also, $\mathbf{Y}^t \boldsymbol{\Sigma}^{-1} \mathbf{Y} \sim \chi_p^2(\theta)$ is a non-central chi-square, with non-centrality parameter $\theta = \boldsymbol{\mu}^t \boldsymbol{\Sigma} \boldsymbol{\mu}$.

(b) The chi distribution

The chi distribution with n dof, denoted by $X \sim \chi_n$, has density

$$f(x) = \frac{2^{-\frac{n}{2}+1}}{\Gamma\left(\frac{n}{2}\right)} x^{n-1} e^{-x^2/2}, \quad x > 0.$$

Hence, $X \sim \chi_n \Rightarrow X^2 \sim \chi_n^2$ and $X \sim N(0,1) \Rightarrow |X| \sim \chi_1$.

Two interesting results on the cdf of χ_2 and χ_2^2 are as follows:

$$\{X \sim \chi_2\} \Rightarrow \left\{\mathbb{P}(X \leq a) = 1 - \exp\left(-\frac{a^2}{2}\right)\right\}, \quad (4.12)$$

$$\{X \sim \chi_2^2\} \Rightarrow \left\{\mathbb{P}(X \leq a) = 1 - \exp\left(-\frac{a}{2}\right)\right\}. \quad (4.13)$$

In $\mathbb{R}^p$, for $X \sim N_p(\boldsymbol{\mu}, \boldsymbol{\Sigma})$, the Mahanabolis distance $\Delta_X \sim \chi_p$ while $\Delta_X^2 \sim \chi_p^2$, with $\Delta_X^2 = (\mathbf{X} - \boldsymbol{\mu})^t \boldsymbol{\Sigma}^{-1}(\mathbf{X} - \boldsymbol{\mu})$. Applications will be found in Chapter 7.

(c) The univariate central beta distribution

The univariate central beta distribution, already seen in Chapter 1, comes in two types:

Type 1: $X \sim \beta_1^I (a, b)$, $a, b > 0 \Rightarrow f(x) = \frac{x^{a-1}(1-x)^{b-1}}{\text{Beta}(a,b)}$, $0 < x < 1$,

where $\text{Beta}(a, b) = \frac{\Gamma(a)\Gamma(b)}{\Gamma(a+b)}$.

Type 2: $X \sim \beta_1^{II} (a, b)$, $a, b > 0 \Rightarrow f(x) = \frac{x^{a-1}(1-x)^{-a-b}}{\text{Beta}(a,b)}$, $x > 0$.

We can see that for $\{Z_1, \ldots, Z_n\} ; \{W_1, \ldots, W_m\} \overset{\text{iid}}{\sim} N(0, 1)$, we have

$$\frac{\sum_{j=1}^{n} Z_j^2}{\sum_{j=1}^{n} Z_j^2 + \sum_{j=1}^{m} W_j^2} \sim \beta_1^I (n/2, m/2)$$

and

$$\frac{\sum_{j=1}^{n} Z_j^2}{\sum_{j=1}^{m} W_j^2} \sim \beta_1^{II} (n/2, m/2).$$

(d) Non-central distributions

The chi-square, gamma, and beta have another form, called non-central, which occurs under sampling when the population mean is not zero, for example. Their densities are more complicated and, for convenience, we have grouped all non-central distributions in Chapter 13, the appropriate section of which can be referred by the reader for more details.

4.10   Conclusion

From what is presented above, we can see that distributions derived from the univariate normal distribution are numerous and versatile, with several of them now having important roles in probability and statistics. There is no doubt that they continue to make the normal distribution a popular statistical model in theory and in practice.

Bibliography

Ahsanullah, M., Golam Kibria, B.M. and Shakil, M. (2014). *Normal and Student's t-Distributions and Their Applications* (Atlantis Press, Amsterdam).

Epstein, B. (1948). Some applications of the Mellin transform in statistics, *Annals of Mathematical Statistics*, **19**, 370–379.

Forchini, G. (2008). The distribution of the sum of a normal and a t random variable with arbitrary degrees of freedom, *Metron — International Journal of Statistics*, **LXVI**(2), 205–208.

Gaunt, R.E. (2016). A probabilistic proof of some integral formulas involving the Meijer G-Function, *arXiV:1602.08143v4[math.CA]*.

Gupta, A.K. and Naradajah, S. (2004). *Handbook of the Beta Distribution and Its Applications* (Marcel Dekker, New York).

Marsaglia, G. (1965). Ratios of normal variables and ratios of sums of uniform variables, *Journal of the American Statistical Association*, **60**, 193–204.

Nason, G.P. (2006). On the sum of t and Gaussian random variables, *Statistics & Probability Letters*, **76**, 1280–1286.

Novosyolov, A. (2006). The Sum of Dependent Normal Variables May be Not Normal, Technical Report, Krasnoyarsk, Russia: Institute of Computational Modelling, Siberian Branch of the Russian Academy of Sciences.

Olsson, S. (2020). The ongoing COVID-19 epidemic curves indicate initial point spread in China with log-normal distribution of new cases per day with a predictable last date of the outbreak, *Preprints*, doi: 10.20944/preprints202003.0077.v1.

Pham-Gia, T. (2008). Exact expression of Wilks's statistic and applications, *Journal of Multivariate Analysis*, **99**, 1698–1716.

Pham-Gia, T. and Dinh, N.T. (2016). Hypergeometric functions, from one scalar variable to several matrix arguments, in statistics and beyond, *Open Journal of Statistics*, **6**(5), 951–994.

Pham-Gia, T. and Turkkan, N. (2005). Distributions of ratios of random variables from the power-quadratic exponential family and applications, *Statistics*, **39**(4), 355–372.

Pham-Gia, T., Turkkan, N. and Marchand, E. (2006). Distribution of the ratio of normal variables and applications, *Communications in Statatistics — Theory and Methods*, **35**, 1569–1591.

Springer, M. (1979). *The Algebra of Random Variables* (Wiley, New York).

Stuart, A. and Ord, J.K. (1987). *Kendall's Advanced Theory of Statistics*, Vol. I, 5th ed. (Oxford University Press, New York).

Chapter 5

Applications of the Univariate Normal Distribution

5.1 Introduction

It is difficult to talk about the normal distribution's applications in various disciplines because they are present almost anywhere quantitative approaches are used with some stochastic arguments. Nevertheless, there are some important applications worth mentioning, and we wish to report them here. Others serve to illustrate approaches, and distributions, we have presented in the previous chapters. One point that should be understood is that applications made outside the exact sciences domains are frequently subject to many approximations, and subjective decisions sometimes have to be made.

5.2 Overview

The normal distribution is a convenient model for quantitative phenomena in natural and behavioral sciences. Psychological test scores and physical phenomena like photon counts have been found to approximately follow a normal distribution. While the underlying causes of these phenomena are often unknown, the use of the normal distribution can be theoretically justified in situations where many small effects are added together into a score, or variable, that can be observed. The normal distribution also arises in many areas of statistics, for example, the sampling distribution of the mean is approximately normal, even if the distribution of the population the sample is taken from is not normal.

In addition, the normal distribution maximizes information entropy among all distributions with known mean and variance, a fact which makes it the natural choice of underlying distribution for data summarized in terms of sample mean and variance. In statistics, tests are often based on the assumption of normality, and the normal distribution arises again as an indispensable tool. In this chapter, applications in science and engineering are presented first, followed by those in education, psychology and sociology. Management science and image processing applications complete the list. The history of the normal distribution is given in Chapter 8.

5.3   Applications in Physics and Statistics

5.3.1   *Measurements and errors*

The oldest application, which also gave rise to the normal, is the theory of errors in physics. The histories of the theory of errors, of probability, and of the normal law are very much intertwined. The theory of errors can be traced to Cotes in 1722. But Simpson, in 1756, first discussed errors systematically and asserted that positive and negative errors are equally probable. Laplace established the first law in 1774, where he stated that the frequency of an error can be expressed as an exponential function of the numerical magnitude of the error. For the second law in 1778, it was changed to exponential of the square of the error.

In 1808, Adrain put this law as $\phi(t) = c \exp(-\alpha^2 x^2)$, which is quite close to the standard normal. Gauss himself gave a proof in 1809. Several well-known mathematicians brought their own contributions in the later part of the 19th century.

5.3.1.1   *Error function*

Error function (also called the Gauss error function) is a special function (non-elementary) of sigmoid shape, which occurs in probability, statistics, and partial differential equations, describing diffusion. There are several versions of the error function in the literature now, but they are all equivalent and are related to the standard normal distribution. The name and abbreviation for the error function (and the error function complement) were developed by Glaisher

(1848–1928) in 1871 on account of its connection with "the theory of Probability, and notably the theory of Errors". Writing on definite integrals, and considering the interval $[x, \infty]$, Glaisher said as follows: "One function, the integral $\int_x^\infty e^{-t^2} dt$, well-known for its use in Physics, that, with the exception of receiving a name and a fixed notation, it may almost be said that to have already become primary... I propose to call it the Error-function, on account of its earliest and still most important use being in connection with the theory of Probability, and the theory of Errors, and to write $\int_x^\infty e^{-t^2} dt = \mathrm{erf}(x)$". Glaisher then used erf(.) to evaluate several definite integrals. But present day erf(.) (and there are several versions) has different integral limits and constant coefficient. Glaisher also cited that for the "law of facility" of errors — the normal distribution — whose density is given by $f(t) = (\frac{a}{\pi})^{1/2} \exp(-at^2)$, with the chance of an error lying between r and s being

$$\left(\frac{a}{\pi}\right)^{1/2} \int_r^s \exp(-at^2) dt = \frac{\mathrm{erf}(s\sqrt{a}) - \mathrm{erf}(r\sqrt{a})}{2}.$$

5.3.1.2 *The complementary error function*

We define the following:

$$\mathrm{erf}(x) = \frac{1}{\sqrt{\pi}} \int_{-x}^x e^{-t^2} dt = \frac{2}{\sqrt{\pi}} \int_0^x e^{-t^2} dt.$$

It has the interpretation that it is the probability that $Y \in [-x, x]$, with $Y \sim N(0, 1/2)$.

The standard normal cdf Φ is used more in probability and statistics while the error function in other branches of math.

The complementary error function, denoted $\mathrm{erfc}(x)$, is defined as

$$\mathrm{erfc}(x) = 1 - \mathrm{erf}(x) = \frac{2}{\sqrt{\pi}} \int_x^\infty e^{-t^2} dt.$$

Another form is known as Craig's formula after its discoverer:

$$\mathrm{erfc}(x) = \frac{2}{\pi} \int_0^{\pi/2} \exp\left(-\frac{x^2}{\sin^2 \varphi} d\varphi\right), \quad x \geq 0.$$

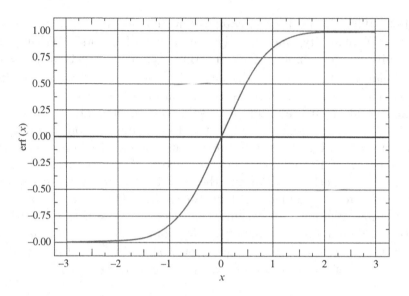

Fig. 5.1   erf function.

The error function is related to the cumulative distribution of the standard normal by

$$\Phi(x) = \frac{1}{2} + \frac{1}{2}\mathrm{erf}\left(\frac{x}{\sqrt{2}}\right) = \frac{1}{2}\mathrm{erfc}\left(-\frac{x}{\sqrt{2}}\right).$$

The error function is an entire function, it has no singularities (except at infinity) and its Taylor expansion always converges (Fig. 5.1).

Glaisher had a long and distinguished career and was Editor of *The Quarterly Journal of Mathematics* for 50 years from 1878 until his death in 1928.

There are several applications of $\mathrm{erf}(x)$ and $\mathrm{erfc}(x)$ to problems familiar to Glaisher and Kelvin, such as the telegrapher equation, estimate of the age of the earth, and short circuit power dissipation in electrical engineering.

In the complex domain, we have Faddeeva's function $w(z) = \exp(-z^2)\,\mathrm{erfc}(-iz)$.

There is a famous remark by Poincaré: "Everybody believes in the law of Errors, the experimenters because they think it is a mathematical theorem, and the mathematicians because they think it is an experimental fact". But, in fact, both parties can be right since mathematical proof shows that under some quite general conditions, it is justified to expect a normal distribution, and statistical experience shows that distributions encountered are quite often approximately normal.

Another important application in physics is the theory of random matrices, often referred to as random matrix theory (RMT).

5.3.2 *Random matrix*

Since the early 1950s, large random matrices with independent standard univariate normal as entries have been studied by theoretical physicists, with important results, not only in theoretical energy physics but also in number theory. We refer the reader to Chapter 13 for further details.

5.3.3 *Uses in statistics*

5.3.3.1 *Importance of the normal*

The normal distribution is considered the most prominent probability distribution in statistics for a variety of reasons. First, the normal distribution is very tractable analytically, i.e. a large number of results involving this distribution can be derived in explicit form. Second, the normal distribution arises as the outcome of the central limit theorem, which states that under mild conditions, the sum of a large number of random variables is distributed approximately normally. This is the case, for example, of a variable observed in some biological investigation which could be the total effect of a large number of independent causes. According to the *theory of elementary errors* by Hagen and Bessel, the total error committed for a physical or astronomical measurement can be regarded as the sum of a large number of mutually independent elementary errors. We can imagine that errors made in technical and socioeconomic problems are of similar nature.

Finally, the "bell" shape of the normal distribution makes it an attractive and convenient choice for modeling a large variety of random variables encountered in practice.

But statistical inference using a normal distribution is not robust in the presence of outliers (data unexpectedly far from the mean, etc.). When outliers are expected, data may be better described using a heavy-tailed distribution, such as Student's *t*-distribution.

5.3.3.2   *Characterizations*

From a technical perspective, alternative characterizations are possible:

- The normal distribution is the only absolutely continuous distribution, all of whose cumulants beyond the first two (i.e. other than the mean and variance) are zero.
- For given mean and variance, the corresponding normal distribution is the continuous distribution with maximum entropy.

Statistical methods can roughly be divided into parametric and non-parametric approaches, and almost all of the parametric approaches are based on the normal. In classical statistical methods, it is supposed first that data collected come from a normal population, and then inferences can be made on parameters. In multivariate models, we have principal component analysis, factor analysis, discrimination and prediction, etc., and the multivariate model is used so that hypothesis testing can be performed at various steps of these methods. Without the normal, we can either restrict ourselves to using descriptive statistics or numerical models. For normal matrix variates, although only few applications are seen at the present time, it is not surprising that they will be used more frequently in image processing and medical image processing in the near future. Some are already present in wireless communication.

5.4   Applications in Stochastic Processes: Gaussian and Brownian Motion Processes

5.4.1   *Stochastic process*

A set of rv's, indexed by a set of numbers, is a stochastic process, or a random process, and in a very general context, that set of indices can be numbers, the real line, the Cartesian plane, or some higher-dimensional Euclidean space (called a random field). Since t can be time, stochastic processes can be thought of as probability models plus time.

(a) **Classification:** A simple classification of random processes can be as follows:

(1) **Stationary processes:** Their joint distribution functions are invariant to time translation.
(2) **Gaussian processes:** These joint distributions are multivariate normal.
(3) **Markov processes:** The future behavior of the process depends only on the present.

There are important properties of random processes, such as *regularity* and *ergodicity*. A process is *regular* when any time average of any one sample function is equal to the corresponding time average of any other sample function. A process is *ergodic* if time averages of sample functions of the process can be used as approximations to the corresponding ensemble averages or expectations.

(b) Tools for stochastic processes: Advanced mathematical methods, in particular Fourier transform methods, are extensively used in the analysis and processing of stochastic processes. These include various notions of stochastic continuity, stochastic differentiation, and integration, all interpreted in the mean square sense. For example, $X(t)$ is continuous in mean square if $\lim_{\varepsilon \to 0} \mathbb{E}\{[X(t+\varepsilon) - X(t)]^2\} = 0$.

Furthermore, the Fourier transform of $R_X(t)$ is called power spectral density (PSD), where $R_X(t) = \mathbb{E}[X(t)X(t+\tau)]$ is the autocorrelation function of the continuous-time random process.

It can be shown that the Fourier transform of continuous-time random process is itself a random process, $\tilde{X}(\omega) = \int_{-\infty}^{\infty} X(t)\exp(-iwt)dt$, this integral being interpreted in the mean square sense.

When we consider two random processes together, $\{X(t), Y(t)\}$, we have similar measures, like the cross-correlation function $R_{XY}(t) = \mathbb{E}[X(t)Y(t+\tau)]$. Other cross-measures are also used.

(c) Gaussian process: The Gaussian process seems to occupy a special position due to its use of the univariate and multivariate normal distribution.

Definition 5.1. A time-continuous stochastic, or random, process is Gaussian if and only if for every finite set of indices $\{t_1, \ldots, t_n\} \subset T$, we have $X_{t_1,\ldots,t_n} = \{X_{t_1}, \ldots, X_{t_n}\}$ as a multivariate Gaussian random variable.

In the more general case, both the state space and the index set can be integers, or the real line, or a p-dimensional Euclidean space. The change of a stochastic process between two index values is called an increment.

Hence, it is a statistical model where observations occur in a continuous domain, e.g. time or space, and every point is associated with a normal variable. Moreover, every finite collection of those random variables has a multivariate normal distribution. Viewed as a *machine-learning algorithm*, a Gaussian process uses lazy learning and a measure of the similarity between points (this is the kernel function) to predict the value for an unseen point from training data. The prediction is not just an estimate for that point, but also has uncertainty information. It is hence related to *Kriging, the science of predicting in a surface domain or in space.*

5.4.1.1 *Brownian motion process or Wiener process*

Definition 5.2. A family of rv's $\{X(t)\}$, indexed by t continuous on $[0, \infty)$, is called a Brownian motion process, or Wiener process, if it satisfies the following:

(1) $X(0) = 0$.
(2) The increments $X(s_i + t_i) - X(s_i)$ over any arbitrary finite set of intervals $\{(s_i, s_i + t_i)\}$ are independent rv's.
(3) For $s > 0, t > 0$, the differences $X(s + t) - X(t) \sim N(0, s)$, and they are stationary and independent.

Then we have $\{X(t), t > 0\}$ as normal variables. A Wiener process is hence a Gaussian process with stationary independent increments.

We can see that the process is obtained by a limiting passage from random walks (Breiman, 1968). Wiener has noted that almost all its paths are continuous, i.e. for almost all ω, the function $t \to X(t, \omega)$ is a continuous function of t on $[0, \infty)$. But he also noted that almost every path is nowhere differentiable, i.e. the curve does not have a tangent anywhere.

The process has a drift coefficient μ and variance parameter σ^2 if, moreover, $X(t) \sim N(\mu t, \sigma^2 t)$ (Fig. 5.2).

Playing a central role in the theory of probability, the Wiener process is often considered to be the most important and the most

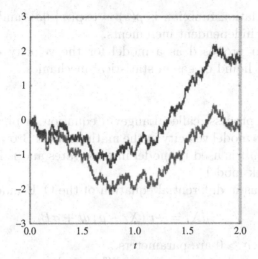

Fig. 5.2 Realizations of the Wiener processes (or the Brownian motion processes) with drift (upper curve) and without drift (lower curve).

studied stochastic process, with connections to other stochastic processes. The first explanation of the Brownian motion is due to Einstein, in 1905, who thought that the immersed particle is continually being subjected to bombardment by the molecules of the surrounding medium. He inspired Wiener to devise a type of measure theory to establish the Wiener process as a mathematical topic. The process now has many applications. It plays a central role in quantitative finance, where it is used, for example, in the Black–Scholes–Merton model. The process is also used in different fields, including the majority of natural sciences as well as some branches of social sciences, as a mathematical model for various random phenomena.

The following process is a particular case of the Gaussian process.

5.4.1.2 *The Ornstein–Uhlenbeck process*

The Ornstein–Uhlenbeck process is defined as a continuous Gaussian Markov process such that for $t > 0$, $\mathbb{E}(U(t)) = u_0 \exp(-\beta t)$, $\mathrm{Var}(U(t)) = \frac{\sigma^2}{2\beta}(1 - \exp(-2\beta t))$, $\beta > 0$. Then

$$Cov(U(t), U(t+s)) = \frac{1}{2}\frac{\sigma^2}{\beta}(1 - \exp(-2\beta t))\exp(-\beta|s|), \quad t > 0.$$

The autocorrelation function is $\rho(s) = \exp(-|\beta|s)$, and the process does not have independent increments.

It has been proposed as a model for the velocity of a particle immersed in a liquid or gas in statistical mechanics.

Notes:

(1) The O–U process, called Langevin equation in physics, was an attempt to model velocity in the mathematical Brownian motion. In finance, it is used to model interest rates and is known under the Vasicek model.

The stochastic differential equation of the O–U model is

$$dX_t = -r\left(X_t - \mu\right)dt + \sigma B_t,$$

where $r, \mu, \sigma > 0$ are parameters.

For $\sigma = 0$, we have an ordinary differential equation, which gives a deterministic solution $X_t = \mu + (X_0 - \mu)e^{-rt}$, where $X_t \to \mu$ as $t \to \infty$. But using Ito's lemma (in stochastic calculus) to solve the above stochastic differential equation, we obtain

$$X_t = \mu + (X_0 - \mu)e^{-rt} + \sigma \int_0^t e^{-r(t-s)} dB_s.$$

If X_0 is constant, X_t is normal, and we have

$$X_t \sim N\left(\mu + (X_0 - \mu)e^{-rt}, \frac{\sigma^2}{2r}\left(1 - e^{-2rt}\right)\right),$$

which gives a normal limiting distribution $N(\mu; \sigma^2/2r)$.

(2) Another definition of the O–U process, denoted $V(t)$, is found in the literature:

$$\mathbb{E}(V(t)) = 0, t > 0, \quad \operatorname{Var}(V(t)) = \frac{\sigma^2}{2\beta}, \quad \beta > 0.$$

$$\operatorname{Cov}(V(t), V(t+s)) = \frac{\sigma}{2\beta}\exp(-\beta|s|), \quad t > 0.$$

The website for the Gaussian processes is as follows: http://www.gaussianprocess.org/. This website aims at providing an overview of resources concerned with probabilistic modeling, inference, and learning based on Gaussian processes. Although Gaussian processes have a long history in the field of statistics, they seem to have been employed extensively only in some areas. With the advent of kernel machines in the machine learning community, models based on Gaussian processes have become commonplace for problems of regression (kriging) and classification, as well as a variety of more specialized applications.

5.5 Applications in Engineering

Applications of the normal distribution in engineering are numerous, and this distribution is pervasive in all levels and aspects of this discipline. We just mention some specific topics.

Definition 5.3. Engineering reliabilty is the probability that a system performs its purpose adequately, for the period of time intended, under operating conditions encountered.

Let T be the lifelength of the system or the time to failure. If it is a rv with density $f(t)$, reliability at time t is $R(t) = \mathbb{P}(T \geq t) = 1 - F(t)$. We have $F'(t) = f(t)$ and the failure rate, or hazard rate, is $\rho(t) = \frac{f(t)}{R(t)}$.

The normal distribution is often used for systems that fail because of some wearing effects and not because of a particular problem. We then have $R(t) = 2 - [\Phi(\frac{t-\mu}{\sigma}) + \Phi(\frac{\mu}{\sigma})]$.

When t is time, reliability has been defined as "quality plus time".

5.5.1 *Static reliability model*

Another approach to reliability is the stress–strength model (Fig. 5.3), mostly used for events not time-dependent. A system has its strength X operating under the stress Y, both considered as random variables, with determined distributions. As long as we have $X > Y$, the system is functioning safely and $R = \mathbb{P}(X > Y)$ is the system reliability, which does not depend on time but on built-in robustness (strength) under outside pressure (stress). This concept, very simplified though, can be easily computed when X and Y are independent. When they are not, however, the distribution of $X - Y$ can be less easily computed.

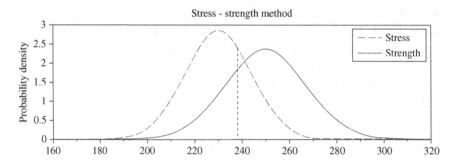

Fig. 5.3   Strength and stress normally distributed.

We can take both distributions of X and Y as normal if they have distributions that can be expressed in same units. We have

$$R = \mathbb{P}(\text{strength}(X) > \text{stress}(Y)) = \mathbb{P}(X - Y > 0) = \mathbb{P}(W > 0),$$

since

$$W \sim N(\mu_X - \mu_Y, \sigma_X^2 + \sigma_Y^2),$$

we have

$$R = \mathbb{P}(W \geq 0) = \mathbb{P}\left(Z \geq \frac{0 - (\mu_X - \mu_Y)}{\sqrt{\sigma_X^2 + \sigma_Y^2}} \right)$$

$$= \mathbb{P}\left(Z \geq \frac{\mu_Y - \mu_X}{\sqrt{\sigma_X^2 + \sigma_Y^2}} \right) \geq 0.5.$$

Alternately, we have reliability $\mathbb{P}(X/Y > 1)$ and if both X and Y are normal, the density of their ratio can be obtained from Pham-Gia *et al.* (1994) (see Chapter 4).

The main difficulty in applying this model is the determination of the distributions of X, and of Y, when they are not normal, which, in engineering, can only be done through tests and subjective judgments. Then the two distributions of $X - Y$ or X/Y have to be determined, while they are readily available in the normal case.

As applications, we offer two cases of stress–strength reliability:

(1) Strength R and stress W follow some theoretical statistical distributions related to the normal, here the log-normal model, i.e $R \sim LN(\mu_x, \sigma_x^2)$, $W \sim LN(\mu_y, \sigma_y^2)$.

(2) X and Y follow the normal model in a Bayesian context.

5.5.1.1 *Application 1: Estimation of the median safety factor in a log-normal stress–strength model (Pham-Gia and Turkkan, 1994)*

As presented in Chapter 4, the log-normal distribution is well suited to model phenomena where multiplicative effects, instead of additive ones, are reported. This frequently happens in engineering. We have $R \sim LN(\mu, \sigma^2)$ if $R = \exp(X)$, where $X \sim N(\mu_x, \sigma_x{}^2)$.

Both classical and Bayesian approaches are used in this example to estimate the median safety factor, and the median coefficient of variation of this stress–strength model, with log-normal distributions for both stress and strength. A numerical example on fire load studies in civil engineering illustrates some of these results.

The density of strength R is

$$f(r) = \frac{1}{r\sigma_x\sqrt{2\pi}} \exp\left\{-\frac{(\ln r - \mu_x)^2}{2\sigma_x^2}\right\}, r > 0, \sigma_x > 0, -\infty < \mu_x < \infty.$$

We then have $\mu_R = \exp(\mu_x + \sigma_x^2/2)$, $\sigma_R^2 = \exp(2\mu_x + \sigma_x^2)[\exp(\sigma_x^2) - 1]$.

The coefficient of variation of R is $V_R = \frac{\sigma_R}{\mu_R} = [\exp(\sigma_x^2) - 1]^{1/2}$. The median of R is $\mathrm{Med}(R) = \exp(\mu_x)$ and the coefficient of variation of R about its median is $V_{\mathrm{Md}}(R) = \frac{\sigma_R}{\mathrm{Md}_R} = [\exp(\sigma_x^2) - 1]^{1/2}\exp(\sigma_x^2/2)$.

Similar relations hold for the stress $W \sim LN(\mu_y, \sigma_y^2)$.

Assuming strength and stress to be independent, the reliability ρ of the system is $\rho = \mathbb{P}(R - W > 0) = \mathbb{P}(R/W > 1)$. It can be shown that $\rho = \Phi(\beta)$, where the safety index is $\beta = \frac{\mu_x - \mu_y}{\sqrt{\sigma_x^2 + \sigma_y^2}}$. However, the estimation of β by $\widehat{\beta} = \frac{\bar{x} - \bar{y}}{\sqrt{s_x^2 + s_y^2}}$ is not satisfactory.

A measure of safety is $\lambda_k = \frac{R_k}{W_{100-k}}$, with R_k and W_{100-k} being the kth and $(100 - k)$th percentiles of their respective distribution.

For $k = 50$, we have $\lambda_{50} = \lambda_{\text{Md}}$, the ratio of two medians. Let $U = \frac{R}{W}$. Then $U \sim LN(\mu_x - \mu_y, \sigma_x^2 + \sigma_y^2)$, and hence,

$$\text{Med}(U) = \frac{\text{Med}(R)}{\text{Med}(W)} = \lambda_{\text{Md}} = \exp(\mu_x - \mu_y),$$

while the coefficient of variation of U about its median is

$$V_{\text{Md}}(U) = [\exp(\sigma_x^2 + \sigma_y^2) - 1]^{1/2} \exp((\sigma_x^2 + \sigma_y^2)/2).$$

We transform the data $x_i = \ln(R_i)$, $y_i = \ln(W_i)$ and obtain the statistics $\bar{x}$, $\bar{y}$, s_x^2, s_y^2 related to two normal populations. From here, two approaches are taken:

(a) Classical approach: We distinguish between the cases as follows:

(i) σ_x^2, σ_y^2 are known.
(ii) $\sigma_x^2 = \sigma_y^2 = \sigma^2$ unknown.
(iii) $\sigma_x^2 \neq \sigma_y^2$ both unknown.

Obtain, in each case, an estimation of $\mu_x - \mu_y$ and of V_{Md}.

(b) Bayesian approach: Now, μ_x, μ_y, σ_x^2, σ_y^2 are considered as random variables and have different priors or a joint prior. Considering again the cases (i)–(iii) above, we can derive the posterior distribution of λ_{Md} and V_{Md}. From these posterior densities, the highest probability density (hpd) interval can be computed by applying the TPG algorithm (see Chapter 6). Details are given in the publication in reference.

5.5.1.2   *Application 2: The joint predictive–posterior method in the Bayesian analysis of stress–strength reliability (Pham-Gia and Turkkan, 2007)*

Indeed, in the stress–strength model, the distributions of X and Y, however obtained, can be considered as their prior distributions, and the associated prior reliability R_{prior} is first computed. Subsequent works on engineering structures would lead to the re-evaluation of the distribution of X, giving its posterior distribution. Similarly, studies and tests related to stress Y would permit one to refine its distribution, leading to its posterior distribution.

Too often, attention is focused on the parameters of the distributions of the variables under study, while neglecting the variables themselves. But, in general, these variables are the only observable entities, and while their predictive distributions can be more easily verified and monitored, they also provide better direct conclusions on the whole problem. Also, there is seldom some probability attached to different posterior reliability values, hence leaving the user with a wealth of information on possible outcomes, but with no information on how likely they could happen.

Here, by combining the predictive and the posterior distributions, and using an appropriate set of parameters for both distributions, we hope to overcome this weakness. Moreover, graphs and charts will efficiently assist us further in relating a posterior reliability value to its probability of occurrence. As usual, we suppose that X and Y are independent.

We will use the following statistical result in the Bayesian analysis.

Proposition 5.1. *Let X and Y be independent normal variables with unknown means and known precisions: $X \sim N(W, \lambda_1)$, with $W \sim N(\mu, \lambda_2)$, and $Y \sim N(V, \lambda_3)$, with $V \sim N(\theta, \lambda_4)$.*

Then, for two fixed samples of sizes n_1, n_2, setting $t_1 = \bar{x}_{n_1}$, $t_2 = \bar{y}_{n_2}$, the predictive density of (t_1, t_2) is the product of two normal densities, while the posterior reliability (Fig. 5.4) of the system is $R_{\text{post}} = \mathbb{P}(Y_{\text{post}} < X_{\text{post}}) = \Phi(\eta^*)$, with

$$\eta^* = \frac{\sqrt{\lambda_1 \lambda_3}}{\sqrt{(\lambda_4 + n_2 \lambda_3)(\lambda_2 + n_1 \lambda_1)}}$$

$$\times \frac{\lambda_2 \lambda_4 (\mu - \theta) + n_2 \lambda_2 \lambda_3 (\mu - t_2) + n_1 \lambda_1 \lambda_4 (t_1 - \theta) + n_1 n_2 \lambda_1 \lambda_3 (t_1 - t_2)}{\sqrt{(\lambda_2 + (n_1 + 1)\lambda_1)\lambda_3(\lambda_4 + n_2 \lambda_3) + (\lambda_4 + (n_2 + 1)\lambda_3)\lambda_1(\lambda_2 + n_1 \lambda_1)}},$$

where $\Phi(.)$ being the cdf of the standard normal.

Numerical example 5.1: Let us consider, for example, the case of a concrete beam, whose strength is normally distributed about a value W, with precision λ_1 estimated as $1/10$. W itself is considered normal, with mean $\mu = 12$ MPa (megapascal) and precision $\lambda_2 = 1/12$ (Fig. 5.5). On the other hand, the stress that acts on this beam is also normal, with mean V and precision λ_3 estimated as $1/8$, where V is normal with mean $\theta = 11$ and precision $\lambda_4 = 1/12$. Let us fix the

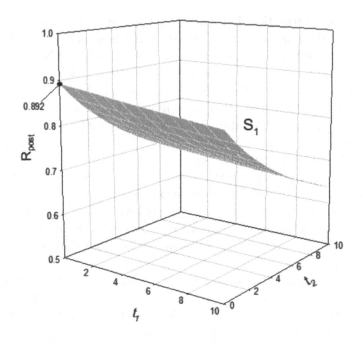

Fig. 5.4   Posterior reliability $R_{\text{post}}(t_1, t_2), 0 \le t_1, t_2 \le 10$, normal case.

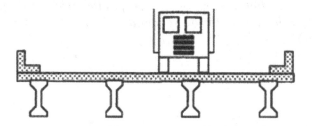

Fig. 5.5   Stress–strength in civil engineering.

two sample sizes $n_1 = n_2 = 20$. We can see that the only variables left in the study of R_{post} are $t_1 = \bar{x}_{n_1}$ and $t_2 = \bar{y}_{n_2}$.

First, we have $R_{\text{prior}} = \mathbb{P}(Y_{\text{prior}} < X_{\text{prior}}) = 0.512$ by direct computation. A certain number of tests are now performed separately on the beam strength and the stress factor that will affect it, leading to the two sample of observations $\{x_1, \dots, x_{20}\}$ and $\{y_1, \dots, y_{20}\}$. We have two surfaces Δ_2, representing the predictive density of (t_1, t_2) and S_2, giving the values of the posterior reliability R_{post}. Figure 5.4

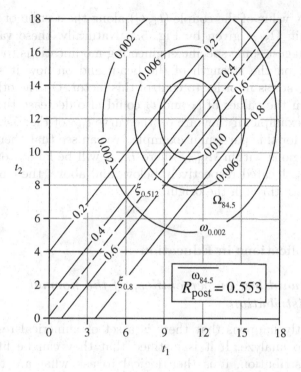

Fig. 5.6 Joint graph for level curves ξ_c of posterior reliability $R_{\text{post}}(x_1, y_2)$ and credible regions $\Omega_{(1-\varepsilon)100}$ of predictive distribution.

illustrates a similar case (see Pham-Gia and Turkkan (2007) for further details).

When values of R_{post} and credible regions are plotted together, as in Fig. 5.6, we have level curves ξ_c, $0 < c < 1$, of S_2 intersecting various confidence regions $\Omega_{(1-\alpha)100}$. For example, we can see that posterior reliability varies approximately between 0.2 and 0.8, within the credible region $\Omega_{84.5}$, determined by the level curve $\omega_{0.002}$ of Δ_2. The mean value of R_{post} within this region is $\bar{R}_{\text{post}}^{\Omega_{84.5}} = 0.553$.

The prior reliability, $R_{\text{prior}} = 0.512$, is quite low due to uncertainties in the estimation of the two means W and V, of strength and stress, respectively, in the prior phase. From this starting value, to increase the value of posterior reliability, while remaining within $\Omega_{84.5}$, either t_1 or t_2 has to increase, or both. For example, to bring this reliability to 0.80, we have to obtain $t_1 = 12$ and $t_2 = 8.25$, or take any value of the couple (t_1, t_2) along the section of the curve $\xi_{0.80}$, within $\Omega_{84.5}$, given by Fig. 5.4. Naturally, these values will, in turn, suggest some specific engineering modifications to be made, separately, on the strength of the beam and on how it should be exposed to stress in order to control this factor. On the other hand, increases in the values of t_1 and t_2 could also decrease the value of R_{post}, for example, if they are on the curve $\xi_{0.2}$ of Fig. 5.4.

Hence, for this particular example, we can see that there are two distinct regions within $\Omega_{84.5}$, where R_{post} will be larger, or smaller, than R_{prior}, located respectively below, and above, the curve $\xi_{0.512}$, which passes through the origin.

5.6   Applications in Education

5.6.1   *Quantiles, or percentiles, of the normal distribution*

It frequently happens that there is a set of numerical results that we wish to analyze. If it is justified that they can be fitted by a normal distribution, it is then logical to ask what are the values of the distribution under which fixed percentages of the scores are located. These are quantiles or percentiles of this population. For the standard normal density Z, as shown here, these values are well known. The median Md has 50% of the total area on each side, and the quartiles Q_1, $Q_2 = $ Md, Q_3 have 25% between them and the same percentage above Q_3 and below Q_1. Quintiles are related to 20% and deciles to 10%.

Stanines are special measures related to the division of the normal into nine parts. But, unlike the quantiles, they represent whole intervals. They are St1, ..., St9, having 4%, 7%, 12%, 17%, 20% in consecutive disjoint intervals, according to Table 5.1 (stanines = standard nine). The idea is to simplify the results and assign to a set of standard scores a number between 1 and 9 using the normal distribution (Comrey, 1975). The sets of standard scores are grouped as in Table 5.1.

We see that there is a symmetry wrt the origin, and the middle interval has the highest probability and the probabilities of each stanine is fixed arbitrarily, according to some argument. We now

Table 5.1 Values of stanines.

Standard scores equivalent	Probability	Stanine scores
Above 1.75	0.0401	9
1.25 to 1.75	0.0655	8
0.75 to 1.25	0.1210	7
0.25 to 0.75	0.1747	6
−0.25 to 0.25	0.1974	5
−0.75 to −0.25	0.1747	4
−1.25 to −0.75	0.1210	3
−1.75 to −1.25	0.0655	2
Below −1.75	0.0401	1
Total	1.00	

conveniently assign scores from 1 to 9 to a set of raw data. Dividing a set of real numbers into nine different categories needs some approximation.

For a general normal distribution $N(\mu, \sigma^2)$, the corresponding quantiles are obtained from those of the standard normal Z by a simple transformation $X = \mu + z\sigma$.

Fitting a normal distribution to a set of data, we can use the moments method, taking the sample mean m as population mean and the sample variance s^2 as population variance. For a score X that has served to compute these parameters, we have its $z - score = z_x = \frac{X-m}{s}$.

5.6.2   *Method to determine values of stanines*

For a distribution of raw scores, we should do the following:

- Trim the outliers.
- Put them in increasing order.
- Divide them in nine intervals, the first and last ones being open-ended, according to the stanine proportions. For equal scores, some decision has to be made on where to put them.
- Take the z scores of these stanines $s_1, s_2, \ldots, s_9$.
- Compute the adjusted raw scores of these stanines according to the formula $X_i = m + Z_i s$.

5.6.3   *An application to the normalization of a set of data*

Background: The final grade X given to a student for a course, with different instructors, can vary so much from one instructor to another. In order to have some uniformity, *and fairness*, and have a grade distribution where 65 is the average, with 10 as standard deviation, we set (with some subjectivity of course), similar to Table 5.1, the following rules, where G is a grade transformed from X. They are reasonable, again, on the average. To build the table of final grades, we do the following:

(a) Fix $A+$ and F: $A+$ if $G \geq 95$, F if $G < 55$.
(b) Divide the interval 55 to 95 into 10 intervals according to $N(65, 10^2)$, (or 10 grades other than F and $A+$), with each interval containing $(0.84/10 = 0.084)$ probability.
(c) Using Z distributed as $N(0,1)$, find the values of $Z(A+)$, $Z(A)$, etc. and transform them into $G(A) = 65 + Z(A) \times 10$. $G(A-)$, etc. These values of G will be used according to Table 5.2 of *final letters*.

So, we have 12 possible grades, leading to 12 intervals (with the first and last being open-ended), based on $N(65, 10^2)$ (Fig. 5.7). The slanted shape is intentional.

Table 5.2   Values of letter grades.

Classification	Range	Score	Probability
F		< 55	0.15865
D	55.00	58.02	0.08400
D+	58.02	60.51	0.08400
C−	60.51	62.74	0.08400
C	62.74	64.87	0.08400
C+	64.87	66.98	0.08400
B−	66.98	69.20	0.08400
B	69.20	71.64	0.08400
B+	71.64	74.57	0.08400
A−	74.57	78.70	0.08400
A	78.70	95.00	0.08400
A+		> 95	0.00135

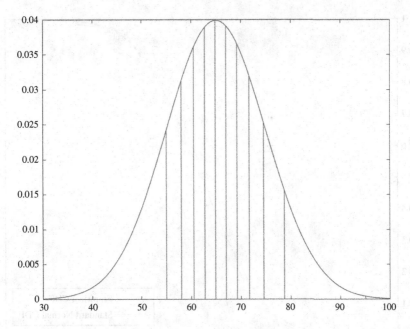

Fig. 5.7 $N(65, 10^2)$ and grade intervals.

5.6.4 *Application*

Three classes of 80, 90, and 100 students have taken the same exam with three different instructors, who used the same tests. Grades were obtained from the department's records. There are differences in the means of the classes and we wish to normalize the whole batch of grades. Using the Kolmogorov–Smirnov test (Fig. 5.8), we have the following:

(1) Kolmogorov–Smirnov statistic is KS_STAT = 0.0252, and the p-value of the test is P-VALUE = 0.8763. So, the set of 270 grades can be considered as coming from a normal population.
(2) Compute the sample mean and variance:

$$\hat{\mu} = \bar{X} = \frac{1}{n} \sum_{i=1}^{n} X_i = 54.65;$$

$$\hat{\sigma} = s^2 = \frac{1}{n-1} \sum_{i=1}^{n} \left(X_i - \bar{X}\right)^2 = 16.2017.$$

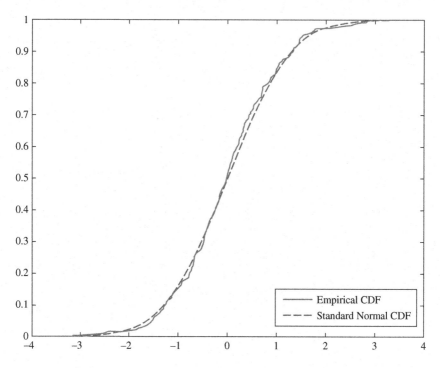

Fig. 5.8   Kolmogorov–Smirnov test.

(3) Obtain the normal distribution $N(54.65, 4.0251^2)$ which can now represent the whole batch of 270 grades. This moment approach to determine the approximating normal distribution here is simple and convenient, but is often not the best. It can be replaced by other methods available in the applied statistics literature (see, e.g. Olkin *et al.* (1994)). It is clear that the normal $N(54.65, 4.0251^2)$ is different from the normal $N(65, 10^2)$ to which it will be transformed.

(4) Make the linear transformation of $N(54.65, 4.0251^2)$ to $N(0, 1)$:

$$z = \frac{X - 54.65}{4.0251},$$

and then $N(0, 1)$ to $N(65, 10^2)$: $Y = 65 + z \times 10$.

Note: We can combine these two transformations into one: $X \sim N(54.65, 4.0251^2) \to Y \sim N(65, 10^2)$: $Y = \alpha X + \beta$, with $\alpha = \frac{\sigma_Y}{\sigma_X}$, $\beta = \mu_Y - \alpha \mu_X$.

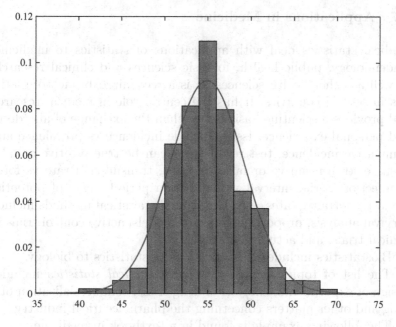

Fig. 5.9 Histogram of X.

(5) Give the grades for three values of X (Fig. 5.9):

- For Jill, $X_1 = 44$, we have $z(X_1) = z(44) = \frac{44-54.65}{4.0251} = -2.6459$ and

$$Y_1 = 65 + (-2.6459) \times 10 = 38.54 < 55.$$

So, Jill has obtained a real grade of 44 in her group, with her instructor. After normalization and using the grade table above, she has the official grade $G = 38.54$, resulting in the letter F according to Table 5.2.

- For Jack, who has obtained 56 in his group, we have $z(X_2) = z(56) = \frac{56-54.65}{4.0251} = 0.33$, $Y_2 = 65 + (0.33) \times 10 = 68.3$ or the letter $B-$.

- Mary has a score, noted as X_3 equal 67, i.e $X_3 = 67$. Its z-score is then $z(67) = \frac{67-54}{4.025} = 3.06$. Her Y score is then $Y_3 = 65 + (3.06 \times 10) = 95.68$. According to the table giving Y score as alphabetical grade, she has $A+$.

5.7   Applications in Medicine

Medical statistics deal with applications of statistics to medicine, epidemiology, public health, forensic sciences and clinical research, as well as other health sciences. It is a recognized branch of statistics in several countries. It has the central role in medical research and provides a scientific basis better than the exchange of anecdotes and personal experience. Issues include incidence vs prevalence and cumulative incidence, test results that can be true positive or false positive, true negative or false negative, transmission rate vs force of infection, serial interval vs incubation period, years of potential life lost, mortality rates, etc. The related statistical methods include survival analysis, proportional hazard models, active control trials in clinical trials, and actuarial statistics.

Biostatistics include all applications of statistics to biology.

The list of topics treated by *pharmaceutical statistics* includes design of experiments, analysis of drug trials, commercialization of a drug and other matters concerning the pharmaceutical industry.

The following example is found in a textbook in medicine.

The level of cholesterol for women in the 20–34 age bracket is approximatively normal, with $\mu = 185$ and $\sigma = 39$.

The level of 240 mg/dL found on a patient requires immediate attention because $\mathbb{P}(X > 240) = 0.079$. Hence, approximately only 8% of the women of this age bracket have a cholesterol level that high.

The normal distribution is well used in medical sciences in practice and research. Just take a look at the *Journal of Medical Statistics and Informatics* and *The New England Journal of Medicine*, you will be impressed by the research results of all kinds, with the normal present somewhere, explicitly, or implicitly via regression, ANOVA, MANOVA, etc.

5.8   Applications in Psychology, Behavioral Science: Intelligence Quotient, Stanford–Binet Test

5.8.1   *Intelligence quotient and the Stanford–Binet Test*

There are a variety of intelligence measurement tests. The most commonly used individual IQ tests in use in the English-speaking world are the Wechsler Adult Intelligence Scale for adults and the Wechsler

Intelligence Scale for Children for school-age test-takers. Other commonly used individual IQ tests (some of which do not label their standard scores as "IQ"scores) include the current versions of the Stanford–Binet Intelligence Scales, the Woodcock–Johnson Tests of Cognitive Abilities, the Kaufman Assessment Battery for Children, the Cognitive Assessment System and the Differential Ability Scales.

The 1996 Task Force investigation on Intelligence sponsored by the American Psychological Association concluded that there are significant variations in IQ across races. The problem of determining the causes underlying this variation relates to the question of the contributions of "nature and nurture". Psychologists such as Kaufman and Brody argue that there is no sufficient data to conclude that this is because of genetic influences. A review article published in 2012 by leading scholars on human intelligence, concluded, after reviewing the prior research literature, that group differences in IQ are best understood as environmental in origin.

IQ scores are not stable over time, but the intellectual development of an individual is highly complex. Twin studies, using the normal IQ range, show that IQ is heritable with IQ concordance between twins raised together or apart. Twins adopted and raised in different family environments provide similar evidence of IQ concordance, which increases with age.

5.8.2 The IQ: How to interpret it?

Table 5.3 provides the most acceptable interpretation of IQs.

About 2% of the population has an IQ score lower than 69 (Fig. 5.10). Such a low IQ score is often hard to measure using a

Table 5.3 IQ and its interpretation.

IQ	Percentage of the population with this IQ	Interpretation
>130	2.1	Very gifted
121–130	6.4	Gifted
111–120	15.7	Above-average intelligence
90–110	51.6	Average intelligence
80–89	15.7	Below-average intelligence
70–79	6.4	Cognitively impaired

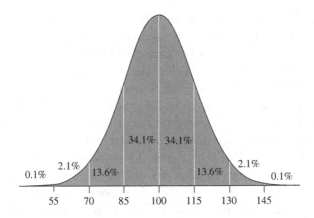

Fig. 5.10   Distribution of normal IQ with mean 100 and standard deviation 15.

regular intelligence test. Very high IQ scores are also hard to determine accurately. This is because you need a lot of reference measurements to determine a specific score reliably. As very high and very low IQ scores simply do not occur often, it is hard to form such a reference group.

5.8.3   *The terms*: *Intelligence and quotient*

The topic of IQ comes up quite frequently in conversations and is a source of debate. The two terms, *intelligence* and *quotient*, both need some clarification. First, what is intelligence? Why *quotient* and not *measure* or *score*? Originally, "IQ" tests were created to be able to identify children who might need special education due to their retarded mental development. Binet's test included varied questions and tasks. The tasks even included unwrapping a piece of candy and comparing the weights of different objects.

   To relate the mental development of a child to the child's chronological age, the IQ was invented. IQ = (MA/CA) × 100. The intelligence quotient was equal to 100 times the mental age divided by the chronological age. For example, if a certain child started reading, etc., at the age of 3 (CA) and average children start reading, etc., at the age of 6 (MA), the child would get an IQ score of 200 (such a score is very rare). Since people wanted to also use IQs for adults, this formula was not very useful since raw scores start to level off around the age of 16.

Thus, the deviation IQ replaced the ratio IQ. It compares people of the same age or age category and assumes that IQ is normally distributed, that the average (mean) is 100 and that the standard deviation is something like 15 (IQ tests sometimes differ in their standard deviations).

Politicians have high IQs according to some popular magazines. *Life Buzz* gave the following IQs to politicians: Trump (156), Kennedy (150), Clinton (149), Carter (145), Obama (144), Eisenhower (132), Nixon (131), Bush Senior (130), Reagan (130), Ford (127), Bush Junior (124). Following IQs were given for deceased past presidents: Lincoln (140), Roosevelt (139), Washington (132).

Concerning ethnic groups, it was found that Chinese, Japanese, South Koreans have an average of above 105, North Americans and Western Europeans have an average between 95 and 100, some South East Asians between 90 and 95, most South East Asians, Middle East, South America between 80 and 90, and Africans between 65 and 70.

5.8.4 *Howard Gardner: Multiple intelligences*

In Howard Gardner's theory of multiple intelligences, instead of analyzing test scores, attention should be put on eight different types of human intelligence, based on skills and abilities valued in different cultures.

The eight kinds of intelligence described are as follows:

(1) Visual–spatial intelligence
(2) Verbal–linguistic intelligence
(3) Bodily–kinesthetic intelligence
(4) Logical–mathematical intelligence
(5) Interpersonal intelligence
(6) Musical intelligence
(7) Intrapersonal intelligence
(8) Naturalistic intelligence

A high score obtained in a test related to one of the above types of intelligence, will certainly provide the motive to the individual to choose studies and a career/occupation in a certain specific domain.

5.9   Applications in Sociology/Political Science: Bell-Shaped Curve and Its Controversies

The book, *The Bell Curve: Intelligence and Class Structure in American Life*, which has caused much reaction at the time of its publication, takes the IQ several steps further to explain and predict various social phenomena in the American society.

5.9.1   *Social classes and IQ*

In this volume of 600 pages, published in 1994, the two authors wish to prove some theories of theirs related to the status and advancement of some ethnic groups in the American society. The statistical tool used is basically regression analysis. The normal distribution was used via this tool.

Herrnstein and Murray's central argument is that human intelligence is substantially influenced by both inherited and environmental factors. It is a better predictor of many personal dynamics, including financial income, job performance, birth out of wedlock, and involvement in crime, than an individual's parental socioeconomic status or education level (Table 5.4).

They also argue that those with high intelligence, the *cognitive elite*, are becoming separated from those with average and below-average intelligence. The book was controversial, especially where the authors wrote about racial differences in intelligence and discussed the implications of those differences.

The book argued the average genetic IQ of the United States is declining, owing to the tendency of the more intelligent to have fewer children than the less intelligent, the generation length to be shorter

Table 5.4   Values are the average earnings (1993 USD) of each IQ sub-population.

Relation between IQ and earnings in the US (USD)				
IQ	<75	75–90	90–110	110–125	>125
Age 18	2,000	5,000	8,000	8,000	3,000
Age 26	3,000	10,000	16,000	20,000	21,000
Age 32	5,000	12,400	20,000	27,000	36,000

for the less intelligent, and the large-scale immigration to the United States of those with low intelligence. We do not like to comment on the validity of the arguments presented in the book and ask our readers to do it if they wish.

Reaction: Because several sensitive questions raised by the book are not "politically correct", there are various reactions to the conclusions made in the book and to the issues raised there too. Maybe the most important reply is the book by a group of scholars from Berkeley University: *Inequality by Design.* They contested the validity of some conclusions and also raised the question on how the system itself has contributed to inequalities in society.

Claude Fisher, Michael Hout, Martín Sánchez Jankowski, Samuel R. Lucas, Ann Swidler, and Kim Voss, in their book, recalculated the effect of socioeconomic status, using the same variables as *The Bell Curve*, but weighting them differently. They found that if IQ scores are adjusted, as Herrnstein and Murray did, to eliminate the effect of education, the ability of IQ to predict poverty can become dramatically larger, by as much as 61% for whites and 74% for blacks. According to the authors, Herrnstein and Murray's finding that IQ predicts poverty much better than socioeconomic status is substantially a result of the way they handled the statistics.

5.10 Applications in Management

5.10.1 *A few business uses*

In the field of operations management, results of many processes fall along the normal distribution curve, which governs many aspects of human performance. Human resource professionals often use the normal distribution to describe employee performance. A diversified portfolio will typically have returns that fall in a normal distribution, which is often a rough substitute for any distribution that is symmetrically distributed about an axis and is unimodal.

5.10.2 *Application in engineering/management quality control: Six Sigma program*

This program is used to control work process and products quality using the normal distribution (Fig. 5.11).

6σ

Fig. 5.11   Six Sigma program logo.

In the USA, Motorola (1986) has applied the general principle in its plants and offices, then General Electric (1990) followed.

Reasoning adopted: In the normal distribution, there is only 0.0013 probability that the variable X is distant of the mean y more than three standard deviations, i.e. 0.26%. If A is the tolerance of produced parts and σ is the standard deviation of produced parts, having $3\sigma = A$, there will be only 0.26% of the parts that will be rejected or 2,600 defectives for one million parts produced. If $6\sigma = A$, it would be two parts for one billion of parts produced. This argument can be based on either the confidence interval, or the tolerance interval, of the Gaussian distribution since these intervals are very close to each other.

5.10.2.1   *Background*

The starting problem began in the late 1970s when the American economy had to compete with emerging economies from the East, most particularly Japan. It was found that American manufacturing products lag behind the Japanese products in quality and reliability. Efforts made to catch up led to the careful study of the Japanese methods, in particular those popularized by Professor Genichi Taguchi, called Taguchi methods. These are essentially experimental design protocols that permit one to tighten some practices, leading to higher quality and reliability. Subsequently, these methods were improved and taught at the North American universities. Another practice from Japan was studied by managers: *the quality circles*, by which managers from a manufacturing area form a group to study and discuss about the quality of their work and the products they put in the market and how to improve their quality.

5.10.2.2   *Design of experiments and Taguchi's approach*

Taguchi developed his experimental theories independently. He read works following Fisher only in 1954. His framework for design of experiments is idiosyncratic and often flawed, but contains much that is of enormous value. He made a number of innovations. Taguchi's designs aimed to allow greater understanding of variation than did many of the traditional designs from the classical analysis of variance (following Fisher). Taguchi contended that conventional sampling is inadequate here, as there is no way of obtaining a random sample of future conditions.

In Fisher's design of experiments and analysis of variance, experiments must aim at reducing the influence of nuisance factors to allow comparisons of the mean treatment effects. Variation becomes even more central in Taguchi's thinking.

Taguchi methods are statistical methods, or sometimes called *robust design methods*, developed to improve the quality of manufactured goods, and more recently, also applied to engineering, biotechnology, marketing, and advertising. His work includes three principal contributions to statistics: a specific loss function, the philosophy of off-line quality control, and innovations in the design of experiments.

However, Taguchi insisted that manufacturers should broaden their horizons to consider cost to society. Though the short-term costs may simply be those of non-conformity, any item manufactured away from nominal would result in some loss to the customer or to the wider community through early wearout.

5.10.2.3 *Education and training according to Six Sigma*

Belonging now to Motorola, this educational program boasts several levels of competency, similar to those in martial arts, all in the field of management and quality. It was developed after quality control processes, including Taguchi's approach, proved to be very popular.

The term *Six Sigma* (capitalized because it was written that way when registered as a Motorola trademark on December 28, 1993). Motorola set a goal of *Six Sigma* for all of its manufacturing operations, and this goal became a byword for the management and engineering practices used to achieve it.

Six Sigma was registered on June 11, 1991 as US Service Mark 1,647,704. In 2005, Motorola attributed over USD 17 billion in savings to Six Sigma.

The methodology has five phases: *Define* the system, listen to the voice of the customer and their requirements, and understand the project goals, specifically. *Measure* key aspects of the current process and collect relevant data, and calculate the "as-is" process capability. *Analyze* the data to investigate and verify cause-and-effect relationships. Determine what the relationships are, and attempt to ensure that all factors have been considered. Seek out root cause of the defect under investigation. *Improve or optimize* the current process, based upon data analysis using techniques, such as design of experiments, mistake proofing, and standard work to create a new, future state process. Set up pilot runs to establish process capability. *Control the future state process* to ensure that any deviation from the target is corrected before it results in defects. Implement control systems such as statistical process control, production boards, visual workplaces, and *continuously monitor* the process. This process is repeated until the desired quality level is obtained. Naturally, there are criticisms of Six Sigma.

5.10.2.4   *Criticisms*

They are essentially as follows:

(a) **Lack of originality:** Quality expert Joseph M. Juran described Six Sigma as "a basic version of quality improvement", stating that "there is nothing new there. They've adopted more flamboyant terms, like belts with different colors. I think that concept has merit to set apart, to create specialists who can be very helpful". The American Society for Quality long ago established certificates, such as reliability engineers. Quality expert Philip B. Crosby pointed out that the use of "Black Belts" as itinerant change agents has fostered an industry of training and certification.

(b) **Overselling:** Critics have argued there is overselling of Six Sigma by too great a number of consulting firms, many of which claim expertise in Six Sigma when they have only a rudimentary understanding of the tools and techniques involved. In most cases, more attention is paid to reducing variation and searching for any significant factor and less attention is paid to developing robustness in the first place (which can altogether eliminate the need for reducing variation).

(c) **Creativity and Six Sigma:** Stifling creativity in research environments: According to an article, the use of Six Sigma is inappropriate in a research and development environment. A critic wrote as follows: "Excessive metrics, steps, measurements and Six Sigma's intense focus on reducing variability water down the discovery process. Under Six Sigma, the free-wheeling nature of brainstorming and the serendipitous side of discovery is stifled". He concludes: "There's general agreement that freedom in basic or pure research is preferable while Six Sigma works best in incremental innovation, when there's an expressed commercial goal".

5.10.3   *An application of the normal project management*

5.10.3.1   *S-shaped curves used in project management*

In project management, senior managers have to first budget the whole project and the total length of time until completion. The budget is then divided into different periods of time, usually months, or quarters. Large construction projects can last a few years and high-tech development projects even longer, with a lot of uncertainties. It is important to note that although science and statistics take an important part in the conduct of these projects through different techniques used in operations research and management science, realistic accounting methods coupled with effective management of human and material resources are the first prerequisites for success.

How do time and money come together? Money allocated to the project is divided into different lumps, according to the manager's best judgment, destined to be spent in consecutive periods. Typically, we have budgets for the first, second, and third year if this is a three-year project. Within each year, this budget is allocated to the first, second, third, and fourth quarters. This is where the S-curves come into play (see Comrey (1975)). S-curves reflect the usual different phases, which are as follows: start-up, picking up, cruising speed for full work and finally slowing down. The more curved they are, the more concentrated are the activities in the corresponding period. However, they are usually stretched out on different periods of time. The cumulative normal is one of them and has a moderate curvature which seems to fit several kinds of activities or projects.

Eight types of curves, with precise mathematical expressions, are used to represent the different ways activities can be planned.

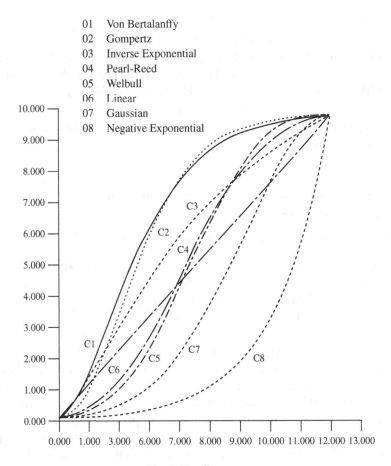

01	Von Bertalanffy
02	Gompertz
03	Inverse Exponential
04	Pearl-Reed
05	Welbull
06	Linear
07	Gaussian
08	Negative Exponential

Fig. 5.12   S-curves.

They are drawn here with the same end points to show the different degrees of curvature. Precisely plotted on a computer terminal, they offer better tools for planning and control, but can never replace a competent manager's experience.

5.10.3.2   *Equations of S-curves*

Figure 5.12 shows the following S-curves:

(1) Von Bertalanffy curve, $y = c - a \exp\left(-\frac{bx}{3}\right)^2$

(2) Gompertz curve, $y = c \exp(-a \exp(-bt))$

(3) Cumulative inverse exponential distribution, $y = a[1 - \exp(-bt)]$

(4) Pearl–Reed logistic curve, $y = \frac{c}{1 + a \exp(-bx)}$

(5) Cumulative Weibull distribution, $y = c \exp[-(\frac{x}{\lambda})^k]$

(6) Linear function, $y = a + bx$

(7) Cumulative Gaussian distribution, $y = \frac{1}{b\sqrt{2\pi}} \int_{-\infty}^{x} \exp(-\frac{(t-a)^2}{2b^2}) dt$

(8) Exponential function, $y = c \exp(x^a)$.

In project cost control, various methods have been developed, applied, and even required. Under the US Department of Defense (DOD) guidelines, within a "Work Breakdown Structure", the "Budgeted Cost of Work Performed" (BCWP) plays a major role in monitoring the cost of a work package. Together with the "Budgeted Cost of Work Scheduled" (BCWS) and the "Actual Cost of Work Performed" (ACWP), the BCWP provides a basis for cost analysis and control. Theoretically, BCWP is an excellent tool for making cost forecasts. However, the estimated amount of "real" work accomplished can be difficult to establish from a laboratory, a plant, or a building site, resulting in inaccurate values for the BCWP. In most research and development projects, the "real" amount of work finished yields the value of the BCWP and different solutions are suggested for its evaluation: (1) subjective evaluation, (2) weighted average of work items, and (3) Using work sub-packages with duration and cost identified at the start.

The project manager has to monitor the project status and ensure that everything is progressing as planned. For example, if at the end of the second quarter, 60% of the work has been done while 55% of the budget is spent, then the project is within time and within budget. On the contrary, if 50% of the budget is spent while only 35% of the work is done, we are obviously overspending and late for project completion. We then have to take corrective measures, such as increasing work time and reducing expenses. But most likely, we have to seek some extra funds to pay for augmented work. At the end of each calendar year, a summary of work completed vs budget unspent has to be made and residual funds, if any, have to be allocated to the following year, while work to be completed is assigned to that year, and another cycle of budgeting, planning and control begins.

5.10.3.3 *Project control*

Besides the BCWP mentioned earlier, other notions can be used for control: These quantities are related to the real amount of work that has been realized at the time of review (see Pham-Gia (1985)). The "Expected Cost of Remaining Work" (ECRW) represents the real value of remaining work to be accomplished, for project completion, according to preset budget, taking inflation and material price increase into consideration.

The notions presented above are used in the planning and control of the start-up phase of a new manufactured product, as presented in the following application.

5.10.3.4   Application: Tooling-up in a high-technology environment: Can it be controlled? (Pham-Gia and Middleton, 1990)

Tooling-up, or start-up, is an important phase in the manufacturing of a new product. In most manufacturing industries, tooling-up begins as soon as the prototype built by the experimental engineering department is judged satisfactory from the standpoint of aesthetics and performance, and full production is decided upon. For some products that could affect public safety, one or several tests have to be performed. For example, for aircraft reactors, the number of continuous satisfactory flight hours has to be above a certain range, and in the bird ingestion test, the reactor would not explode when a small bird is caught by its blades while in the air. The prototype is usually not in its final form yet and is generally composed of a combination of old and new components and sub-assemblies. The components may be new, as may be their mode of assembly or testing. Tooling-up is often conducted under its own conditions, possibly with a separate management team, an accounting sub-system, and a separate budget (Baloff, 1970). It involves everything from the assembly of tools, equipment, and machines required to manufacture and inspect all new parts to the production and inspection of a limited number of each of these parts to ensure that no obstacle or undetected problem remains. Ultimately, the project is turned over to production engineering, which manufactures the new product for market, according to blueprints provided. Depending on the industry, tooling-up can take anywhere from a few days to several months and can involve many people (Fig. 5.13).

Here, we deal with a hypothetical large company in a high-technology context, such as aerospace or electronics, with special reference to the former. Tooling-up for a new, modified, or improved

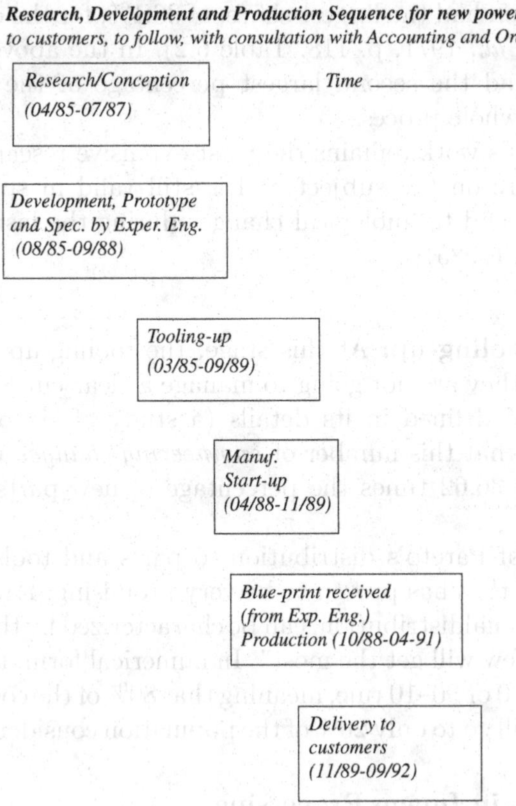

Research, Development and Production Sequence for new powerplant WP100. *Delivery to customers, to follow, with consultation with Accounting and Order Dept.*

Research/Conception
(04/85-07/87) ———— Time ————

Development, Prototype
and Spec. by Exper. Eng.
(08/85-09/88)

Tooling-up
(03/85-09/89)

Manuf.
Start-up
(04/88-11/89)

Blue-print received
(from Exp. Eng.)
Production (10/88-04-91)

Delivery to
customers
(11/89-09/92)

Fig. 5.13   Tooling-up within production Gantt Chart.

aerospace product is usually a project (or sub-project) by itself, with a budget easily running into the millions of dollars. When there are innovative concepts and designs, the similarity between aerospace and electronics has been shown by Mansfield *et al.* (1971) and his students during the early 1970s. Tooling-up corresponds to Stage 4 of their "five-stage breakdown of an innovative process". Based on a sample of cost data and time data from 29 large companies classified under the three general categories of Chemicals, Machinery, and Electronics, they found that tooling-up had the second largest share of the total cost of the whole process. This process goes from applied research to development and ends with manufacturing start-up. Tooling-up came immediately after prototype or pilot plant building, with respective percentages of 37.1% and 30.4% for Machinery and Electronics. It even had the largest share, of 36.1%, for the entire sample (Mansfield *et al.*, 1971, p. 118, Table 6.2). In the above two fields, tool-up also had the second largest percentage of the total elapsed time for the whole process.

To date, Mansfield's work remains the most extensive research of an econometric nature on the subject and is still valid in spite of rapid socio-economic and technological changes during the last four decades (see Clawson (1985)).

Remarks

(a) **Planning for tooling-up:** At this stage, the tooling-up team must accept the fact they are not going to manage a clear-cut, stable project, which is well defined in its details (a study of six of our past projects reveals that this number of *engineering changes* varies widely from 20.65 to 36.62 times the percentage of new parts (see Quinn (1985))).

(b) The application of Pareto's distribution to parts and tools is a "short-cut" approach that has proven to be very promising. Pareto's distribution, or rather maldistribution, can be characterized by the following rule: "A vital few will get the most". In numerical form, this is also known as the 80–20 or 90–10 rule, meaning that 80% of the cost (or effort, or problems) will go to only 20% of the population considered.

5.11   Applications in Image Processing

5.11.1   *Blur and noise*

Image processing, a relatively new domain, has revolutionized our approach in taking pictures. While at the turn of the 20th century photography was based on the action of sunlight on a sensitive film, modern photography is based on *pixels* (picture cells) and the intensity of sunlight on these pixels. It is now possible to perform a wide variety of operations on these pixels, making digital photography a very exciting discipline and hobby (see also an application in imagery in Chapter 10). Satellite images, transmitted from space by satellites such as Landsat, have contributed to the exploration of various corners of the earth.

Image processing has become an applied science using quite advanced mathematical and statistical tools. Medical imaging is the technique and process of creating visual representations of the interior of a body for clinical analysis and medical intervention, as well as visual representation of the function of some organs or tissues (physiology). Medical imaging seeks to reveal internal structures hidden by the skin and bones, as well as to diagnose and treat disease. It also establishes a database of normal anatomy and physiology to make it possible to identify abnormalities. It is the set of techniques that non-invasively produce images of the internal aspect of the body. In this restricted sense, medical imaging can be seen as the solution of mathematical inverse problems. This means that cause (the properties of living tissue) is inferred from effect (the observed signal). The term "non-invasive" is used to denote a procedure where no instrument is introduced into a patient's body, which is the case for most imaging techniques used.

The normal distribution has been applied to two major operations: blurring and noise creation.

Gaussian distribution appears everywhere and is ubiquitous in image processing, given that we usually would want to model noise present in an image, or the image itself, and this is the most natural of distributions.

Blurring an image by a Gaussian distribution is called Gaussian smoothing or blurring. It is used to reduce image detail with the visual effect of a smooth blur, like looking at the image through a translucent screen. In the frequency domain, we convolve the image with a normal distribution and reduce the image's high-frequency components. The $N(0, \sigma^2)$ distribution is used at each pixel. In two dimensions, we use the product of two independent densities $N(0, \sigma^2)$, i.e. the density (see Chapter 7);

$$ f(x, y) = \frac{1}{2\pi\sigma^2} \exp\left(-\frac{x^2 + y^2}{2\sigma^2}\right). $$

We need to discretize this continuous Gaussian function and store it as discrete pixels (Gonzalez *et al.*, 2004). For example, an integer-valued 5×5 convolution kernel with $\sigma = 1$ is given as follows:

$$ \frac{1}{273} \begin{bmatrix} 1 & 4 & 7 & 4 & 1 \\ 4 & 16 & 26 & 16 & 4 \\ 7 & 26 & 41 & 26 & 7 \\ 4 & 16 & 26 & 16 & 4 \\ 1 & 4 & 7 & 4 & 1 \end{bmatrix}. $$

5.11.2   *Filter*

A filter is a combination of a mask (a rectangle with sides of odd lengths) and a function.

By moving the mask over the image, using the function to compute the new value at each pixel, the new image is the filtered image. Following are a 3×5 mask M, with center at $m(0,0)$ and a neighborhood $P(3 \times 5)$, of pixel $p(i,j)$:

$$M = \begin{bmatrix} m(-1,-2) & m(-1,-1) & m(-1,0) & m(-1,1) & m(-1,2) \\ m(0,-2) & m(0,-1) & m(0,0) & m(0,1) & m(0,2) \\ m(1,-2) & m(1,-1) & m(1,0) & m(1,1) & m(1,2) \end{bmatrix},$$

$$P = \begin{bmatrix} p(i-1,j-2) & p(i-1,j-1) & \cdots & p(i-1,j) & p(i-1,j+1) & p(i-1,j+2) \\ p(i,j-2) & p(i,j-1) & \cdots & p(i,j) & p(i,j+1) & p(i,j+2) \\ p(i+1,j-2) & p(i+1,j-1) & \cdots & p(i+1,j) & p(i+1,j+1) & p(i+1,j+2) \end{bmatrix}.$$

We apply the formula for the new value of the pixel:

$$p^*(i,j) = \sum_{s=-1}^{1} \sum_{t=-2}^{2} m(s,t)p(i+s,j+t).$$

Each pixel value becomes a weighted average of pixels in its neighborhood, with the original pixel value having the most important weight. Gaussian smoothing is frequently used for edge detection since its application reduces the noise level in the image and improves the result in subsequent edge detection.

5.11.3   *Gaussian filtering*

The Gaussian filter works by using the $2D$ distribution as a point-spread function. This is achieved by convolving the $2D$ Gaussian distribution function with the image. We need to produce a discrete approximation to the Gaussian function. This theoretically requires an infinitely large convolution kernel, as the Gaussian distribution is non-zero everywhere. Fortunately, this distribution is very close to zero at about three standard deviations from the mean. Around 99% of the distribution falls within three standard deviations. This means we can normally limit the kernel size.

Gaussian filtering is used to remove noise and detail. It is not particularly effective at removing salt and pepper noise. Gaussian filtering is more effective at smoothing images. It has its basis in the human visual perception system. It has been found that neurons create a similar filter when processing visual images.

Digital cameras generally include specialized digital image processing hardware — either dedicated chips or added circuitry on other chips — to convert the raw data from their image sensor into a color-corrected image in a standard image file format.

5.11.4   *White noise*

5.11.4.1   *White noise in image processing*

In imagery, a random vector (i.e. a partially indeterminate process that produces vectors of real numbers) is said to be a *white noise vector* or *white random vector* if its components each have a distribution with zero mean and finite variance, and are statistically uncorrelated, i.e. their joint probability distribution must be the product of the distributions of the individual components (see Chapter 9 on whitening). White noise is often a nuisance without being a real disturbance.

It is often incorrectly assumed that Gaussian noise (i.e. noise with a Gaussian amplitude distribution) necessarily refers to white noise, yet neither property implies the other. Gaussian refers to the probability distribution with respect to the value of the signal falling within any particular range of amplitudes, while the term "white" refers to the way the signal power is distributed (i.e. independently) over time or among frequencies.

Whether the unwanted *sounds* are snoring, traffic, or noisy neighbors, white noise offers the *sleeping* solution. White noise is also excellent for sound *masking* in an office setting. Because real white noise allows the brain to ignore distracting ambient *sounds*, white noise is often used in the workplace.

In signal processing, *noise* is a general term for unwanted (and, in general, unknown) modifications that a signal may suffer during capture, storage, transmission, processing, or conversion.

Sometimes the word is also used to mean signals that are random (unpredictable) and carry no useful information, even if they are not interfering with other signals or may have been introduced intentionally, as in comfort noise.

Noise reduction, the recovery of the original signal from the noise-corrupted one, is a very common goal in the design of signal processing systems, especially filters. The mathematical limits for noise removal are set by information theory, namely the *Nyquist–Shannon* sampling theorem.

5.11.4.2   *White noise stochastic process*

We define first $X(t)$ as a wide-sense stationary (WSS) process if

$$\mathbb{E}[X(t)] = \mu,$$

$$R_X(t, s) = \mathbb{E}[X(t)X(s)] = R_X(|s - t|).$$

A WSS process $W(t)$ is a continuous-time white noise process if its autocorrelation function is $R_W(t) = \sigma^2 \delta(\tau)$, where δ is the Dirac function.

White noise is hence the generalized mean-square derivative of the Wiener process.

5.12   Conclusion

We have presented some applications of the univariate normal in various disciplines, but we are barely scratching the surface. The normal distribution intervenes deep inside many fields (Fig. 5.14). More

What is the Normal Distribution?
Statistician and Chemist Jack Youden's Graphic:

THE
NORMAL
LAW OF ERROR
STANDS OUT IN THE
EXPERIENCE OF MANKIND
AS ONE OF THE BROADEST
GENERALIZATIONS OF NATURAL
PHILOSOPHY ♦ IT SERVES AS THE
GUIDING INSTRUMENT IN RESEARCHES
IN THE PHYSICAL AND SOCIAL SCIENCES AND
IN MEDICINE AGRICULTURE AND ENGINEERING ♦
IT IS AN INDISPENBABLE TOOL FOR THE ANALYSIS AND THE
INTERPRETATION OF THE BASIC DATA OBTAINED BY OBSERVATION AND EXPERIMENT

Fig. 5.14   What is the Normal Distribution? Statistician and Chemist Jack Youden's Graphic.

advanced applications exist but will demand, from the reader, more knowledge on the field concerned. But we have not mentioned the Applications of Probabilities in everyday life, in risk management and in strategic modeling. These duties are often the responsibilities of ministries, or governmental agencies, which do not want to expose publicly any weakness. Insurance industry, for example, through their actuaries, determine pricing of insurance premiums of all kinds. Governments apply probabilistic methods for environmental regulation, as well as to evaluate ripple effects of political happenings on the economy. The probabilities are never assessed independently or very rationally, and never subjected to an enquiry. Casinos use probabilistic methods to provide payouts to players, while keeping an advantage in their games. Finally, industrial manufacturers use reliability to plan for their products to assure profit. The interested reader should consult directly the articles presented in order to obtain other details.

Bibliography

Baloff, N. (1970). Start-up management, *IEEE Transactions on Engineering Management*, **EM-17**(4), 132–141.

Clawson, R.T. (1985). Controlling the manufacturing start-up, *Harvard Business Review*, **63**(3), 6–20.

Comrey, A.L. (1975). *Elementary Statistics: A Problem Solving Approach* (The Dorsey Press, Homewood, Illinois).

Fischer, C.S., Hout, M., Jankowski, M.S., Lucas, S.M., Swidler, A. and Voss, K. (1996). *Inequality by Design, Cracking the Bell Curve Myth* (Princeton University Press, New Jersey).

Gonzalez, R.C., Woods, R.E. and Eddins, S.L. (2004). *Digital Image Processing Using MATLAB* (Publishing House of Electronics Industry, New York).

Herrnstein, R.J. and Murray, C. (1994). *The Bell Curve: Intelligence and Class Structure in American Life* (Free Press, New York).

Kotz, S., Lumelskii, Y. and Pensky, M. (2003). *The Stress–Strength Model and Its Generalizations* (World Scientific, Singapore).

Mansfield, E. Rapaport, J., Schnee, J., Wagner, S. and Hamburger, M. (1971). *Research and Innovation in the Modern Corporation* (W.W. Norton, New York).

Olkin, I., Gleser, L.J. and Derman, C. (1994). *Probability Models and Applications* (Macmillan, New York).

Pham-Gia, T. (1985). The elusive budgeted cost of work performed, *Project Management Journal*, **16**, 76–79.

Pham-Gia, T. and Middleton, G. (1990). Tooling up in a high tech environment: can it be controlled?, *SAM Advanced Management Journal*, **55**(1), (Winter 1990): 14–19.

Pham-Gia, T. and Turkkan, N. (1994). Estimation of the median safety factor in a lognormal stress–strength model, *International Journal of Reliability, Quality and Safety Engineering*, **1**(4), 417–431.

Pham-Gia, T. and Turkkan, N. (2007). The joint predictive posterior method in the Bayesian analysis of stress–strength reliability, *International Journal of Reliability, Quality and Safety Engineering*, **14**(1), 29–48.

Quinn, J.B. (1985). Managing innovation: controlled chaos, *Harvard Business Review*, **63**(3), 73–84.

Chapter 6

Bayes Approach in Applied Probability

6.1 Introduction

In a Bayesian statistics research and development project, the three main parts are as follows: the prior, the likelihood (or sampling), and the posterior (and the predictive). They should be consecutively monitored after a particular model is chosen and followed through until completion (see, e.g. Pham-Gia (1989)).

The plan of this chapter is as follows. In Section 6.2, we present the general *Bayesian paradigm*. In Section 6.3, various questions related to the prior normal distribution are discussed: *elicitation*, *non-informative*, *least informative* and *reference priors*. The sampling phase is treated in Section 6.4, with the *conjugacy* property highlighted. The *posterior distribution* is dealt with in Section 6.5 and the *predictive distribution* in Section 6.6. Finally, Section 6.7 deals with various sub-domains of Bayesian statistics, including *Bayesian Decision Theory*, *Sequential Bayes*, and *Empirical Bayes*. In particular, Section 6.8 is in the form of an appendix and presents the known results in the conjugate Bayesian theory, where the prior and the likelihood have similar mathematical expressions, resulting in a posterior distribution expressible in closed form. Throughout the chapter, our focus is on the normal.

6.2   The Bayesian Paradigm

Classical, or *frequentist*, statistics is based on the basic principle that the parameters of a distribution are constants, and a statistical model is used, on which we base our analysis and inference. Usually, it is the normal model, and the most popular topics are *hypothesis testing*, *regression, estimation*, and *analysis of variance*. The meaning of the probability of an event is never clearly stated, but it could be a mixture of *Laplace symmetry principle* and *von Mises limit proportion*. In the 20th century, other meanings of probability have been suggested, in particular, subjective notions associated with betting (see also Chapter 8). Works by Savage, and by De Finetti on exchangeable sequences, have provided a solid basis to the *subjective approach* to probability. These should be at least three kinds of probability: physical, logical, and subjective (Good, 1965), leading to three areas of Bayesian statistics: *Logical Bayes* (Keynes, Carnap), *Subjectivist Bayes* (Ramsey, Savage, de Finetti), and others, including *Empirical Bayes* (developed by Robbins). The interested reader can consult the thorough treatise on the foundations of different probability theories by Fine (1973), who concluded that "of all the theories considered, subjective probability holds the best position with respect to the value of probability conclusions, however arrived at" (p. 240).

Subjective probability concepts constitute the main engine that permitted the development of the Bayesian approach to statistics, which is based on Reverend Thomas Bayes (1702–1761) simple formula:

$$\mathbb{P}(A|B) = \frac{\mathbb{P}(A \cap B)}{\mathbb{P}(B)}, \quad \mathbb{P}(B) > 0,$$

when considering single events A and B, and

$$\mathbb{P}(A_k|B) = \frac{\mathbb{P}(B|A_k)}{\sum\limits_{j=1}^{n} \mathbb{P}(B|A_j)}$$

for the set of exhaustive events $\{A_j\}_{j=1}^{n}$. In the formal *Bayesian paradigm*, the parameter θ under consideration is a random variable with a *prior distribution* $g(\theta)$, defined in $\Omega \subset \mathbb{R}^n$. In the statistical continuous model $f(x|\theta)$, a value of θ serves to determine the distribution, from which a sample $X = (X_1, \ldots, X_n)$ will be taken.

The *likelihood function* $\prod_{i=1}^{n} f(x_i|\theta)$, *where θ is now considered as the variable*, can be combined with the prior to obtain the *posterior distribution* of θ, according to the Bayes theorem. This is the *conditional distribution* of θ, given X, and we have

$$g(\theta|X_1,\ldots,X_n) = \frac{g(\theta)\prod_{i=1}^{n} f(x_i|\theta)}{\int_{\Omega} g(\theta)\prod_{i=1}^{n} f(x_i|\theta)d\theta}.$$

We can see that the posterior is proportional to the product (prior by likelihood), denoted as

$$\textbf{Posterior} \propto (\textbf{Prior} \times \textbf{Likelihood}).$$

Often, a *sufficient statistic* of θ, $\tau_n = k(X_1,\ldots,X_n)$ is considered, with density $h(\tau_n|\theta)$, and we have

$$g(\theta|X_1,\ldots,X_n) = \frac{g(\theta)h(\tau_n|\theta)}{\int_{\Omega} g(\theta)h(\tau_n|\theta)d\theta}. \tag{6.1}$$

All information on the parameter θ, i.e. the prior information, as well the sampling results, being contained in the posterior distribution, it is natural that any Bayesian inference should be based on this distribution.

6.3 The Prior Distribution

The prior is an indispensable component of the Bayesian paradigm, and, strictly speaking, it should be based on subjective probability. However, in practice, its derivation can come from personal beliefs, historical data, or other means of information, or a combination of these sources. The prior distribution is the *weakest link* in the Bayesian chain, a situation that, in many cases, has seriously prevented the application of Bayesian statistical methods in real-life projects. Concerning this problem, Fine (1973, p. 240) wrote as follows: "The measurement problem in subjective probability is sizable and conceivably insurmountable".

One of the goals of the determination of the posterior is to obtain a tighter posterior distribution of the parameter, resulting in a shorter credible interval for the estimation of its value, as shown in Fig. 6.1, where the curve Obs. Likelihood refers to the normalized likelihood,

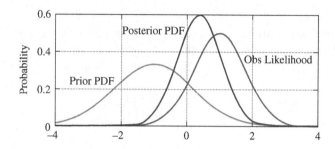

Fig. 6.1   Distributions of the parameter θ in a Bayesian approach.

with the parameter as variable. Besides the posterior, the posterior predictive distribution of the variable and the posterior marginal distribution of the parameter, if they can be derived, can serve for estimation as well.

6.3.1   *The normal as a prior: Its elicitation*

Elicitation of the prior, and in fact, the establishment of a probability distribution for an event, or a parameter, has been the subject of much study and debate (Winkler, 1967). Since the probability is taken here as a *degree of belief*, it has a subjective meaning and is closely associated with the *anchoring phenomenon* in human quantitative evaluations (Kahneman *et al.*, 1982, p. 14),[1] with the consequence that different initial values suggested by the formulation of the problem may yield different estimates. Various biases inherent to human judgments, and evidenced by experiences in psychology, are also causes for concern in elicitation, and less subjective information, such as the histogram of relative frequencies (Raiffa and Schlaifer, 1961), can also serve as input to assist in making quantitative judgments. In general, this is a common area between statistics and psychology that still needs much attention, and close cooperation between experts in these two disciplines has been called for to alleviate serious difficulties that plague human assessment of probability. Hogarth (1975) should be consulted for a general view of the whole question. For the normal, it is highly convenient that we only need to determine its mean and its standard deviation. We mention

---

[1]Nobel prize in Economics in 2002.

below a few practical approaches used, being fully aware of their shortcomings.

6.3.1.1 *Methods based on interviews*

In practice, quantitative information is obtained from the *assessor or expert*, who can be different from the applied statistician in charge of obtaining the prior distribution. One important remark to be made is that the normal is theoretically a variable defined on $(-\infty, +\infty)$, so taking possibly negative values. In applications, we have, often, only positive values to consider. Hence, the normal can be restricted to $(\mu - 3\sigma, \mu + 3\sigma)$, where there is 99.74% of its probability. The assessor could use the following:

(a) The mean absolute deviation method: This method is based on the idea that the *mean absolute deviation* about the mean of the normal, denoted δ_1, should be an information easier to obtain for the assessor, since it reflects an error independent of the sign. It is easier for most people interviewed to understand the meaning of δ_1 than to grasp the meaning of the variance or of the standard deviation.

From the assessed values of δ_1 and the assessed value of the mean obtained first, we can deduce the values of the normal parameters (see Pham-Gia *et al.* (1992)) since we have $\delta_1(X) = \sqrt{\frac{2}{\pi}}\sigma$.

(b) The percentiles method: Methods to directly determine the density, or the cumulative distribution of the prior distribution, are generally based on estimated percentiles to be obtained from the assessor. Examples of questionnaires to be used in interviews for that purpose are given in Winkler (1967). For example, the assessor is asked to provide the numerical values of two percentiles of his prior density, say $C_{.90}$ and $C_{.10}$. A computer program will provide values of the parameters by numerical search. A variation of this method asks for the values of the mean and the 5th (or 95th) percentile. It is worth noting that Kahneman *et al.* (1982, p. 17) has reported that "actual values of the assessed quantities are either smaller than $C_{.01}$ or larger than $C_{.99}$, reflecting an over-confidence of the expert in his judgment". Corrective measures have been suggested. For example, Bunn (1975) suggested a hysteresis-based interview strategy, to anchor some of the biases, but recognized that other more important biases could still be present in the interviewing process.

(c) The Pearson–Tukey approach: In this approach to estimate densities in general, a fixed set of weights is used on three percentiles $C_{.05}$, $C_{.50}$, $C_{.95}$, provided by the assessor, to obtain the *estimated mean*, $\hat{\mu} = 0.185(C_{.05} + C_{.95}) + 0.63C_{.50}$, and the estimated variance in two steps:

$$\hat{\sigma}_0^2 = \left[\frac{(C_{.95} + C_{.05})}{3.25} \right]^2$$

and

$$\hat{\sigma}^2 = \frac{(C_{.95} + C_{.05})}{\left[3.29 - 0.1(C_{.95} + C_{.05} - 2C_{.50})/\sigma_0^2\right]}.$$

In their investigation, Keefer and Bodily (1983) found that the above method is the most accurate among three-point estimation method for a probability density. Zaino and d'Enrico (1989), however, claimed that their method, based on the 4th, 50th, 96th percentiles, provides an improvement over the above method.

(d) The PERT approach: Finally, in classical Operations Research PERT Time (Program Evaluation and Review Technique), a simple procedure is used to elicit the general beta, or the normal: the max, min, and most likely (M) values of the distribution are obtained from the expert, and we take as approximate values,

$$\hat{\mu} = \frac{\max + 4M + \min}{6}$$

and

$$\hat{\sigma} = \frac{\max - \min}{6}.$$

This method has been highly criticized, however, because of its inaccuracy.

6.3.1.2 *Methods based on historical data*

These classical methods (histogram, curve fitting, moments method, etc.) well known in applied statistical estimation, strictly speaking, do not belong to Bayesian statistics and neither do their variations/improvements. They are, at best, associated with *parametric empirical Bayes*. They also suppose that the past has a significant input into the future and that the long-range frequency interpretation of probability can be reconciled with its personal degree of belief interpretation. However, they can provide useful information to the assessor, on which he/she will build his/her prior as some other similar methods, such as the estimation from data, of the αth, and $(1 - \alpha)$th percentiles.

6.3.1.3 *Working with the normal*

Several points are raised when working with the normal in the Computational Appendix in Chapter 7. Some of the recommendations made there can be applied here to derive the best normal prior.

6.3.2 *Special priors*

6.3.2.1 *Non-informative priors*

Since the prior introduces subjective information into the Bayesian process, it is thought that a prior distribution that brings in no information at all would provide a good starting reference point. Short of this complete absence of information, the least information possible brought in would be acceptable. The general idea of a prior distribution in Bayesian statistics, that would behave neutrally enough so that the *data can speak for itself*, is a complex question that still has not found a satisfactory answer. Several methods have been suggested, the most accepted one is probably Jeffrey's, based on data invariance, which leads to a density proportional to the square root of Fisher's information (Berger, 1985, p. 87).

$$I(\theta|X) = -\mathbb{E}_{X|\theta}(\partial^2 L/\partial\theta^2), \qquad (6.2)$$

where $L(\theta|X)$ is the log-likelihood function of the parameter. The basic argument used here is that if there is no information about θ, then there should be no information about a transformed ϕ of

θ as well. More precisely, to define the non-informative prior of θ, we consider a transform ϕ of θ, such that the likelihood function of ϕ is *data-translated*, i.e. its form is completely determined, except for its location that depends only on X. We now assign a (locally) uniform distribution to ϕ, and by the reverse transformation, obtain the non-informative prior for θ.

Considering Fisher's information in this transformation, we can show that the non-informative distribution for θ is proportional to $\sqrt{I(\theta|X)}$ as in (6.2).

An exhaustive list of non-informative prior distributions has been compiled by Yang and Berger (1998), and it is still growing.

Vague priors and improper priors have their definitions different from non-informative priors. They are not real distributions (do not integrate to one), but would function like ones because the corresponding posteriors are real distributions.

6.3.2.2   *Reference prior*

The *reference prior* is based on the amount of *entropy information* measure on the parameter of interest, that an experiment is expected to provide, and on the missing information about the parameter as a function of the prior $\pi(\theta)$. The reference prior is then the prior that maximizes the missing information functional. It is obtained usually via a limiting process by applying the Bayes theorem to the posterior (see Bernardo and Smith (1994, p. 306)). For example, let

$$K_n(\pi) = \int \pi(\theta|\{x_j\}_{j=1}^n) \log(\pi(\theta|\{x_j\}_{j=1}^n))/\pi(\theta)d\theta$$

be the Kullback–Liebler divergence between the prior $\pi(.)$ and its posterior, with $\{x_j\}_{j=1}^n$ being the sample. We use $\mathbb{E}[K_n(\pi)]$, where the expectation is taken under the marginal distribution of $\{x_j\}_{j=1}^n$ as a measure of missing information. We define the reference prior as the distribution π maximizing $K^*(\pi) = \lim_{n\to\infty} \mathbb{E}[K_n(\pi)]$.

Under some regularity conditions, the reference prior can be characterized in terms of the parametric model, and it agrees with Jeffrey's prior, both being proportional to the square root of Fisher's information. The reference prior can be shown to be independent of the sample size, invariant under one-to-one transformation, and is compatible with sufficient statistics (see Bernardo and Smith (1994, p. 309)).

6.4 Sampling Phase

6.4.1 *The sampling phase*

The *sampling phase* is associated with the likelihood function of classical statistics. If $X_1, \ldots, X_n$ is a sample from a distribution, with density $f(x|\theta)$ indexed by θ, then likelihood function is $\prod_{j=1}^{n} f(x_j|\theta)$. The *likelihood principle* (see Press (1989, p. 56)) states that only information contained in the sample already obtained should be considered, not the one contained in a potential sample. The Bayesian approach is hence in conformity with the *likelihood principle*, unlike several important concepts in frequentist statistics.

The *likelihood principle* itself is the result of two other principles: the *weak conditionality principle* and the *weak sufficiency principle*. Naturally, associated with the normal prior can be any sampling model. We will only discuss normal sampling here, i.e. the sample of observations come from a normal population.

6.4.2 *Conjugacy*

In the general case, when likelihood functions and prior distribution bear no mathematical similarities, the integral (6.1) has to be computed numerically, and no general mathematical expression can be given to the posterior distribution. In some cases, we have the posterior in closed form, thanks to the above-mentioned similarity. The posterior is then obtained by merely updating the parameters of the prior with sampling results. The prior is then called *natural conjugate to the likelihood function* or more simply *conjugate to sampling*. For the exponential family of distributions, of the form

$$f(\theta|X) = h(\theta)g(x)\exp\{\psi(\theta)t(x)\},$$

we can see that the likelihood is

$$l(\theta|X) = h(\theta)^n \exp\left\{\sum t(x_i)\psi(\theta)\right\},$$

and hence, the conjugate prior has the form

$$h(\theta)^v \exp(\tau\psi(\theta)).$$

The notion of conjugacy is important because it allows the convenient handling of mathematical computations. Also, since most densities with complicated expressions can be adequately approximated by a beta, or a normal, in practical situations, approximate posteriors can then be handily obtained.

In order to simplify equation expressions, we will use precision τ, instead of variance, with $\tau = \frac{1}{\sigma^2}$. The normal model $N(\mu, \sigma^2)$ becomes $N(\mu, \tau)$,

$$f(x; \mu, \tau) = \sqrt{\frac{\tau}{2\pi}} \exp\left(-\frac{\tau(x - \mu)^2}{2}\right), \quad -\infty < x < \infty.$$

6.4.3 *Case of the normal:* $N(\mu, \tau)$

There are three cases usually considered:

(1) Precision known, $\tau = \tau_0$, and μ unknown, with prior $N(\mu_0, \tau_0)$.
(2) $\mu = \mu_0$ known, while precision τ unknown, has prior $\text{Gam}(\alpha, \beta)$, a gamma distribution, and
(3) Both μ and τ unknown, with a joint prior distribution, the normal–gamma compound distribution.

A sample $\{X_1, \ldots, X_n\}$ is taken from the population $N(\mu, \tau)$ and permits to derive, by conjugacy, the posterior distributions of μ, of τ and of (μ, τ), respectively. Prior and posterior marginal distributions can also be derived (see Section 6.8 on conjugate Bayesian analysis table for complete expressions).

6.4.4 *Case of the multinormal:* $\mathbf{X} \sim N_k(\mathbf{M}, \mathbf{r} = \mathbf{\Sigma}^{-1})$

$$f(\mathbf{X}) \propto \exp\left(-\frac{(\mathbf{X} - \mathbf{M})^t \mathbf{\Sigma}^{-1}(\mathbf{X} - \mathbf{M})}{2}\right).$$

(a) Mean vector $\mathbf{M}$ variable, precision matrix $\mathbf{\Sigma}_0^{-1}$ given

Prior: $\mathbf{M} \sim N_k(\boldsymbol{\mu}_0, \boldsymbol{\tau}_0)$.

(b) Mean $\mathbf{M} = \mathbf{m}_0$ given, precision $\mathbf{R}$ variable

Prior: $\mathbf{R} \sim Wis(\alpha_0, \boldsymbol{\tau}_0)$, $\alpha_0 > k - 1$ (Wishart distribution).

(c) Both mean $\mathbf{M}$ and precision $\mathbf{R}$ are variable

Multinormal-Wishart Prior: $\mathbf{R} \sim Wis(\alpha_0, \boldsymbol{\tau}_0), \alpha_0 > k - 1$,

$$\mathbf{M} \,|\, \mathbf{R} = \mathbf{r}_0 \sim N_k(\boldsymbol{\mu}_0, \nu \mathbf{r}_0), \quad \nu > 0.$$

Prior and posterior predictive distributions of the parameters, and of the variable, can be derived in closed form.

6.5 The Posterior Distribution

The following sections describe the points that characterize the posterior distribution.

6.5.1 *Computation of the posterior*

Other sampling schemes frequently encountered in the literature include *gamma sampling, Poisson sampling, exponential* and *uniform sampling*. When the beta or normal is taken as a prior for these sampling schemes, numerical methods have to be used to derive the posterior and predictive distributions. This fact could be a hindrance for the establishment of further properties of the posterior or for variables depending on it. In the last few decades, however, advances in computer technology have allowed the application of *computer intensive methods* in the determination of the posterior. Techniques such as the *Gibbs sampler*, the *Metropolis–Hastings algorithm* and the *Monte Carlo Markov Chains* simulation approach certainly helped derive, numerically, the posterior distribution under very extreme conditions. The free software WINBUGS was very popular at one time. Still, the analytic form of the posterior, if available, would be more useful.

6.5.2 *Dominance*

It is immediate that if one of the two factors dominates in this product *prior* × *likelihood*, the posterior will inherit more features from that factor. Dominance here can be thought of as having more peakedness in the region of interest. Hence, for a small sample, the prior will dominate while as n increases, the likelihood function takes over, and, at the limit, the posterior will be close to a *normalized likelihood function*. By virtue of the central limit theorem, that limit is also close to a normal distribution (see Bernardo and Smith (1994, Section 5.3)).

Also, the form of this posterior depends on the sampling result. If there is agreement between the two sources, sampling and prior, the posterior will be tighter than the prior (there will be less dispersion) while if the two information sources are conflicting, the posterior will be more spread out.

A similar interpretation applies for the posterior mean. Let's take the first case of the univariate normal, where τ is known, $\tau = r_0$, and the mean μ has a normal prior $N(\mu_0, \tau_0)$. We know that the posterior distribution of μ is also normal and

$$\mu_{\text{post}} = \mu \,|\, Data \sim N\left(\frac{\tau_0\mu_0 + nr_0\bar{x}}{\tau_0 + nr_0}, \tau_0 + nr_0\right)$$

(see Section 6.8). Writing

$$\mathbb{E}(\mu_{\text{post}}) = \frac{nr_0}{\tau_0 + nr_0}\bar{x} + \frac{\tau_0}{\tau_0 + nr_0}\mu_0, \tag{6.3}$$

we can see that the posterior mean, or mean of the posterior distribution of μ, is a weighted average of the sample mean and the mean of the prior distribution. For $n = 0$, $\mathbb{E}(\mu_{\text{post}})$ is the mean of the prior and for $n \to \infty$, it is the mean of the sample. Between these two extremes, the larger the sample size n gets, the greater will be the weight given to $\bar{x}$.

6.5.3 *The highest posterior density region*

In relation to the posterior distribution of a parameter, the interval with $(1-\gamma)$ probability is called the $(1-\gamma)100\%$ *credible region*. This really means that the probability is $(1 - \gamma)$ that the parameter lies within that region, unlike the $(1-\gamma)100\%$ confidence region in classical statistics. Furthermore, if any point inside the region has a higher probability than any point outside it, we have the *highest posterior density* (hpd) region, which can consist of a single interval or a set of disjoint intervals. Only for symmetrical densities, this interval is the same as the equal-tailed ($\gamma/2$ area on each side) interval. Otherwise, we have to solve the set of equations:

$$\int_{\theta_a}^{\theta_b} f(\theta|x_1, \ldots, x_n)d\theta = 1 - \alpha$$

and

$$\int f(\theta_a|x_1, \ldots, x_n)d\theta = \int f(\theta_b|x_1, \ldots, x_n)d\theta,$$

with $\alpha = \gamma$ or $1 - \gamma$. Turkkan and Pham-Gia (1993) provided an algorithm, based on another approach, to compute the hpd region for general distributions, and an extension of it to the bivariate case is given by Turkkan and Pham-Gia (1997).

6.5.4 *Generalization of the hpd interval*

Our problem is related to Bayesian statistics, but can be dealt with in a more general setting. Let $f(x)$ be the density of the rv X and α an adopted significance level. We look for the interval(s) $\{I_k\}_{k=1}^n$ such that $\sum_{k=1}^n \int_{I_k} f(x)dx = 1 - \alpha$ and also such that any point inside these intervals has higher probability than any point outside them. $(1 - \alpha)$ is the area below the density, i.e. the sum of the areas between the density curve and the horizontal line L at height h to be determined, added to the areas of rectangles below L, determined by the intersections of L and the density. It is called the highest posterior density set if $f(x)$ is the posterior density. Naturally, if $f(x)$ is unimodal and symmetric, like the case of a single normal univariate density, the solution is simpler and we have only to take the interval $[-k, +k]$, where $k = 100(1 - \alpha/2)$th percentile of the density. However, when $f(x)$ becomes multimodal, as is the case of discrete mixtures of normal variables, an appropriate approach has to be taken. The hpd intervals, or areas, or volumes (for multidimensional spaces), are also the intervals, or areas, volumes with minimum lengths, or surfaces, or volumes, which satisfy the probability conditions.

6.6 The Predictive Distribution

The predictive distribution of X is also the continuous mixture of the sampling distribution with the prior, or, for some authors, with the posterior. When we consider the mixing of the binomial probability of each value of the sample (of fixed size n), with the prior distribution beta(α, β) of the parameter π, we obtain the prior predictive distribution of X. In other terms, if $X|\pi \sim \text{Bin}(n, \pi)$ and $\pi \sim \text{beta}(\alpha, \beta)$, then $X \sim \text{Bbin}(\alpha, \beta; n)$ called the *beta-binomial or Polya* distribution with density

$$\mathbb{P}(R = r) = \frac{\binom{n}{r} B(\alpha + r, \beta + n - r)}{B(\alpha, \beta)}, \quad 0 \leq r \leq n.$$

The predictive distribution, which is the marginal distribution of X, is used to compute the expected value wrt data, in a sample of size n, of any random variable which is the function of the posterior parameters. For the normal case $N(\mu, \sigma^2)$, it is a normal distribution in the case μ is unknown and has a normal prior and a Gamma–gamma (Gg) distribution in the case $\tau = 1/\sigma^2$ is unknown and has a gamma prior (see Section 6.8).

Note: The distribution of any variable obtained from the prior is called prior marginal distribution of that variable, except the distribution of X, called prior predictive. Similarly, when considering the posterior, we have the posterior marginal and the posterior predictive distributions. Since X is often observable, there is a call to use it in the Bayesian analysis, instead of the parameters, which are not likely to be observable.

Winkler stated as follows: "The major advantage of asking about predictive distributions rather than asking directly about prior distributions is that potentially observable statistics are often easier to understand and relate more directly to an expert's knowledge than unobservable parameters". (see Winkler (1980)).

Moreover, information obtained from an expert on the predictive distribution can also provide some flexibility in terms of model building since we can contemplate different combinations of likelihood functions and prior distributions, such that the statistics given by corresponding predictive distribution would best fit the expert's given values. As an example, Chaloner and Duncan (1983) also used the beta-binomial, but were essentially concerned with the mode. See also Chapter 5 for an example of the joint posterior–predictive approach in Bayesian stress–strength reliability.

6.7 Bayesian Sub-domains

Although the Bayesian paradigm applies in the whole domain of Bayesian statistics, there are distinct sub-domains, where some techniques and approaches are preferred.

6.7.1 *Sequential Bayes*

The sequential approach in classical statistics can be applied to Bayesian statistics and the *stopping rule principle*, which follows from the likelihood principle, is a logical consequence of the Bayesian approach. The beta again plays an important role when dichotomous variables are considered. The normal can also be used in many situations. Basically, here, we take one observation at a time and continuously compare the expected cost of taking the next observation (or its Bayes risk) with the cost of making an immediate decision. We only proceed further if the former is less than the latter. An example is provided by Pham-Gia (1998).

6.7.2 *Empirical Bayes*

Empirical Bayes is concerned with the case where the prior is not known explicitly, only as possibly belonging to a family of distributions, and has to be obtained from data or by another mean. Due to Robbins (1955), it originally dealt with the compound decision problem, and data could be either current or past. It is further divided into parametric empirical Bayes, where the family of distributions is known but not its hyperparameters, and non-parametric empirical Bayes, where no other information is available. In this case, it constitutes a class of decision theoretic procedures, that use past or current data to bypass the necessity of identifying the unknown prior, which has a frequency interpretation however (see Waller and Martz (1982, Chapter 13)).

Although empirical Bayes has certain similarities with the Bayesian methods in general and has relations with important topics in statistics, such as the Stein estimator, it has been frequently argued that the empirical Bayes methods should not be considered as Bayesian.

It is often recognized that the empirical Bayes theory blurs the distinction between estimation and testing as well as between frequentist and Bayesian methods.

6.7.3   *Bayes operator*

The Bayes paradigm hence depends on two components: the prior and the sampling model. The prior and the sampling scheme could belong to mathematically similar, or completely different, families of parametric distributions, i.e. there could be conjugacy or not. However, in all cases, we have a prior distribution mapped into a posterior one, and, under some restrictive hypotheses on the sampling phase, we can define an operator from the space of prior densities to the one of posterior densities. Cuevas and Sanz (1988), for example, have considered such an approach. Although this approach could be highly theoretical, based, for example, on Frechet derivatives, the L^1-metric can be used to measure distances between the priors and posteriors. Also, the use of conjugacy can also help make the study simpler. Under some hypotheses, the very important topic of Bayesian robustness can be tackled with this approach (see also Pham-Gia *et al.* (2006)).

6.7.4   *Bayesian decision theory*

In this section, we recall some basic notions in statistical decision theory first. We then discuss the Bayesian decision theory and important related problems, such as James–Stein estimator and the St. Petersburg paradox. We conclude by considering the cases of loss functions frequently encountered, such as quadratic and absolute values.

(i) Statistical decision theory

Statistical decision theory is a well-developed sub-field of statistics and a *decision-theoretic* approach can be taken in most topics of classical statistics. Ferguson (1967) provides a concrete example. We consider a *parameter space* Ω (also called *state of nature*) and a *decision space* D. For any value $\theta \in \Omega$, which determines a probability distribution $\mathbb{P}$, and any decision $d \in D$, let $\gamma(d, \theta)$ be the *consequence* of choosing d, when the parameter has value θ. The decision

d is chosen on the basis of an observation $x \in X$, subsequent to an experience E. The subjective notion of *utility* is usually used to deal with that consequence. The real-valued function $U(\theta, d)$ is called an *utility function* if, for two distributions $\{\mathbb{P}_1, \mathbb{P}_2\}$, $\mathbb{P}_1$ is not preferred to $\mathbb{P}_2$ if and only if $\mathbb{E}(U|\mathbb{P}_1) \leq \mathbb{E}(U|\mathbb{P}_2)$. Hence, between two distributions, we would prefer the distribution for which the expected utility of the consequence, or $\mathbb{E}(U|\mathbb{P}, d)$, is the larger one. If $\mathbb{P}$ is fixed, we then choose the decision d^*, called the Bayes decision, such that $\mathbb{E}(U|\mathbb{P}, d^*)$ would be maximum. *Bayesian decision theory is the Bayesian approach to statistical decision theory* and introduces a prior distribution on θ. The basic decision problem can be presented as a decision graph, where there are two decision nodes and two random nodes and is worked on backward to maximize the utility function (see Lindley (1972)) as follows:

$$\max_e \left\{ \int_X \left[\max_d \int_\Theta U(d, \theta, e, x) p(\theta|x, e) p(x|e) d\theta \right] dx \right\}, \qquad (6.4)$$

where U is the utility function.

Alternately, the *loss function* $L(\theta, d)$ is the penalty of making decision d when the value of the parameter is θ. The expected loss is called the risk $\mathbb{E}(L|P, d)$, and we should minimize that risk for a given value of $\mathbb{P}$. This duality between utility and loss functions is very convenient and permits looking at any decision problem from two complementary points of view. A comprehensive table is provided by Raiffa and Schlaifer (1961).

(ii) Admissibility

This notion plays an important role since it selects the best decision, according to an effectiveness criterion. Loosely speaking, a decision is admissible only if there is no decision strictly more effective than itself. *Bayes decisions* can be proved to be admissible when the prior is proper, although admissibility is often not accepted as part of the Bayesian approach since it violates the *likelihood principle*. However, the converse, that every admissible rule is the Bayes rule for some prior distribution, is generally not true. One important result is the inadmissibility of the sample mean as an estimator of the population mean for a multivariate normal population. This is known as the James–Stein theorem.

(iii) James–Stein estimator

For a multivariate normal distribution, $\bar{\mathbf{x}}$ is most efficient for each component where each element $\bar{x}_j$ of $\bar{\mathbf{x}}$ is the minimum variance unbiased estimator of μ_j. However, using the loss function $(\mathbf{t} - \boldsymbol{\mu})^t \boldsymbol{\Sigma}^{-1} (\mathbf{t} - \boldsymbol{\mu})$, James and Stein showed that for a certain c, $\mathbf{t} = (1 - \frac{c}{\bar{\mathbf{x}}^t \mathbf{S}^{-1} \bar{\mathbf{x}}}) \bar{\mathbf{x}}$ has a smaller risk than $\bar{\mathbf{x}}$ for all $\boldsymbol{\mu}$, using empirical Bayes methods (see also Muirhead (1983, Chapter 4)).

This estimator has been in use in several important applications, since for $N \geq 3$, the James–Stein estimator everywhere dominates the MLE in terms of expected total squared error.

(iv) Utility function and the St. Petersburg paradox

Money has a face value clearly indicated and is taken into consideration in cold accounting activities. However, in people's mind, when making a decision related to some financial problem, this value might not reflect the face value. For example, 5\$ can mean to a decision maker five times the value of 1\$, but the second million earned might be less valued than the first million. Hence, we usually have a function of utility vs value, which is an increasing function, with graph located below the first diagonal and not necessarily linear. It rather has the form of a tempered exponential, which shows that utility decreases when the face value increases.

This function is used in decision theory, although it could be hard to determine it in real applications. Its elicitation is similar to the personal elicitation of probability.

In the St. Petersburg paradox, a person is offered to pay a sum and double his/her winning every time. If we compute the math expectation of his/her gain, it is infinite so that he/she should pay an infinite amount of money to play the game. Since this is impossible, there is a paradox, that can be solved subjectively, by using his/her utility function.

6.7.5   *Common loss functions in Bayesian decision theory*

(a) General definitions

In the Bayesian decision theory, an estimation, denoted here by $\mathbf{a}$, of the parameter θ is made, based on our knowledge of that parameter, represented by the prior distribution $F(\theta)$, and on a loss function $L(\theta, \mathbf{a})$, which gives the penalty of taking as estimate while the real value is θ. Naturally, the best estimate of θ is the one that minimizes the average value of the loss function, $R(\theta) = \mathbb{E}_\theta\left[L(\theta, \mathbf{a})\right]$, also called the risk function. With that value of the estimate, denoted $\mathbf{a}^*$, and prior to taking any sample, the corresponding value of the minimum risk is the Expected Value of Perfect Information (EVPI), since this is the value we should pay, on the average at the start, for information to acquire perfect knowledge and hence to eliminate all risk. But, in general, θ is the parameter of the distribution of an observable random variable X, of density $f(x|\theta)$, and a random sample of observations $X = (X_1, \ldots, X_n)$ gives rise to the posterior distribution of θ, denoted $g(\theta|X)$, and a revised estimation of θ. Again, the optimal point estimation $\mathbf{a}^{**}$ of θ minimizes the average value of the loss function over the posterior distribution, or $R_{\text{post}}(\theta) = \mathbb{E}_{\theta|X}\left[L(\theta, \mathbf{a}^{**})\right]$. Alternately, we can consider the interval with credibility $(1 - \alpha)$, $0 \le \alpha \le 1$, denoted $I_{(1-\alpha)}^{hpd}$, already encountered. It gives the interval estimation of θ, where $\mathbb{P}(\theta \in I_{(1-\alpha)}^{hpd}) = 1 - \alpha$, with the probability of any point inside $I_{(1-\alpha)}^{hpd}$ larger than that of any point outside it.

Since the posterior risk depends on X, which is unknown before sampling, we have to consider its average with respect to sampling outcomes, i.e. over the marginal distributions of X, called the Bayes risk, denoted by $\rho(n)$, which depends only on the sample size n. Prior to taking a sample, in *preposterior analysis*, we are interested in knowing what would be the average worth of a sample of size n. The reduction in risk, represented by the difference between the prior risk and the above-average posterior risk, or EVPI-$\rho(n)$, is that average worth and is called the Expected Value of Sample Information, EVSI(n). If we consider the cost of sampling, $C(X_1, \ldots, X_n)$, for a sample of size n, the Expected Net Gain in Sampling (denoted by ENGS(n)) is just the difference between EVSI(n) and the expected cost $\mathbb{E}\left[C(X_1, \ldots, X_n)\right] = C(n)$. As long as ENGS($n$) is positive, it is advantageous to sample. ENGS(n) is usually a concave function presenting a maximum at n_0, which is the optimal sample size, at which point we would gain the most out of sampling (see Raiffa and Schlaifer (1961) for an in-depth treatment of these topics).

(b) Quadratic loss

The quadratic loss function $Q(\theta, \mathbf{a}) = A(\theta - \mathbf{a})^2$ is well accepted in statistics and has several important applications. Here, it provides a correspondence with central moments of the distribution of θ. Taking $A = 1$, we have the following:

$\mathbf{a}^* =$ mean of prior distribution,
EVPI = variance of prior distribution.

Let $f(\theta \,|\, X)$ be the posterior distribution of θ. We have the following:
$\mathbf{a}^{**} =$ mean of posterior distribution, or μ_{post},
Posterior risk = variance of posterior distribution, or Var_{post},
Bayes risk = expectation wrt X of posterior risk = $E_X(\text{Var}_{\text{post}}) = \rho(n)$,
EVSI$(n) =$ EVPI$-\rho(n) = \text{Var}(\mu_{\text{post}})$, and ENGS$(n) =$ EVSI$(n) - C(X_1, \ldots, X_n)$.

(c) Absolute value loss

For the *general linear loss function*, defined by $L(\theta, \mathbf{a}) = k_u(\theta - \mathbf{a})$ for $\mathbf{a} \leq \theta$ and $L(\theta, \mathbf{a}) = k_0(\mathbf{a} - \theta)$ for $\mathbf{a} > \theta$, where $\mathbf{a}$ is an estimate to the parameter θ and k_u and k_0 are positive constants, results are complex, but this loss function is a realistic one in applications. However, left and right linear loss integrals have to be considered separately here, and several difficulties in computation have led to the use of numerical methods to evaluate the associated decision criteria (Pratt *et al.*, 1995). A simpler case is obtained by choosing $k_0 = k_u = K$, and we will consider it from now on. We then have an *absolute value loss function* of the form $L(\theta, \mathbf{a}) = K\,|\theta - \mathbf{a}|$, which is intimately related to the absolute distance and the L_1-approach in statistics and has the clear advantage of being more robust than the quadratic form.

Let us consider a random variable with distribution $F(x)$ and finite mean μ. Its median $\text{Md}(X)$, which does not need to be unique, is defined as the value such that

$$\int_{-\infty}^{\text{Md}(X)} dF(t) = 0.5 \quad \text{or} \quad F^{-1}(\text{Md}(x)) = 0.5,$$

where F^{-1} is the quantile function.

We define the mean absolute deviation of X about its mean as

$$\delta_1(X) = \int_{-\infty}^{\infty} |x - \mu|\, dF(x) \tag{6.5}$$

and about its median as

$$\delta_2(X) = \int_{-\infty}^{\infty} |x - \mathrm{Md}(X)|\, dF(x)$$

$$= \int_{0}^{1/2} \left[F^{-1}(1 - \alpha) - F^{-1}(\alpha) \right] d\alpha. \tag{6.6}$$

Using Liapunov's inequality (see Pham-Gia and Turkkan (1994)), we always have $\delta_2 \leq \delta_1 \leq \sigma$, where σ is the standard deviation. In particular, for the normal $N(\mu, \sigma^2)$, we have a simple relation $\delta_1(X) = \delta_2(X) = \sqrt{\frac{2}{\pi}}\sigma$.

Since the median of a distribution rarely comes in closed form, δ_2 does not either, unlike δ_1, Table 6.1 gives the expressions of $\delta_1(X)$ for some distributions.

Table 6.1   Expressions of $\delta_1(X)$ for some common distributions.

No.	Model	$\delta_1(X)$
1	Exponential(λ) $f(x\|\lambda) = \lambda \exp(-\lambda x)$, $x, \lambda > 0$.	$\frac{2}{\lambda e}$
2	Gamma(α, β) $f(x\|\alpha, \beta) = \frac{\beta^\alpha x^{\alpha-1} \exp(-\beta x)}{\Gamma(\alpha)}$, $\alpha, \beta > 0, x > 0$. Shape: α = shape parameter, β = scale parameter.	$\frac{2\alpha^\alpha}{\beta e^\alpha \Gamma(\alpha)}$
3	Beta(α, β) $f(x\|\alpha, \beta) = \frac{x^{\alpha-1}(1-x)^{\beta-1}}{B(\alpha,\beta)}$, $\alpha, \beta > 0, 0 \leq x \leq 1$.	$\frac{2\alpha^\alpha \beta^\beta}{(\alpha+\beta)^{\alpha+\beta+1} B(\alpha,\beta)}$
4	Student(μ, λ, α) $f(x\|\mu, \alpha, \beta) = \sqrt{\frac{\lambda}{\alpha\pi}} \frac{\Gamma(\frac{\alpha+1}{2})}{\Gamma(\frac{\alpha}{2})} \left[1 + \frac{\lambda}{\alpha}(x-\mu)^2\right]$, $\lambda, \alpha > 0, \mu, x \in \mathbb{R}$.	$\sqrt{\frac{\alpha}{\lambda\pi}} \cdot \frac{\Gamma(\frac{\alpha-1}{2})}{\Gamma(\frac{\alpha}{2})}$
5	Normal(μ, R) $f(x\|\mu, R) = \sqrt{\frac{R}{2\pi}} \exp\left(-\frac{R(x-\mu)^2}{2}\right)$.	$\sqrt{\frac{2}{\pi R}}$

Note: For $Y = \alpha + \beta X$, we have $\delta_1(Y) = |\beta| \delta_1(X)$, $\delta_2(Y) = |\beta| \delta_2(X)$ and $\sigma(Y) = |\beta| \sigma(X)$.

Like the standard deviation in the L_2-norm, the two dispersion measures δ_l and δ_2 are, respectively, the L_1-norms of the two "centered variables" $X_1 = X - \mathbb{E}(X)$ and $X_2 = X - \text{Md}(X)$, with the use of either the mean or the median as the "center" of a distribution. Depending on the distance between the median and the mean of the distribution considered, values of δ_l and δ_2 can be very close together or significantly apart. They are associated with the study of Lorenz curves (see Pham-Gia and Turkkan (1994)). See Pham-Gia and Tranloc (2001) for a survey of up-to-date results available on δ_l and δ_2, and their sample estimations.

Let us now consider a distribution $f(x|\theta)$, with its parameter θ having prior distribution $F(\theta)$ and loss function $L(\theta, \mathbf{a}) = K |\theta - \mathbf{a}|$. We can take $K = 1$ without loss of generality. The prior median, denoted Md_{prior}, minimizes the risk $\int_{-\infty}^{\infty} |\theta - \mathbf{a}| \, dF(\theta)$ and hence, $\text{EVPI} = \delta_2(\theta_{\text{prior}})$, the mean absolute deviation of θ about its prior median. Similarly, the median of the posterior distribution, denoted Md_{post}, minimizes the posterior risk, with value

$$R_{\text{post}}(\theta) = \int\limits_{-\infty}^{\infty} |\theta - \text{Md}_{\text{post}}| \, dF(\theta|X) = \delta_2(\theta_{\text{post}}).$$

Its average wrt X is the Bayes risk, $\rho(n) = E_X(\delta_2(\theta_{\text{post}}))$, independent of X. We now have $\text{EVSI}(n) = \delta_2(\theta_{\text{prior}}) - \rho(n)$, but, unlike the case of the quadratic loss function, $\text{EVSI}(n) \neq \delta_2(\text{Md}_{\text{post}})$. For $\text{ENGS}(n)$, we have, as before, $\text{ENGS}(n) = \text{EVSI}(n) - C(n)$ (see Pham-Gia and Bekker (2005)).

Table 6.2 gives a comparative view of the expressions of different quantities under the absolute value and square error loss functions (we have chosen the notation $\mathbb{E}(\theta_{\text{post}})$ and $\text{Var}(\theta_{\text{post}})$ for clarity).

(d) Loss function using L^P-norm

For the loss function of the general form $L(\theta, \mathbf{a}) = |\theta - \mathbf{a}|^p, p > 1$, as presented in Bar-Lev *et al.* (1999), a solution $\mathbf{a}^*$ exists for p taking integral values larger than 1. For $p > 4$, as well-known in algebraic Galois theory, no algebraic solution can be obtained and we have to resort to numerical methods to find these values that would minimize $L(\theta, \mathbf{a})$. Also, when $\mathbf{a}^*$ is not unique, the computation of

Table 6.2 Decision criteria associated with the two main loss functions.

Distribution of parameter	Absolute value loss function	Quadratic loss function
1 Prior Distribution		
(a) Optimal estimate =	Prior median (Md_{prior})	Prior mean (or μ_{prior})
(b) Risk $R(a^*) = \text{EVPI} =$	$E_\theta\left[|\theta - \text{Md}_{\text{prior}}|\right] = \delta_2(\theta_{\text{prior}})$	$E_\theta\left[(\theta - \mu_{\text{prior}})^2\right]$
		$= \text{Var}(\theta_{\text{prior}})$
2 Posterior Distribution		
(a) Optimal estimate =	Posterior median (or Md_{post})	Posterior mean (or μ_{post})
(b) Post. Risk $R(a^{**}|X) =$	$\delta_2(\theta_{\text{post}})$	$\text{Var}(\theta_{\text{post}})$
(c) Bayes risk, $\rho(n) =$	$E_X(\delta_2(\theta_{\text{post}}))$	$E_X(\text{Var}(\theta_{\text{post}}))$
(d) EVSI(n)	$\delta_2(\theta_{\text{prior}}) - E_X(\delta_2(\theta_{\text{post}}))$	$\text{Var}(\theta_{\text{prior}})$
$= \text{EVPI} - \rho(n) =$	$\neq \delta_2(\text{Md}_{\text{post}})$	$-E_X(\text{Var}(\theta_{\text{post}}))$
		$= \text{Var}(\mu(\theta_{\text{post}}))$

$$\delta_1(\theta) = \delta_2(\theta) \text{ for a symmetrical distribution of } \theta$$

the associated decision criteria (EVPI, posterior risk and EVSI) has to be done numerically first and then by comparing the values of the risk associated with each solution. For $p = 4$, $\mathbf{a}^*$ can be computed as follows:

$$\mathbf{a}^* = \mu_1 + \left[a + (b^3 + a^2)^{1/2}\right]^{1/3} + \left[a - (b^3 + a^2)^{1/2}\right]^{1/3},$$

with $\mu_j = \mathbb{E}(X^j)$ finite, $a = (\mu_3 - \mu_1\mu_2)/2 - \mu b$ and $b = \mu_2 - \mu_1^2$.

Additional hypotheses have to be made on the distribution to compute $\mathbf{a}^*$ for other values of p.

(e) The normal loss integral in Bayesian decision theory
Suppose that the production volume has a normal prior distribution $N(\mu_0, \sigma_0^2)$ determined by subjective means. The payoff of the action considered is a linear function of the production volume x, with equation $V_1 = \alpha x - \beta$. It passes through point b on the horizontal axis as the break-even quantity. We can prove (see Chou (1972)) that

the EVPI, which is the expected opportunity loss of the best possible decision under uncertainty, is given by $\text{EVPI} = b\sigma_0 L(z_b)$, where $z_b = \left|\frac{b-\mu_0}{\sigma_0}\right|$ and $L(z) = \varphi(z) - (1 - \Phi(z))$, where φ and Φ are the pdf and cdf of the standard normal $Z \sim N(0,1)$. is called the *standard normal loss integral*, with tabulated values in some textbooks. It is also called *Rederre* integral in the gambling theory (see Epstein (2009)).

6.8   Conjugate Bayesian Analysis Table

Section 6.4 deals with conjugacy in the Bayesian analysis. Here, for each case, we identify the model, the variable, the parameter, its prior (and non-informative prior whenever available), the sample, its sufficient statistic and the likelihood function. Then we give the posterior and, when possible, the prior and posterior marginals for the parameters, and the prior and posterior predictive distributions of the variable.

Note: *To use this table, for the likelihood, use $f(x; w)$ if there is only one observation x. Otherwise, use $f(u|w)$, where u is the sufficient statistic (Suff. Stat.).*
   Post = Posterior distribution,
   Pred = Predictive distribution,
   N.I. = "non-informative" prior.

(A) Non-normal models
(1) Model: Poisson, $X \sim Poi(w)$

$$f(x; w) = e^{-x}\frac{w^x}{x!}, x = 0, 1, 2, \ldots$$

   Prior: $W \sim Gam(\alpha, \beta), w > 0, (N.I. = \pi(w) = w^{-1/2})$,
   Sample: $\{x_1, \ldots, x_n\}$,
   Suff. Stat: $U = \sum x_i$,
   Likelihood: $f(u|w) \propto Poi(u|nw)$,
   Post.: $W \sim Gam(\alpha + u, \beta + n)$,
   Pred.: $U \sim Pg(\alpha, \beta, n)$,
   Pg = Poisson–gamma distribution,
   Poi = Poisson distribution.

(2) Model: Bernoulli, $X \sim Be(p)$

$$f(x;p) = 1, \text{ or } 0,$$

Prior: $p \sim \text{beta}(\alpha, \beta), 0 \leq p \leq 1$ (N.I.: beta(1/2, 1/2)),
Sample: $\{x_1, \ldots, x_n\}$,
Suff. Stat.: $U = \sum x_i$,
Likelihood: $f(u \mid n, p) = \binom{n}{u} p^u (1-p)^{n-u}, u = 0, 1, \ldots, n,$
Post.: $p \sim \text{beta}(\alpha + u, \beta + n - u)$,
Pred.: $U \sim \text{Bbin}(n, \alpha, \beta)$.
Bbin = beta-binomial distribution.

(3) Model: Negative binomial, $X \sim NB(w, r), r > 0, x = 0, 1, 2, \ldots$

$$f(x; w, r) = \binom{r + x + 1}{r - 1} w^r (1 - w)^x.$$

Prior: $W \sim \text{beta}(\alpha, \beta), 0 \leq w \leq 1$, (N.I.: beta(0, 1/2)),
Sample: $\{x_1, \ldots, x_n\}$,
Suff. Stat.: $U = \sum x_i$,
Likelihood:

$$f(u \mid r, w) \propto \binom{r + u - 1}{r - 1} w^{nr} (1 - w)^u, \ u = 1, 2, \ldots,$$

Post.: $W \sim \text{beta}(\alpha + nr, \beta + u)$,
Pred.: $U \sim \text{Nbb}(\alpha, \beta, nr)$,
Nbb: Negative beta-binomial.

(4) Model: Continuous uniform, $X \sim Un(0, w), 0 \leq x \leq w$

$$f(x) = \frac{1}{w}.$$

Prior: $W \sim Pa(\alpha, \beta)$, (N.I.: $\pi(w) = w^{-1}$),
Sample: $\{x_1, \ldots, x_n\}$,
Suff. Stat: $U = \max\{x_1, \ldots, x_n\}$,
Likelihood: $f(u \mid w) \propto IP(n, 1/w)$ for $u < w$, $f(u \mid w) = 0$ otherwise,
Post.: $W \sim Pa(\alpha + n, \max\{\beta, x_1, \ldots, x_n\})$,
Pred.: $U \sim \frac{n}{\alpha+n} Pa(\alpha, \beta)$ if $u > \beta$, and $U \sim \frac{n}{\alpha+n} IP(n, \beta^{-1}), u \leq \beta$,
Pa = Pascal distribution,
IP = Inverse Pascal.

(5) Model: Exponential, $X \sim \text{Exp}(w), 0 < x$

$$f(x; w) = w \exp(-wx).$$

Prior: $W \sim \text{Gam}(\alpha, \beta)$, (N.I.: $\pi(w) = w^{-1}$),
Sample: $\{x_1, \ldots, x_n\}$,
Suff. Stat.: $U = \sum x_i$,
Likelihood: $f(u \mid w) \propto u^{n-1} \exp(-wu)$,
Post.: $W \sim \text{Gam}(\alpha + n, \beta + u)$,
Pred.: $U \sim Gg(\alpha, \beta, n)$,
Gg = Gamma–gamma distribution.

(B) Univariate normal models

(6) Model: Normal, $X \sim N(\mu, r = 1/\sigma^2)$

$$f(x; \mu, r) = \sqrt{\frac{r}{2\pi}} \exp\left(-\frac{r(x - \mu)^2}{2}\right), -\infty < x < \infty.$$

(a) $R = r_0$ known, μ unknown,
Prior: $\mu \sim N(\mu_0, \tau_0)$ (N.I.: $\pi(\mu) = $ const.),
Sample: $\{x_1, \ldots, x_n\}$,
Suff. Stat.: $U = \sum x_i/n = \bar{x}$,
Likelihood: $f(x \mid r_0, \mu) \propto \exp(-\frac{nr_0}{2}(u - \mu)^2)$,
Post.: $\mu \sim N(\frac{\tau_0\mu_0 + nr_0\bar{x}}{\tau_0 + nr_0}, \tau_0 + nr_0)$,
Pred: $U \sim N(\mu_0, \frac{nr_0\tau_0}{\tau_0 + nr_0})$.

(b) $\mu = \mu_0$ known, R unknown with prior, $R \sim \text{Gam}(\alpha, \beta)$, (N.I.: $\pi(R) = r^{-1}$),
Sample: $\{x_1, \ldots, x_n\}$,
Suff. Stat.: $U = \sum (x_i - \mu_0)^2$,
Likelihood: $f(u \mid \mu_0, r) \propto \text{Gam}(n/2, r/2)$,
Post.: $R \sim \text{Gam}(\alpha + \frac{n}{2}, \beta + \frac{\sum(x_i - \mu_0)^2}{2})$,
Pred.: $U \sim Gg(\alpha, 2\beta, n/2)$,
Ga = Gamma distribution,
Gg = Gamma–gamma distribution,
(c) Both unknown. Prior: $R = \text{Gam}(\alpha, \beta), (\mu \mid R = r) \sim N(\mu_0, \tau_0 r), \tau_0 > 0$, (N.I.: $\pi(\mu, R) = r^{-1}$),

(Normal–gamma prior. The conditional of μ is normal, but the marginal is *not* and the two parameters μ and R are dependent.)

Sample: $\{x_1, \ldots, x_n\}$,

Post.: $R \sim \text{Gam}(\alpha* = \alpha + \frac{n}{2}, \beta* = \beta + \frac{\sum(x_i - \bar{x})^2}{2} + \frac{\tau_0 n(\bar{x} - \mu_0)^2}{2(n + \tau_0)})$,

$(\mu \mid R = r) \sim N(\mu* = \frac{\tau_0 \mu_0 + n\bar{x}}{\tau_0 + n}, \tau* = (\tau_0 + n)r)$,

Prior marginal: $\mu \sim St_{2\alpha}(\mu, \frac{\alpha\tau_0}{\beta})$, $R \sim \text{Gam}(\alpha, \beta)$.

Prior predictive: $X \sim St_{2\alpha}(\mu_0, \frac{\tau_0}{\tau_0 + 1} \frac{\alpha}{\beta})$.

For posterior marginal and predictive, replace $(\alpha, \beta, \mu, \tau)$ by $(\alpha*, \beta*, \mu*, \tau*)$, with St = Student t-distribution.

(C) Multivariate normal models

(7) Model: Multinormal $\{X_1, \ldots, X_n\} = $ sample from $X \sim N_k$ $(M, r_0 = \Sigma_0^{-1})$

$$f(X) \propto \exp\left(-\frac{(X - M)^t \Sigma^{-1}(X - M)}{2}\right).$$

(a) Mean vector M variable, precision matrix Σ_0^{-1} given

Prior: $M \sim N_k(\mu_0, \tau_0)$,

Post.: $M \sim N_k(\mu^*, \tau^*)$, $\mu^* = \frac{\tau_0 \mu_0 + n r_0 \bar{X}}{\tau_0 + n r_0}$, $\tau^* = \tau_0 + n r_0$.

(8) Mean $M = m_0$ given, precision R variable

Prior: $R \sim Wis(\alpha_0, \tau_0)$, $\alpha_0 > k - 1$ (Wishart distribution),

Post.: $R \sim Wis(\alpha^*, \tau^*)$, $\alpha^* = \alpha_0 + n$,

$$\tau^* = \tau_0 + \sum_{i=1}^n (X_i - m_0)(X_i - m_0)^t.$$

(9) Both mean M and precision R are variable

Multinormal-Wishart prior: $R \sim Wis(\alpha_0, \tau_0)$, $\alpha_0 > k - 1$,

$$M \mid R = r_0 \sim N_k(\mu_0, \nu r_0), \ \nu > 0,$$

Post.: $R \sim Wis(\alpha^* = \alpha_0 + n, \tau^* = \tau_0 + s + \frac{\nu n}{\nu + n}(\mu_0 - \bar{x})(\mu_0 - \bar{x})^t)$,

$$s = \sum (x_i - \bar{x})(x_i - \bar{x})^t,$$

and

$$(M \mid R = r_0) \sim N_k\left(\mu^* = \frac{\nu \mu_0 + n\bar{x}}{\nu + n}, (\nu + n)r_0\right),$$

Marginal for **M**: prior marginal

$$\mathbf{M} \sim St_{\alpha_0-k+1}(\boldsymbol{\mu}_0, \boldsymbol{\Sigma}_0 = \nu(\alpha_0 - k + 1)\tau_0^{-1}),$$

Posterior marginal of **M**: replace (α_0, τ_0) by (α^*, τ^*).

6.9   Conclusion

The domain of Bayesian statistics is a very vast one and is getting larger every day. In what precedes, we could just give a glimpse at the domain, covering some of its important aspects. We also concentrated on the normal, as prior and as sampling model, in the discussion of conjugacy in Bayesian statistics, and in the Bayesian decision theory. Needless to say, there are other distribution families conjugate to other sampling models. The philosophy of the Bayesian approach, of updating the model with newly acquired data, is also getting wide acceptance in many applied fields and contributes further to the development of the domain of Bayesian statistics.

Bibliography

Bar-Lev, S.K., Boukai, B. and Enis, P. (1999). On the mean square error, the mean absolute error and the like, *Communications in Statistics — Theory and Methods*, **28**(81), 1813–1822.

Berger, J.O. (1985). *Statistical Decision Theory and Bayesian Statistics* (Springer, New York).

Bernardo, J.M. and Smith, A.F.M. (1994). *Bayesian Theory* (John Wiley and Sons, New York).

Bunn, D.W. (1975). Anchoring bias in the assessment of subjective probability, *Operations Research*, **26**, 449–454.

Chaloner, K. and Duncan, G.T. (1983). Assessment of a beta prior distribution: PM elicitation, *The Statistician*, **32**, 174–180.

Chou, Y.L. (1972). *Statistical Analysis for Business and Economics* (Hold, Rinehart and Winston, New York).

Cuevas, A. and Sanz, P. (1988). On differentiability properties of Bayes operators, in *Bayesian Statistics 3* (J.M. Bernardo, M.H., DeGroot, D.V., Lindley and A.F.M., Smith, eds.), pp. 569–570 (Oxford University Press, London).

Epstein, R. (2009). *Theory of Gambling and Statistical Logic*, 2nd ed. (Academic Press, London).

Ferguson, T. (1967). *Mathematical Statistics: A Decision Theoretic Approach* (Academic Press, New York).

Fine, T. (1973). *Theories of Probability* (Academic Press, New York).

Good, I.J. (1965). *The estimation of probabilities: An essay on modern Bayesian methods*, Research Monograph No. 30 (The MIT Press, Cambridge, MA).

Hogarth, R.M. (1975). Cognitive processes and the assessment of subjective probabilities, *Journal of the American Statistical Association*, **70**, 271–289.

Kahneman, D., Slovic, P. and Tversky, A. (1982). *Judgement Under Uncertainty, Heuristics and Biases* (Cambridge University Press, London).

Keefer, D.L. and Bodily, S.E. (1983). Three point approximation for continuous random variables, *Management Science*, **29**, 595–609.

Lindley, D. (1972). *Bayesian Statistics: A Review* (Society for Industrial and Applied Mathematics, Philadelphia).

Muirhead, R. (1983). *Aspects of Multivariate Statistical Theory* (Wiley, New York).

Pham-Gia, T. (1989). WORKSAMP (software survey section), *Mathematical and Computer Modelling*, **12**(2), I.

Pham-Gia, T. (1998). Distribution of the stopping time in Bayesian sequential analysis, *Australian and New Zealand Journal of Statistics*, **40**, 221–227.

Pham-Gia, T. and Bekker, A. (2005). Sample size determination using Bayesian decision criteria under absolute value loss function, *American Journal of Mathematical and Management Sciences*, **25**, 259–291.

Pham-Gia, T. and Tranloc, H. (2001). The mean and median absolute deviations, *Mathematical and Computer Modelling*, **34**, 921–936.

Pham-Gia, T. and Turkkan, N. (1994). The Lorentz and T4-curves: A unified approach, *IEEE Transactions on Reliability*, **43**, 76–84.

Pham-Gia, T., Turkkan, N. and Bekker, A. (2006). A Bayesian analysis in the L_1-norm of the mixing proportion, *Metrika*, **64**, 1–22.

Pham-Gia, T., Turkkan, N. and Duong, Q.P. (1992). Using the mean absolute deviation to determine the beta prior distribution, *Statistics & Probability Letters*, **13**(5), 373–381.

Pratt, J.W., Raiffa, H. and Schlaifer, R. (1995). *An Introduction to Statistical Decision Theory* (The MIT Press, Cambridge, MA).

Press, J. (1989). *Bayesian Statistics: Principles, Models and Applications* (Wiley, New York).

Raiffa, H. and Schlaifer, R. (1961). *Applied Statistical Decision Theory* (Harvard University Press, Cambridge, MA).

Robbins, H. (1955). An empirical Bayes approach to statistics, *in Proceedings of the Third Berkeley Symposium on Mathematical Statistics and Probability*, 1, pp. 157–164.

Turkkan, N. and Pham-Gia, T. (1993). Computation of the highest posterior density region in Bayesian analysis, *Journal of Statistical Computation and Simulation*, **44**, 243–250.

Turkkan, N. and Pham-Gia, T. (1997). Highest posterior density and minimum volume confidence region: The bivariate case, *Journal of the royal statistical Society*, C, **46**, 131–140.

Waller, R. and Martz, H. (1982). *Bayesian reliability analysis* (John Wiley and Sons, New York).

Winkler, R.L. (1967). The assessment of prior distribution in Bayesian analyses, *Journal of the American Statistical Association*, **62**, 776–800.

Winkler, R.L. (1980). Prior information, predictive distributions and Bayesian model building, *in Bayesian Analysis in Econometrics and Statistics*, A. Zellner, ed., pp. 95–109 (North-Holland, New York).

Yang, R. and Berger, J.O. (1998). A catalog of non-informative priors, ISDS Discussion Paper, 97-42.

Zaino, N.A. and d'Enrico, J. (1989). Optimal discrete approximations for continuous outcomes with applications in decision and risk analysis, *Journal of the Operations Research Society*, **40**, 379–388.

Chapter 7

Bivariate and Trivariate Normal Distributions

7.1 Introduction

In this chapter, we look into the cases of $p = 2$ and $p = 3$, the bivariate and trivariate normal distributions. These cases are important because the number of dimensions, or variables, is still small and permits the derivation of many analytic closed-form results, which become quite elaborate in the general case. The operation on axis rotation, for example, is clearly presented in the bivariate case. Its equivalent in $\mathbb{R}^p$, the whitening of a normal vector, can only be presented using mathematical expressions and equations.

We start by reviewing basic notions on conics. We study the bivariate normal distribution first: its shape, the standard form of its density, its characteristic and moment-generating functions. We then discuss the transformations that would bring a dependent binormal couple to an uncorrelated one. The use of the binormal tables is mentioned and several other topics are treated. The trivariate normal distribution is then considered and several important related questions are examined. The chapter concludes with a short introduction to normal copulas. Finally, some published results and tools, available from industrial companies for private use, are brought here to the attention of readers.

7.2 Conics

They are given by the intersections of a plane with a double-summit cone (Fig. 7.1). They can be ellipses, parabolas, or hyperbolas, or, in degenerate cases, intersecting straight lines, depending on the position of the plane wrt to the axis of the cone. We might have these shapes.

Let $\mathbf{x} = (x_1, x_2)$ be a bidimensional variable where the variances are σ_1^2 and σ_2^2. Taking the square statistical distance from the origin as $d^2 = x_1^2/\sigma_1^2 + x_2^2/\sigma_2^2$, expression $x_1^2/\sigma_1^2 + x_2^2/\sigma_2^2 = a^2$ is the locus of all points (x_1, x_2) whose square statistical distance to zero is $a^2 > 0$. This is the equation of an ellipse centered at zero

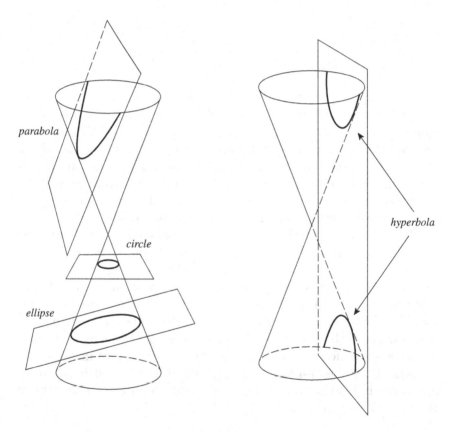

Fig. 7.1   Conics and their intersections with a plane.

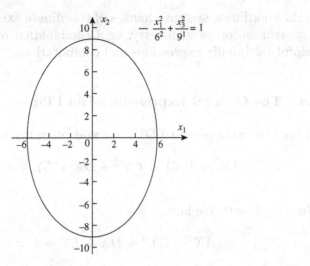

Fig. 7.2   Ellipse $x_1^2/6^2 + x_2^2/9^2 = 1$.

with half-axes $a\sigma_1$ and $a\sigma_2$ along the two coordinate axes, namely $x_1^2/(a\sigma_1)^2 + x_2^2/(a\sigma_2)^2 = 1$.

Figure 7.2 shows the ellipse $x_1^2/6^2 + x_2^2/9^2 = 1$ in the square system of coordinates $x_1 O x_2$ with half-axis lengths 6 and 9.

7.3   Quadratic Expressions and Quadratic Forms

A mathematical expression of the form $Q_2(x, y) = ax^2 + bxy + cy^2$ is called a quadratic form in two variables, and an expression of the form $Q_2(x, y) = gx^2 + hy^2$ is its canonical form.

We will be concerned about going from one form to another, and vice versa, in the study of the binormal distribution: Let

$$Q(x, y) = ax^2 + bxy + cy^2 + dx + ey + f, \quad a, b, c, d, e, f \in \mathbb{R}^1.$$
$$(7.1)$$

It is called a general quadratic expression. It can be brought to the canonical form by changes of variables or rotation of axes. It is a quadratic form if it is homogeneous in the two variables.

In the coordinate system, changes of coordinate axes, either by rotation, translation or symmetry, or a combination of these, are very helpful to simplify expressions and computations.

7.4   The General Expression of an Ellipse

In the Cartesian plane (XOY), a conic curve has general equation:

$$AX^2 + BXY + CY^2 + DX + EY + F = 0. \qquad (7.2)$$

Case 1: $B = 0$. We have

$$AX^2 + CY^2 + DX + EY + F = 0. \qquad (7.3)$$

It is an ellipse if $AC > 0$, a hyperbola if $AC < 0$ and a parabola if $AC = 0$.

Case 2: $B \neq 0$. We have

$$AX^2 + BXY + CY^2 + DX + EY + F = 0. \qquad (7.4)$$

Although this expression can be well handled by present-day computers, the *rectangular* term in XY should be eliminated to have more clarity and convenience. We perform a rotation of angle θ *counterclockwise* to get rid of the rectangular term BXY. Matrixwise, we have the relations between the old coordinates (X, Y) of a point M and its new rotated coordinates (x, y) from XOY to xOy:

$$\begin{pmatrix} x \\ y \end{pmatrix} = \begin{bmatrix} \cos\theta & \sin\theta \\ -\sin\theta & \cos\theta \end{bmatrix} \begin{pmatrix} X \\ Y \end{pmatrix} \quad \text{or} \quad \begin{cases} x = X\cos\theta + Y\sin\theta, \\ y = -X\sin\theta + Y\cos\theta, \end{cases}$$

and conversely,

$$\begin{pmatrix} X \\ Y \end{pmatrix} = \begin{bmatrix} \cos\theta & -\sin\theta \\ \sin\theta & \cos\theta \end{bmatrix} \begin{pmatrix} x \\ y \end{pmatrix}. \qquad (7.5)$$

The new coefficient of xy is now

$$b = -2A\cos\theta\sin\theta + B\{\cos^2\theta - \sin^2\theta\} + 2C\cos\theta\sin\theta.$$

Similarly, we can have the expressions of a, c, d, e and f as

$$a = A\cos^2\theta + B\sin\theta\cos\theta + C\sin^2\theta,$$
$$c = A\sin^2\theta + B(-\sin\theta\cos\theta) + C\cos^2\theta,$$
$$d = D\cos\theta + E\sin\theta,$$
$$e = -D\sin\theta + E\cos\theta,$$
$$f = F.$$

Equation (7.4) becomes

$$ax^2 + bxy + cy^2 + dx + ey + f = 0, \tag{7.6}$$

where $b^2 - 4ac = B^2 - 4AC$ is called the *invariant discriminant* for any angle θ, $0 \leq \theta \leq \pi$.

Using Eq. (7.4), for the angle θ^*, s.t.

$$\tan(2\theta^*) = \frac{B}{A - C},$$

we have $b = 0$.

Adopting this angle θ^* for rotation, since $b = 0$, (7.6) becomes

$$ax^2 + cy^2 + dx + ey + f = 0.$$

In the new system, we have the following:

(a) an ellipse if $ac > 0$;
(b) a hyperbola if $ac < 0$;
(c) a parabola if $ac = 0$.

7.5   Some Examples

(a) Figure 7.3 shows the curve with equation $X^2 + 4Y^2 + 2X - 24Y + 33 = 0$. To have its canonical form, we complete the squares to obtain

$$(X^2 + 2X) + 4(Y^2 - 6Y) = -33,$$
$$(X + 1)^2 + 4(Y - 3)^2 = 4,$$
$$\frac{(X + 1)^2}{2^2} + \frac{(Y - 3)^2}{1^2} = 1.$$

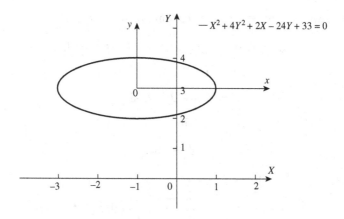

Fig. 7.3   Ellipse $X^2 + 4Y^2 + 2X - 24Y + 33 = 0$.

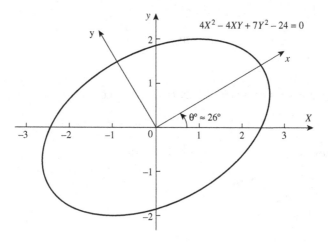

Fig. 7.4   Ellipse $f(X,Y) = 4X^2 - 4XY + 7Y^2 - 24 = 0$.

This is an ellipse centered at $(-1,3)$, with half-axis lengths 2 and 1. This case does not need an axis rotation since the rectangular term is absent in the original equation. In the new axis system xOy, we have equation: $\frac{x^2}{2^2} + \frac{y^2}{1^2} = 1$.

(b) Figure 7.4 shows the curve $f(X,Y) = 4X^2 - 4XY + 7Y^2 - 24 = 0$. We see that the rectangular term is now present and needs to be eliminated. We set $A = 4$, $B = -4$, $C = 7$, $F = -24$ in Eq. (7.4).

To get rid of the rectangular term, we rotate the axes through the angle θ^*, given by $\tan(2\theta^*) = B/(A-C) = 4/3$, which gives $\cos(2\theta^*) = 3/5$. Hence,

$$\cos(\theta^*) = \sqrt{\frac{1 + \frac{3}{5}}{2}} = \frac{2\sqrt{5}}{5} \to \sin(\theta^*) = \frac{\sqrt{5}}{5},$$

or $\theta^* \approx 26°$ approximately. Our rotation of axes using θ^* has old coordinates in function of *new ones* as

$$\begin{cases} X = \dfrac{2\sqrt{5}}{5}x - \dfrac{\sqrt{5}}{5}y, \\[2mm] Y = \dfrac{\sqrt{5}}{5}x + \dfrac{2\sqrt{5}}{5}y. \end{cases}$$

Substituting these expressions into (7.4), we obtain the new equation

$$3x^2 + 8y^2 = 24 \quad \text{or}$$

$$\frac{x^2}{8} + \frac{y^2}{3} = 1, \quad \text{or}$$

$$\frac{x^2}{\left(2\sqrt{2}\right)^2} + \frac{y^2}{\left(\sqrt{3}\right)^2} = 1.$$

This is a standard ellipse with half-axis lengths $2\sqrt{2} = 2.828$ and $\sqrt{3} = 1.732$ in the new system of axes, obtained by rotation of the old axes system, of angle $\theta^* \approx 26°$ counterclockwise.

We can see, in the rotated system, that the area of this ellipse is $\pi\sqrt{3}(2\sqrt{2}) \approx 15.39$. This result is much harder to derive without axis rotation. We will see that this transformation is very useful when dealing with dependent components of a bivariate normal distribution.

7.6 Correspondence Between Coefficients When the Ellipse Center is Not at the Origin

We consider Eq. (7.2) and the canonical form

$$\frac{(x - u_0)^2}{a^2} + \frac{(y - v_0)^2}{b^2} = 1, \tag{7.7}$$

where the ellipse center is now located at (u_0, v_0).

(a) We can eliminate B first by axis rotation, so that ellipse axes are parallel to coordinate axis and obtain $ax^2 + \gamma y^2 + \delta x + \eta y + \xi = 0$. Then, we have $\alpha = 1/a^2$, $\gamma = 1/b^2$, $\delta = -2u_0/a^2$, $\eta = -2v_0/b^2$, $\xi = u_0^2/a^2 + v_0^2/b^2 - 1$.

(b) Conversely, an equation of the type $Ax^2 + Cy^2 + Dx + Ey + F = 0$, where A, C and $K = D^2/(4A) + E^2/(4C) - F$ have the same sign, can be written under its canonical form (7.7), where $a = \sqrt{K/A}$, $b = \sqrt{K/C}$, $u_0 = -D/(2A)$, $v_0 = -E/(2C)$.

(c) Also, Eq. (7.2) can be put under matrix form as follows: $\mathbf{X}^t \Lambda \mathbf{X} + \varsigma^t \mathbf{X} + \rho = 0$, where

$$\mathbf{X} = \begin{pmatrix} X \\ Y \end{pmatrix}, \quad \Lambda = \begin{bmatrix} A & B/2 \\ B/2 & C \end{bmatrix}, \quad \varsigma = \begin{pmatrix} D \\ E \end{pmatrix}, \quad \rho = F,$$

and various matrix operations can now be performed.

In the remainder of this chapter, we will treat the binormal distribution first. *This is the case where most graphics can be drawn in 2D or 3D, even in animation on our website, and can shed a lot of light into operations performed on the bivariate vector.*

7.7   Bivariate Normal Distribution

We have, for any bivariate distribution, the following notions concerning its moments:

(1) Non-centered moments, $\alpha_{i,j}, 0 \leq i, j, \ldots$

$$\alpha_{ik} = \mathbb{E}(X^i Y^k) = \int x^i y^k f(x, y) dx dy,$$

$$\alpha_{10} = \mathbb{E}(X) = \mu_1,$$

$$\alpha_{01} = \mathbb{E}(Y) = \mu_2.$$

(2) Centered moments

$$\sigma_{12} = \mathbb{E}[(X - \mu_1)(Y - \mu_2)],$$

$$\sigma_{11} = \sigma_1^2 = \alpha_{20} - \mu_1^2,$$

$$cov(X_1, X_2) = cov(X_2, X_1) = \sigma_{12} = \sigma_{21} = \alpha_{11} - \mu_1 \mu_2,$$

$$\sigma_{22} = \sigma_2^2 = \alpha_{02} - \mu_2^2,$$

and

$$\rho = \frac{cov(X_1, X_2)}{\sigma_1 \sigma_2} = \frac{\sigma_{12}}{\sigma_1 \sigma_2}$$

is the correlation coefficient.

7.7.1 *Normal random vector* $\mathbf{V} = (X, Y) \sim N_2(\boldsymbol{\mu}, \boldsymbol{\Sigma})$

We now consider a variable in two dimensions or a vector $\mathbf{V} = (X, Y)$ in $\mathbb{R}^2$. The density of $\mathbf{V}$ is then an expression containing x, y:

$$f(x, y) = \frac{1}{2\pi\sqrt{|\boldsymbol{\Sigma}|}} \exp\left[-\frac{1}{2}(\mathbf{V} - \boldsymbol{\mu})^t \boldsymbol{\Sigma}^{-1}(\mathbf{V} - \boldsymbol{\mu})\right]. \qquad (7.8)$$

Let $\boldsymbol{\Sigma} = \begin{bmatrix} \sigma_X^2 & \sigma_{XY} \\ \sigma_{YX} & \sigma_Y^2 \end{bmatrix}$, we have

$$\boldsymbol{\Sigma}^{-1} = \begin{bmatrix} \dfrac{1}{\sigma_X^2(1 - \rho^2)} & -\dfrac{\rho}{\sigma_X \sigma_Y (1 - \rho^2)} \\ -\dfrac{\rho}{\sigma_X \sigma_Y (1 - \rho^2)} & \dfrac{1}{\sigma_Y^2(1 - \rho^2)} \end{bmatrix}.$$

In this case, results are clearer and we do not use vectors and matrices, and Eq. (7.8) becomes

$$f(x, y) = \frac{1}{2\pi \sigma_X \sigma_Y \sqrt{1 - \rho^2}}$$

$$\times \exp\left\{-\frac{1}{2(1 - \rho^2)}\left[\left(\frac{x - \mu_X}{\sigma_X}\right)^2 + \left(\frac{y - \mu_Y}{\sigma_Y}\right)^2\right.\right.$$

$$\left.\left. -2\rho\left(\frac{x - \mu_X}{\sigma_X}\right)\left(\frac{y - \mu_Y}{\sigma_Y}\right)\right]\right\}$$

$$-\infty < x, y < \infty. \qquad (7.9)$$

We can see that if $\rho = 0$, the joint density factors into the product of the two marginal densities, with disjoint domains of definition. *Hence, non-correlation and independence are equivalent for the binormal.* But if two variables are separately univariate normal, with zero correlation, it is not certain they are independent, and their definition domains may not be disjoint. They might not be jointly binormal either.

(1) The marginals are both normal: $X \sim N(\mu_X, \sigma_X^2)$, $Y \sim N(\mu_Y, \sigma_Y^2)$, with ρ playing no role, which can be proved by considering the joint moment-generating function of X and Y.

We have the conditional distribution also normal:

$$(X \mid Y = y_0) \sim N(\mu_X + \Sigma_{XY}\Sigma_{YY}^{-1}(y_0 - \mu_Y), \Sigma_{XX.Y})$$

(see Chapter 9 for a general result). Here, we find the conditional mean

$$\mu_{X|y_0} = \mu_X + \rho\frac{\sigma_X}{\sigma_y}(y_0 - \mu_Y),$$

and the conditional covariance, independent of y_0,

$$\Sigma_{XX.Y=y_0} = \sigma_X^2 - \rho\sigma_X\sigma_Y\frac{\rho\sigma_X\sigma_Y}{\sigma_Y^2} = \sigma_X^2(1 - \rho^2).$$

Hence, the conditional distribution is also normal, i.e.

$$(X \mid Y = y_0) \sim N_1\left(\mu_X + \rho\frac{\sigma_X}{\sigma_y}(y_0 - \mu_Y); \ \sigma_X^2(1 - \rho^2)\right).$$

(2) Equation (7.9) can be written as

$$f(x, y) = \left[\frac{1}{\sigma_X\sqrt{2\pi}}\exp\left(-\frac{(x - \mu_X)^2}{2\sigma_X^2}\right)\right]$$

$$\times \left[\frac{1}{\sqrt{2\pi}\sigma_Y\sqrt{1 - \rho^2}}\exp\left(-\frac{(y - A)^2}{2\sigma_Y^2(1 - \rho^2)}\right)\right],$$

with $A = \mu_Y + \rho\frac{\sigma_Y}{\sigma_X}(x - \mu_X)$ and we have $f(x, y) = f_1(x)\,f_{2|1}(y\,|x)$ in the above expression.

Hence, both marginal and conditional distributions are normal, $X \sim N(\mu_X, \sigma_X^2)$ and

$$Y \mid X \sim N\left(\mu_Y + \rho\frac{\sigma_Y}{\sigma_X}(x - \mu_X); \sigma_Y^2\left(1 - \rho^2\right)\right).$$

This result permits us to conclude that most of the probability for the distribution of X and Y (99%) lies in the band:

$$\mu_Y + \rho\frac{\sigma_Y}{\sigma_X}(x - \mu_X) \pm 2.576 \times \sigma_Y\sqrt{1 - \rho^2}.$$

We can see that

$$\{X, Y \ independent\} \Leftrightarrow \{f(x,y) = f_1(x)f_2(y)\} \Leftrightarrow \{\rho = 0\}.$$

(3) Equation (7.9) can also be obtained by considering the density $g(z_1, z_2) = \frac{1}{2\pi} \exp\left(-\frac{z_1^2 + z_2^2}{2}\right)$ of $Z_1, Z_2 \overset{iid}{\sim} N_1(0, 1)$ and performing the change of variables:

$$X_1 = \sigma_1 Z_1 + \mu_1,$$

$$X_2 = \sigma_2 \left(\rho Z_1 + \sqrt{1 - \rho^2} Z_2\right) + \mu_2.$$

This approach can be used to generate a random observation from $\mathbf{X} \sim N_2(\mu_1, \mu_2; \sigma_1^2, \sigma_2^2; \rho)$, using only $N_1(0, 1)$ (see Section 7.10.2).

7.7.2 *Iso-density ellipses*

The theoretical expression for the density of the binormal is

$$f(x_1, x_2) = C \exp\left(-\frac{1}{2}Q_2(x_1, x_2)\right),$$

with $Q_2(x_1, x_2) = \sum_{i=1}^{2} \sum_{j=1}^{2} A_{ij} (x_i - \mu_i)(x_j - \mu_j)$, with $[A_{ij}] = \boldsymbol{\Sigma}^{-1}$. Here,

$$Q_2(x_1, x_2) = \frac{1}{(1-\rho^2)}\left[\left(\frac{x - \mu_X}{\sigma_X}\right)^2 + \left(\frac{y - \mu_Y}{\sigma_Y}\right)^2\right.$$

$$\left. - 2\rho\left(\frac{x - \mu_X}{\sigma_X}\right)\left(\frac{y - \mu_Y}{\sigma_Y}\right)\right]. \quad (7.10)$$

(a) **Shapes of the contours:** Expressed in terms of first and second moments of X and Y, we have the density under the five-parameter form (7.9). Naturally, each parameter will bring in its contribution to the shape of the normal surface (an animated graph on our web page shows the contributions of σ_X, of σ_Y, and of ρ separately). But we have to limit their number in the study of that shape. With $\mu_X = \mu_Y = 0$, $\sigma_X = \sigma_Y = 1$, the only parameter left is $\rho, -1 \leq \rho \leq 1$. This is the standard form of the binormal, which permits one to see the effect of ρ alone.

Here, we consider Eq. (7.10), with $\sigma_1 > \sigma_2$.

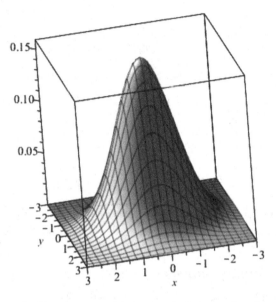

Fig. 7.5   Shape of a bivariate normal density (volume under the normal surface is unity).

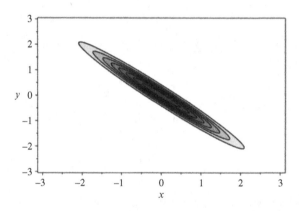

Fig. 7.6   Shapes of the contours ($\mu_1 = \mu_2 = 0$), $\rho = -0.98$.

Figures 7.5–7.14 show contours, or same density value curves, which are ellipses, when using Eq. (7.9), with $\mu_1 = \mu_2 = 0$ and $\sigma_1 > \sigma_2$.

The sign of ρ gives the direction of the main axis of ellipses. If $\sigma_1 \geq \sigma_2$, $\rho \geq 0$ the main axis is in the direction of the first diagonal. For $\rho < 0$, it follows the second diagonal. $|\rho|$ determines how the

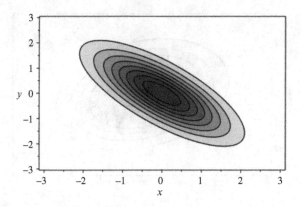

Fig. 7.7 Shapes of the contours ($\mu_1 = \mu_2 = 0$), $\rho = -0.75$.

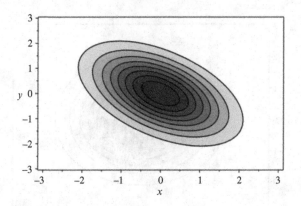

Fig. 7.8 Shapes of the contours ($\mu_1 = \mu_2 = 0$), $\rho = -0.50$.

shape of the ellipse is, wrt to its principal axis: $|\rho| \approx 1$ the surface is thin, concentrated about its principal axis. $|\rho| \approx 0$ the surface is spread out about its principal axis.

(b) **Determination of bivariate normal density contours:**
We have, in linear algebra, the following result: If the matrix $\boldsymbol{\Sigma}$ is positive definite, then so is $\boldsymbol{\Sigma}^{-1}$, and for $\{\lambda_i, e_i\}$ = eigenvalue–eigenvector of $\boldsymbol{\Sigma}$, we have $\{1/\lambda_i, e_i\}$ = eigenvalue–eigenvector of $\boldsymbol{\Sigma}^{-1}$.

(1) Eigenvalues and eigenvectors of $\boldsymbol{\Sigma}$ are obtained by solving $|\boldsymbol{\Sigma} - \lambda \boldsymbol{I}_2| = 0$. This gives

$$\begin{vmatrix} \sigma_{11} - \lambda & \rho\sqrt{\sigma_{11}}\sqrt{\sigma_{22}} \\ \rho\sqrt{\sigma_{11}}\sqrt{\sigma_{22}} & \sigma_{22} - \lambda \end{vmatrix} = 0$$

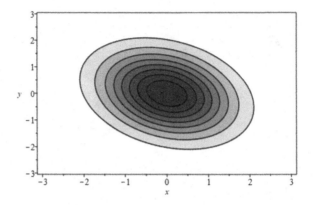

Fig. 7.9 Shapes of the contours ($\mu_1 = \mu_2 = 0$), $\rho = -0.25$.

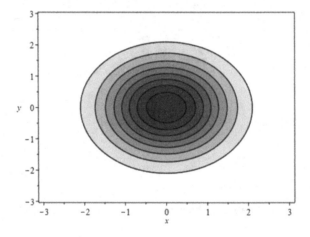

Fig. 7.10 Shapes of the contours ($\mu_1 = \mu_2 = 0$), $\rho = 0$.

or the quadratic equation

$$\lambda^2 - (\sigma_{11} + \sigma_{22})\,\lambda + \left(1 - \rho^2\right)\sigma_{11}\sigma_{22} = 0,$$

which has two roots

$$\begin{cases} \lambda_1 = \dfrac{1}{2}\left\{\sigma_{11} + \sigma_{22} + \sqrt{(\sigma_{11} + \sigma_{22})^2 - 4\left(1 - \rho^2\right)\sigma_{11}\sigma_{22}}\right\}, \\[2mm] \lambda_2 = \dfrac{1}{2}\left\{\sigma_{11} + \sigma_{22} - \sqrt{(\sigma_{11} + \sigma_{22})^2 - 4\left(1 - \rho^2\right)\sigma_{11}\sigma_{22}}\right\}. \end{cases}$$

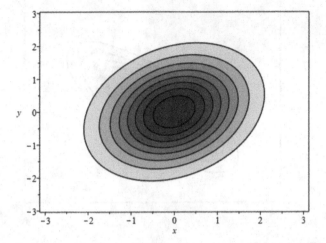

Fig. 7.11   Shapes of the contours ($\mu_1 = \mu_2 = 0$), $\rho = 0.25$.

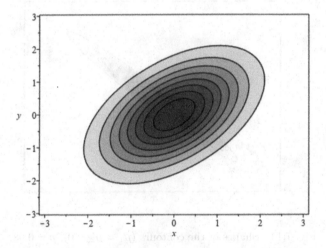

Fig. 7.12   Shapes of the contours ($\mu_1 = \mu_2 = 0$), $\rho = 0.5$.

(2) The eigenvector $E_1 = \begin{pmatrix} e_{11} \\ e_{12} \end{pmatrix}$ associated with λ_1 satisfies

$$\begin{pmatrix} \sigma_{11} & \sigma_{12} \\ \sigma_{12} & \sigma_{22} \end{pmatrix} \begin{pmatrix} e_{11} \\ e_{12} \end{pmatrix} = \lambda_1 \begin{pmatrix} e_{11} \\ e_{12} \end{pmatrix}.$$

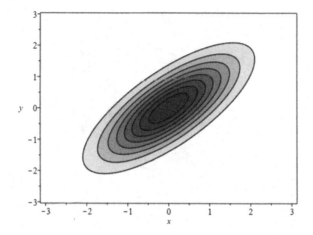

Fig. 7.13   Shapes of the contours ($\mu_1 = \mu_2 = 0$), $\rho = 0.75$.

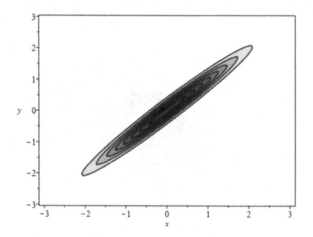

Fig. 7.14   Shapes of the contours ($\mu_1 = \mu_2 = 0$), $\rho = 0.98$.

Solving this system, we have

$$E_1 = \begin{pmatrix} e_{11} \\ e_{12} \end{pmatrix} = \begin{pmatrix} d \\ \alpha d \end{pmatrix},$$

$$\alpha = \frac{(\sigma_{22} - \sigma_{11}) + \sqrt{(\sigma_{22} + \sigma_{11})^2 - 4(1 - \rho^2)\sigma_{22}\sigma_{11}}}{2\rho\sqrt{\sigma_{22}}\sqrt{\sigma_{11}}}.$$

(3) We choose the value d so that vector E_1 has norm 1, i.e. $E_1{}^t E_1 = d^2(1 + \alpha^2) = 1$. Hence,

$$d = \frac{1}{\sqrt{1 + \alpha^2}} \quad \text{or} \quad d = -\frac{1}{\sqrt{1 + \alpha^2}},$$

giving

$$E_1 = \begin{pmatrix} \dfrac{1}{\sqrt{1 + \alpha^2}} \\ \dfrac{\alpha}{\sqrt{1 + \alpha^2}} \end{pmatrix} \quad \text{or} \quad \begin{pmatrix} -\dfrac{1}{\sqrt{1 + \alpha^2}} \\ -\dfrac{\alpha}{\sqrt{1 + \alpha^2}} \end{pmatrix}.$$

(4) Similarly, for the second eigenvector, we have

$$E_2 = \begin{pmatrix} \dfrac{1}{\sqrt{1 + \beta^2}} \\ \dfrac{\beta}{\sqrt{1 + \beta^2}} \end{pmatrix} \quad \text{or} \quad \begin{pmatrix} -\dfrac{1}{\sqrt{1 + \beta^2}} \\ -\dfrac{\beta}{\sqrt{1 + \beta^2}} \end{pmatrix},$$

with

$$\beta = \frac{(\sigma_{22} - \sigma_{11}) - \sqrt{(\sigma_{22} + \sigma_{11})^2 - 4(1 - \rho^2)\sigma_{22}\sigma_{11}}}{2\rho\sqrt{\sigma_{22}}\sqrt{\sigma_{11}}}.$$

(5) The last requirement that they form an orthonormal basis gives $E_1^t E_2 = 0$.

(c) **A particular case:** We look into the $\sigma_{11} = \sigma_{22}$ case, where the two variances are equal, and we have

$$\lambda_1 = \sigma_{11} + \sigma_{12}, \quad \lambda_2 = \sigma_{11} - \sigma_{12},$$

$$E_1 = \begin{pmatrix} 1/\sqrt{2} \\ 1/\sqrt{2} \end{pmatrix}, \quad E_2 = \begin{pmatrix} 1/\sqrt{2} \\ -1/\sqrt{2} \end{pmatrix}.$$

The main axis will follow the 45° line, with the actual value of σ_{12} having no influence. If $\sigma_{12} < 0$, that is the line of 135°.

Remark: This argument can also be made by looking at the exponential part of the equation, i.e. $x^2 - 2\rho xy + y^2$, which is now permutation symmetrical in x and y, resulting in the fact that the density is symmetrical about the line $x = y$ or about $x = -y$ if $\rho < 0$.

7.7.3   *Standard form of the binormal*

Like in the case of the univariate normal, where we work mostly with $Z \sim N_1(0,1)$, instead of the general $X \sim N_1(\mu, \sigma^2)$, it is very convenient to work with the standard bivariate normal, although in some cases, it can be too restrictive (and will leave out some important properties, e.g. when the variances are different, $\sigma_1^2 \neq \sigma_2^2$). It can be obtained using the following steps:

(a) In the new coordinate system, with origin at the ellipse center, an alternate expression of the density, where $\mu_X = \mu_Y = 0$, is

$$f(x,y) = \frac{1}{2\pi\sigma_X\sigma_Y\sqrt{1-\rho^2}}$$

$$\times \exp\left[-\frac{1}{2(1-\rho^2)}\left\{\frac{x^2}{\sigma_X^2} - 2\rho\frac{xy}{\sigma_X\sigma_Y} + \frac{y^2}{\sigma_Y^2}\right\}\right]. \quad (7.11)$$

(b) Taking new scaled variables $\xi = x/\sigma_X$, $\eta = y/\sigma_Y$, we have the standard form already seen:

$$f(\xi,\eta) = \frac{1}{2\pi\sqrt{1-\rho^2}}\exp\left[-\frac{1}{2(1-\rho^2)}\left\{\xi^2 - 2\rho\xi\eta + \eta^2\right\}\right], \quad (7.12)$$

with $-\infty < \xi, \eta < \infty$, where the only parameter left is ρ. This is the simplest form for the bivariate normal, with any other general binormal density obtained by translation of the ellipse center, and by stretching, or compressing, by using variances. We have, in this case, the conditional distribution of η: $(\eta\,|\,\xi = a) \sim N_1\left(\rho a; (1-\rho^2)\right)$. But the standard ellipse axes are not parallel to the coordinate axes when $\rho \neq 0$.

(c) In the new axis system $(\xi O \eta)$, with the origin taken as the ellipse center, and axes of the ellipse parallel to coordinate axes after rotating the axes by the angle θ^*, we have the density

$$f(\xi,\eta) = \frac{1}{2\pi\sigma_\xi\sigma_\eta}\exp\left[-\frac{1}{2}\left\{\left(\frac{\xi}{\sigma_\xi}\right)^2 + \left(\frac{\eta}{\sigma_\eta}\right)^2\right\}\right],$$

where

$$\sigma_\xi = \sqrt{\sigma_X^2\cos^2\theta^* + \rho\sigma_X\sigma_Y\sin 2\theta^* + \sigma_Y^2\sin^2\theta^*},$$

$$\sigma_\eta = \sqrt{\sigma_X^2\sin^2\theta^* - \rho\sigma_X\sigma_Y\sin 2\theta^* + \sigma_Y^2\cos^2\theta^*}. \quad (7.13)$$

Notes: (1) Although the standard form of the binormal is sufficient for most purposes, it can also lead to a loss of information. Following are the cases where we consider the exact values of the variances in the general form and the standard binormal where these variances are 1. Ellipses are fuller in the first case (Fig. 7.15).

(2) For $\sigma_1 > \sigma_2$, the major axis of the ellipse makes an angle

$$\theta^* = \frac{1}{2}\tan^{-1}\left(\frac{2\rho\sigma_1\sigma_2}{\sigma_1^2 - \sigma_2^2}\right), \tag{7.14}$$

with the x axis. Conversely, we have $\rho = \frac{(\sigma_1^2 - \sigma_2^2)\tan(2\theta^*)}{2\sigma_1\sigma_2}$.

We have θ^* varying from $0°$ to $135°$ according to Table 7.1.

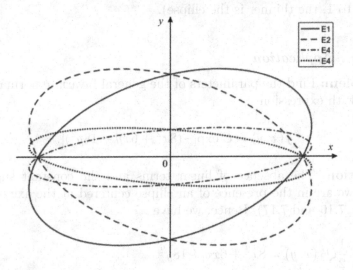

Fig. 7.15   Effects of variances, where $f(x, y; \sigma_X, \sigma_Y; \mu_X, \mu_Y; \rho) = E_1 = f_1(x, y; 1, 3; 0, 0; 0.35)$, $E_2 = f_2(x, y; 1, 3; 0, 0; -0.35)$, $E_3 = f_3(x, y; 1, 1; 0, 0; 0.35)$, and $E_4 = f_4(x, y; 1, 1; 0, 0; -0.35)$.

Table 7.1   Rotation angle θ^* wrt variances and ρ.

	$\sigma_1 > \sigma_2$	$\sigma_1 = \sigma_2$	$\sigma_1 < \sigma_2$
$\rho > 0$	$0 < \theta < 45°$	$\theta = 45°$	$45° < \theta < 90°$
$\rho = 0$	$\theta = 0°$	$\theta = 0°$ $\theta = 0°$	$\theta = 0°$
$\rho < 0$	$135° < \theta < 180°$	$\theta = 135°$	$90° < \theta < 135°$

(3) We see that although σ_1, σ_2, and ρ contribute to determine the angle θ^*, when $\sigma_1 = \sigma_2$, the angle is $45°$ or $135°$, depending on ρ being positive, or negative, but independent of its value.

The standard binormal corresponds to the middle column where $\sigma_1 = \sigma_2 = \sigma = 1$. The ellipse has equation $(x^2 - 2\rho xy + y^2)/(1 - \rho^2) = c$ and has its major axis at $\theta = 45°$ if $\rho > 0$ and at $\theta = 135°$ if $\rho < 0$. The length of the major axis is $2\sqrt{c(1 + |\rho|)}$, and the length of the minor axis is $2\sqrt{c(1 - |\rho|)}$. If $\rho = 0$, we have a circle of radius $\sqrt{c}$. When $\rho \to \pm 1$, the length of the major axis $\to 2\sqrt{2c}$ and the length of the minor axis $\to 0$ and the ellipse collapses into the major axis. Hence, the sign of ρ determines first the major axis, and the value of $|\rho|$ determines how large the minor axis is (the closer $|\rho|$ is to 1, the thinner is the ellipse).

7.7.4   *Application*

Problem: Find the parameters of the general bivariate normal density, with expression

$$f(x, y) = C \exp[-(8x^2 + 6xy + 18y^2)].$$

Solution: The absence of linear terms and the constant suggests that we are in the presence of an ellipse centered at the axes origin (Figs. 7.16 and 7.17). Hence, we have

$$\frac{1}{2}Q_2(x, y) = 8x^2 + 6xy + 18y^2$$

$$= \frac{1}{2\pi\sigma_X\sigma_Y(1 - \rho^2)}\left(\frac{x^2}{\sigma_X^2} - 2\rho\frac{xy}{\sigma_X\sigma_Y} + \frac{y^2}{\sigma_Y^2}\right).$$

Identifying different factors, we have

$$\sigma_X\sqrt{1 - \rho^2} = \frac{1}{4}, \quad \sigma_Y\sqrt{1 - \rho^2} = \frac{1}{6}, \quad -\frac{\rho}{(1 - \rho^2)\sigma_Y\sigma_X} = 6.$$

The first two equalities give $\sigma_X\sigma_Y(1 - \rho^2) = 1/24$. The third equality then gives $\rho = -1/4$, and we have $\sigma_X^2 = 1/15$, $\sigma_Y^2 = 4/135$.

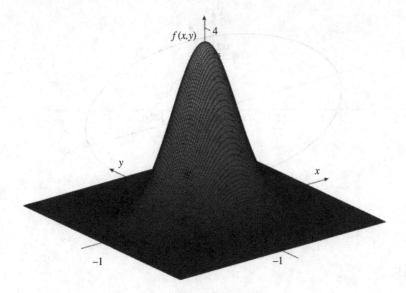

Fig. 7.16   Binormal density $N(0, 0, \frac{1}{15}, \frac{4}{135}; -0.25)$.

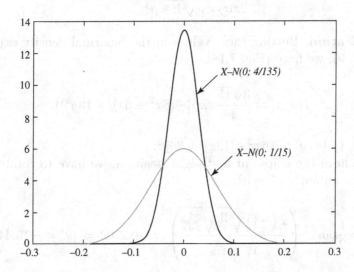

Fig. 7.17   Marginal densities: $X \sim N(0; 1/15)$, $Y \sim N(0; 4/135)$.

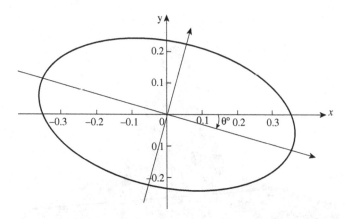

Fig. 7.18   Ellipse $8x^2 + 6xy + 18y^2 = 1$.

Also,

$$C = \frac{1}{2\pi\sigma_X\sigma_Y\sqrt{1-\rho^2}} = \frac{3\sqrt{15}}{\pi}.$$

Verification: Putting these values in the binormal density expression (7.9), we have (Fig. 7.18)

$$f(x,y) = \frac{3\sqrt{15}}{\pi}\exp\left[-(8x^2 + 6xy + 18y^2)\right].$$

Hence, $Q_2(x,y) = 16x^2 + 12xy + 36y^2$.

To have the ellipse at a straight position, we have to rotate the axes by an angle θ^*, with

$$\theta^* = \frac{1}{2}\tan^{-1}\left(\frac{2\left(-\frac{1}{4}\right)\sqrt{\frac{1}{15}}\sqrt{\frac{4}{135}}}{\frac{1}{15} - \frac{4}{135}}\right) = -0.2702 \Rightarrow (\theta^* = -15.44°).$$

Equation of ellipse in rotated system is

$$f(\xi,\eta) = \frac{1}{2\pi\sigma_\xi\sigma_\eta}\exp\left[-\frac{1}{2}\left\{\left(\frac{\xi}{\sigma_\xi}\right)^2 + \left(\frac{\eta}{\sigma_\eta}\right)^2\right\}\right],$$

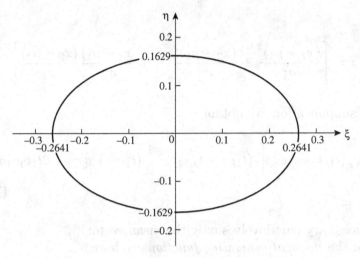

Fig. 7.19   Ellipse $f(\xi, \eta) = \left(\frac{\xi}{0.2641}\right)^2 + \left(\frac{\eta}{0.1629}\right)^2 = 1$.

where

$$\sigma_\xi = \sqrt{\sigma_X^2 \cos^2\theta^* + \rho\sigma_X\sigma_Y \sin 2\theta^* + \sigma_Y^2 \sin^2\theta^*} = 0.2641,$$

$$\sigma_\eta = \sqrt{\sigma_X^2 \sin^2\theta^* - \rho\sigma_X\sigma_Y \sin 2\theta^* + \sigma_Y^2 \cos^2\theta^*} = 0.1629.$$

Hence, as shown in Fig. 7.19,

$$f(\xi, \eta) = \left(\frac{\xi}{0.2641}\right)^2 + \left(\frac{\eta}{0.1629}\right)^2 = 1.$$

7.8   Characteristic Function of the Bivariate Normal

$$\varphi_{X_1,X_2}(t_1, t_2) = \frac{1}{2\pi\sigma_1\sigma_2\sqrt{1-\rho^2}}$$

$$\times \int_{-\infty}^{\infty} \int_{-\infty}^{\infty} \exp[i(t_1 x_1 + t_2 x_2)]$$

$$\times \exp\left(-\frac{z}{2(1-\rho^2)}\right) dx_1 dx_2,$$

with

$$z = \left[\frac{(x_1 - \mu_1)^2}{\sigma_1^2} + \frac{(x_2 - \mu_2)^2}{\sigma_2^2} - 2\rho \frac{(x_1 - \mu_1)}{\sigma_1} \frac{(x_2 - \mu_2)}{\sigma_2} \right].$$

After simplification, we obtain

$$\varphi_{X_1, X_2}(t_1, t_2) = \exp \left[i(t_1\mu_1 + t_2\mu_2) - \frac{1}{2}\left(t_1^2\sigma_1^2 + t_2^2\sigma_2^2 + 2t_1t_2\sigma_1\sigma_2 \right) \right],$$

(7.15)

the imaginary part involves only the mean vector.

For the *moment-generating function*, we have

$$M_{X_1, X_2}(t_1, t_2) = \mathbb{E}[\exp(\mathbf{tX})]$$

$$= \exp \left[(t_1\mu_1 + t_2\mu_2) + \frac{1}{2}\left(t_1^2\sigma_1^2 + t_2^2\sigma_2^2 + 2t_1t_2\sigma_1\sigma_2 \right) \right].$$

7.9   Cdf of the Bivariate Normal

(a) An important point, not much treated in theoretical texts, is the cumulative distribution function of a multivariate normal variable. In two dimensions, the explanation is clear: when the correlation is different from zero, the ellipse axis is inclined on the horizontal axis by an angle θ^* and we can compute $\int_c^d \int_a^b f(x,y) dx dy$ easily only after performing an axis rotation to get $(X, Y) \to (W_1, W_2)$. If (W_1, W_2) are jointly normal, with independent components (after a rotation of axes, for example), we can write

$$\int_c^d \int_a^b f(x,y) dx dy = \int_{a^*}^{b^*} f_1(w_1) dw_1 \int_{c^*}^{d^*} f_2(w_2) dw_2,$$

with the integration bounds transformed from (a, b, c, d). Here, f_1 and f_2 are both normal densities.

(b) The standard binormal density can be expressed as

$$f(x_1, x_2) = Z(x_1) \frac{1}{\sqrt{1-\rho^2}} Z\left(\frac{x_2 - \rho x_1}{\sqrt{1-\rho^2}}\right),$$

$$\text{with } Z(t) = \frac{1}{\sqrt{2\pi}} \exp\left(-\frac{t^2}{2}\right), \qquad (7.16)$$

which shows, again, that for (X_1, X_2) binormal (not X_1 and X_2 separately normal), X_1 and X_2 are independent iff $\rho = 0$.

For the general case, we have

$$f(x_1, x_2) = \frac{1}{\sigma_1} Z\left(\frac{x_1 - \mu_1}{\sigma_1}\right) \frac{1}{\sigma_2\sqrt{1-\rho^2}}$$

$$\times Z\left(\frac{x_2 - \mu_2 - \rho\frac{\sigma_2}{\sigma_1}(x_1 - \mu_1)}{\sigma_2\sqrt{1-\rho^2}}\right).$$

(c) For computing, we have

$$F(x_1, x_2) = \int_{-\infty}^{\infty} \Phi\left(\frac{\sqrt{|\rho|}z + a_1}{\sqrt{1-|\rho|}}\right) \Phi\left(\frac{\delta_\rho\sqrt{|\rho|}z + a_2}{\sqrt{1-|\rho|}}\right)$$

$$\times \phi(z)\, dz, \ \rho \in (-1, 1), \qquad (7.17)$$

with $\delta_\rho = 1$ if $\rho \geq 0$, $\delta_\rho = -1$ if $\rho \leq 0$, which reduces to the product of two Φ's when $\rho = 0$.

But two normal variables X_1 and X_2 can have zero correlation and be dependent.

Note: Let $F(x, y) = \mathbb{P}(X \leq x, Y \leq y)$ with the density

$$f(x, y) = \frac{\partial^2 F(x, y)}{\partial x \partial y}.$$

We then have the following:

(a) $F(\infty, \infty) = 1$, $F(-\infty, y) = F(x, -\infty) = 0$.
(b) $F(x, y)$ is monotone non-decreasing in each variable separately, continuous on the right in each variable.

(c) The following relation holds:

$$\mathbb{P}(x_1 < X \le x_2, y_1 < Y \le y_2) = F(x_1, y_1) + F(x_2, y_2) - F(x_1, y_2)$$
$$- F(x_2, y_1) \ge 0. \tag{7.18}$$

Because of the connections between the normal and the Student t-distributions, several computing techniques are shared by these two random variables (see Genz and Bretz (2009)).

7.10   Dependence and Independence of Components

Let (X, Y) be a binormal couple, with independent components $(\rho = 0)$. One of the most interesting operations under this topic is the change of variables that transforms this independent couple into a dependent one. First, let us consider the change.

7.10.1   *From dependent components to uncorrelated ones*

Let's consider again expression (7.9).

This transformation is very important to compute the cumulative probability of a dependent couple (with $\rho \ne 0$), where the axes of its ellipse are not parallel to the two coordinate axes. The main axis makes an angle θ^* with the horizontal axis, which can be determined by $\tan(2\theta^*) = 2\rho\sigma_X\sigma_Y/(\sigma_X{}^2 - \sigma_Y{}^2)$, as we have seen before. A rotation of the axes system will make the ellipse axes parallel to coordinate axes and the new variables with no correlation between them.

Let's consider the following transformation:

$$(X_1, X_2) \sim N_2 (0, 0, \sigma_1, \sigma_2; \rho) \rightarrow (Z_1, Z_2)$$
$$\sim D_2 \left(0, 0, \left(1 - \rho^2\right) \sigma_1^2, \sigma_2^2; 0\right)$$

(see Thomas (1971)). Let Eq. (7.9) be the density of the centered binormal couple $(X_1, X_2) \sim N_2 (0, 0, \sigma_1, \sigma_2; \rho)$. We make the linear change of variable

$$\begin{cases} Z_1 = X_1 - \rho\dfrac{\sigma_1}{\sigma_2}X_2, \\ Z_2 = X_2. \end{cases} \tag{7.19}$$

This transformation is, in fact, a rotation since $|J| = 1$, but affecting only the first axis. We then have

$$\begin{cases} E(Z_1) = E(Z_2) = 0, \\ \text{cov}(Z_1, Z_2) = 0. \end{cases}$$

But the resulting distribution of (Z_1, Z_2) is not necessarily binormal.

Note: Another transformation from dependent components to uncorrelated ones is the following. It is presented in Stuart (1987, p. 26)

$$(X_1, X_2) \sim N_2 (0, 0, \sigma_1, \sigma_2; \rho) \to (Y_1, Y_2) \sim D_2 (0, 0, 1, 1; 0)$$

by using the change of variables,

$$\begin{cases} y_1 = \dfrac{1}{\sqrt{1 - \rho^2}} \left(\dfrac{x_1}{\sigma_1} - \dfrac{\rho x_2}{\sigma_2} \right), \\ y_2 = \dfrac{x_2}{\sigma_2}, \end{cases} \tag{7.20}$$

with Jacobian

$$|J| = \frac{1}{\sigma_1 \sigma_2 \sqrt{1 - \rho^2}}.$$

In spite of the differences in changes of variables, these two transformations (7.19) and (7.20) give the same couple, which is not necessarily binormal.

7.10.2 *Going from independent components to dependent ones*

This reverse path is also useful to introduce the general binormal distribution, or to generate observations from a dependent couple, but using an independent one.

(a) Let $Z_i \sim N_1(0, 1)$, $i = 1, 2$ be two independent variables and ρ be a constant, with $|\rho| \leq 1$.

To have the transformation

$$(Z_1, Z_2) \sim N_2\left(0, 0, 1, 1; 0\right) \rightarrow (X_1, X_2) \sim N_2(0, 0, 1, 1; \rho),$$

we just have to set

$$X_1 = \sqrt{\frac{1+\rho}{2}} Z_1 + \sqrt{\frac{1-\rho}{2}} Z_2, \quad X_2 = \sqrt{\frac{1+\rho}{2}} Z_1 - \sqrt{\frac{1-\rho}{2}} Z_2.$$

$$(7.21)$$

(b) Another transformation from independent $(Z_1, Z_2) \sim N_2\left(0, 0, 1, 1; 0\right)$ to dependent $(Y_1, Y_2) \sim N_2\left(0, 0, 1, 1, \rho\right)$ is

$$Y_1 = Z_1, \quad Y_2 = \rho Z_1 + \sqrt{1 - \rho^2} Z_2. \tag{7.22}$$

(c) The following transformation leads to a dependent couple with all prescribed parameters:

$$(X_1, X_2) \sim N_2\left(0, 0, 1, 1; 0\right) \rightarrow (Y_1, Y_2) \sim N_2\left(\mu_1, \mu_2, \sigma_1^2, \sigma_2^2; \rho\right). \tag{7.23}$$

By choosing the values of the coefficients of the linear transformations from (X_1, X_2) to (Y_1, Y_2), we can have predetermined values for the parameters of (Y_1, Y_2).

(1) Set

$$\begin{cases} Y_1 = m_1 + \alpha_1 X_1 + \beta_1 X_2, \\ Y_2 = m_2 + \alpha_2 X_1 + \beta_2 X_2. \end{cases}$$

(2) We have

$$\mu_1 = m_1, \qquad\qquad \mu_2 = m_2,$$
$$\sigma_1^2 = \alpha_1^2 + \beta_1^2, \qquad\qquad \sigma_2^2 = \alpha_2^2 + \beta_2^2,$$
$$\text{cov}\,(Y_1, Y_2) = \alpha_1\alpha_2 + \beta_1\beta_2, \quad \rho^2 = \frac{(\alpha_1\alpha_2 + \beta_1\beta_2)^2}{(\alpha_1^2 + \beta_1^2)(\alpha_2^2 + \beta_2^2)}. \tag{7.24}$$

(3) If the determinant

$$\begin{vmatrix} \alpha_1 & \beta_1 \\ \alpha_2 & \beta_2 \end{vmatrix} \neq 0,$$

we have the density of (Y_1, Y_2):

$$f(y_1, y_2) = \frac{1}{2\pi\sigma_1\sigma_2\sqrt{1-\rho^2}}$$

$$\times \exp\left[\frac{-1}{2(1-\rho^2)}\left\{\frac{(y_1-\mu_1)^2}{\sigma_1^2}\right.\right.$$

$$\left.\left. -\frac{2\rho^2(y_1-\mu_1)(y_2-\mu_2)}{\mathrm{cov}(Y_1,Y_2)}+\frac{(y_2-\mu_2)^2}{\sigma_2^2}\right\}\right].$$

(4) If the determinant

$$\begin{vmatrix} \alpha_1 & \beta_1 \\ \alpha_2 & \beta_2 \end{vmatrix} = 0,$$

we can express Y_1 and Y_2 as linear functions of a third variable $X_3 = \alpha_1 X_1 + \beta_1 X_2$. Hence,

$$\begin{cases} Y_1 = m_1 + X_3, \\ Y_2 = m_2 + \dfrac{\alpha_2}{\alpha_1}X_3. \end{cases}$$

The singular normal distribution is then concentrated along the line $Y_2 = m_2 + (\alpha_2/\alpha_1)(Y_1 - m_1)$.

Note: We can generate random observations from a dependent couple $(Y_1, Y_2) \sim N_2(0, 0, 1, 1; \rho)$ by using the standard normal $N_1(0, 1)$. It suffices to draw two observations from the univariate normal since if (U_1, U_2) is an independent couple, with values from $N_1(0, 1)$, setting $Y_1 = U_1$, $Y_2 = \rho U_1 + \sqrt{1 - \rho^2}U_2$, we have $(Y_1, Y_2) \sim N_2(0, 0, 1, 1; \rho)$. See also Section 7.7.1.

The unit normal table can hence be used to construct a table of random values for a correlated normal couple (Wold, 1948).

7.10.3   *Going from dependent components*
to independent ones and back

(a) Let $\mathbf{X} \sim N_2\,(\mu_1, \mu_2; \sigma_1, \sigma_2; \rho)$. It can be transformed into $\mathbf{Z} \sim$ $N_2\,(0, 0; 1, 1; 0)$ by using the following matrix transformations:

$$\begin{pmatrix} Z_1 \\ Z_2 \end{pmatrix} = \begin{pmatrix} 1/\sqrt{1-\rho} & 0 \\ 0 & 1/\sqrt{1+\rho} \end{pmatrix} \begin{pmatrix} 1/\sqrt{2} & -1/\sqrt{2} \\ 1/\sqrt{2} & 1/\sqrt{2} \end{pmatrix}$$

$$\times \begin{pmatrix} 1/\sigma_1 & 0 \\ 0 & 1/\sigma_2 \end{pmatrix} \begin{pmatrix} X_1 - \mu_1 \\ X_2 - \mu_2 \end{pmatrix}. \tag{7.25}$$

We begin from the right and note that the third transformation is orthogonal, giving a system rotation of axes (see Tong (1989, p. 11)).

(b) Conversely, let $\mathbf{Z} \sim N_2\,(0, 0; 1, 1; 0)$. We obtain $\mathbf{X} \sim$ $N_2\,(\mu_1, \mu_2; \sigma_1, \sigma_2; \rho)$ by reversing the orders of the transformations:

$$\begin{pmatrix} X_1 \\ X_2 \end{pmatrix} = \begin{pmatrix} \sigma_1 & 0 \\ 0 & \sigma_2 \end{pmatrix} \begin{pmatrix} 1/\sqrt{2} & 1/\sqrt{2} \\ -1/\sqrt{2} & 1/\sqrt{2} \end{pmatrix}$$

$$\times \begin{pmatrix} \sqrt{1-\rho} & 0 \\ 0 & \sqrt{1+\rho} \end{pmatrix} + \begin{pmatrix} \mu_1 \\ \mu_2 \end{pmatrix}. \tag{7.26}$$

7.11   Orthant Probabilities for the Standard Binormal

These probabilities still present some interest in their computation, and they have found several applications in statistics and allied fields. Let

$$I = \mathbb{P}(X_1 > 0, X_2 > 0) = \int\limits_0^\infty \int\limits_0^\infty f(x, y)\,dx\,dy.$$

If we start from equation of a standard binormal, we can proceed as follows: Let

$$\begin{cases} u = x, \\ v = \dfrac{y - \rho x}{\sqrt{1 - \rho^2}}. \end{cases} \tag{7.27}$$

With these new variables, we evaluate

$$I = \frac{1}{2\pi} \int\limits_{0}^{\infty} \int\limits_{-\rho u/\sqrt{(1-\rho^2)}}^{\infty} \exp\left(-\frac{u^2 + v^2}{2}\right) dv du.$$

This is the integral of the circular normal probability density

$$f(u, v) = \frac{\exp\left(-\frac{u^2 + v^2}{2}\right)}{2\pi}$$

over a region determined by the integration bounds, with the angle between the lower boundary and the vertical axis being

$$\cot^{-1}\left(\frac{-\rho}{\sqrt{1 - \rho^2}}\right) = \frac{\pi}{2} + \sin^{-1}\rho. \tag{7.28}$$

We can hence find that

$$I = \frac{1}{2\pi}\left(\frac{\pi}{2} + \sin^{-1}\rho\right) = \frac{1}{4} + \frac{1}{2\pi}\sin^{-1}\rho = \frac{1}{2} - \frac{1}{2\pi}\cos^{-1}\rho.$$

This result was obtained by Sheppard (1898).

Similarly, we have

$$\mathbb{P}(X \leq 0, Y \geq 0) = \mathbb{P}(X \geq 0, Y \leq 0)$$

$$= \int\limits_{-\infty}^{0} \int\limits_{0}^{\infty} f(x, y) dx dy = \frac{\cos^{-1}\rho}{2\pi}.$$

7.12 Ellipses of Prediction (Dispersion or Prevision Ellipses)

We denote by E_k the intersection of the density of the general binormal $f(x, y)$ with means at the origin, with the horizontal plane at height $h > 0$. This is an ellipse of dispersion of order k, determined from h, with k being the common ratio between standard deviations and the ellipse half-axes. In the case of independent components, we have $Z_1^2 + Z_2^2 \sim \chi_2^2$, a chi-square with 2 dof, and its square root $\sqrt{Z_1^2 + Z_2^2} \sim \chi_2$, is a chi variable, also with 2 dof. We have

$$\mathbb{P}(\chi_2^2 < k^2) = 1 - \exp\left(-\frac{k^2}{2}\right)$$

for the chi-square distribution and, similarly,

$$\mathbb{P}(\chi_2 < k) = 1 - \exp\left(-\frac{k^2}{2}\right)$$

for the chi distribution. Both relations have been used by various authors, but the chi distribution will be used in this book. The probability for a point to belong to the ellipse E_k determined by

$$\frac{(x - \mu_X)^2}{(k\sigma_X)^2} + \frac{(y - \mu_Y)^2}{(k\sigma_Y)^2} = 1$$

is

$$\beta = \mathbb{P}(\chi_2 < k) = \mathbb{P}(M \in E_k) = 1 - \exp\left(-\frac{k^2}{2}\right).$$

We also recognize the square Mahalanobis distance in $\mathbb{R}^2$, Δ_x^2, from (x, y) to $\boldsymbol{\mu}$, here $\Delta_x^2 = k^2$.

The relation between k and β: $k = \sqrt{-2\ln(1 - p)}$, obtained by solving

$$\beta = 1 - \exp\left(\frac{-k^2}{2}\right) \tag{7.29}$$

(see Section 9.3 for a relationship between height and width of the normal).

We have, for $k = 1$, $\mathbb{P}(M \in E_1) = 1 - \exp(-1/2) = 0.393$, and for $k = 2, 3$, $\mathbb{P}(M \in E_2) = 1 - \exp(-2) = 0.865$ and $\mathbb{P}(M \in E_3) = 1 - \exp(-9/2) = 0.989$.

A more informative table would relate the probability content of the ellipse to the intersection height h at which the ellipse is determined, where $h = 1/(2\pi\sigma_X\sigma_Y)\exp(-k^2/2)$.

Note: Several aspects of the problem are treated numerically in Section A.3 of the Computational Appendix.

For the ellipse, in general, we have the following.

Theorem 7.1. *For the bivariate normal, taking* $k^2 = \chi^2_{2,\alpha}$, *the* $(\alpha \times 100)$*th percentile, we have the smallest ellipse* E_α, *with probability* α *of containing a randomly selected observation of the population.*

This is the ellipse with half-axis lengths of $\sqrt{\chi^2_{2,\alpha}}\sqrt{\lambda_j}$ and hence with area

$$\mathrm{mes}(E_\alpha) = \pi\chi^2_{2,\alpha}\sqrt{\lambda_1\lambda_2} = \pi\chi^2_{2,\alpha}|\Sigma|^{1/2}, \qquad (7.30)$$

using the formula of the ellipse area. We note that

$$|\Sigma| = \begin{vmatrix} \sigma_{11} & \rho\sigma_1\sigma_2 \\ \rho\sigma_1\sigma_2 & \sigma_{22} \end{vmatrix} = (1-\rho^2)\sigma_{11}\sigma_{22}$$

and that the values of λ_1 and λ_2 have been computed earlier.

For the more general case of p-normal density (see Chapter 9), we have a p-dimensional ellipsoid instead of an ellipse, with hypervolume

$$\mathrm{mes}(E_\alpha^{(p)}) = \frac{2\pi^{p/2}}{p\Gamma(p/2)}[\chi^2_{p,\alpha}]^{p/2}|\Sigma|^{1/2}. \qquad (7.31)$$

In $\mathbb{R}^p, p \geq 3$, this ellipsoid is determined by $(X-\mu)^t\Sigma^{-1}(X-\mu) = c^2$, which defines a constant probability density contour centered at μ. The axes of the ellipsoid are in the directions of the eigenvectors of Σ. The length of the kth longest axis is proportional to $\sqrt{\lambda_k}$, the square root of the eigenvalue associated with the kth eigenvector λ_k. This ellipsoid has α probability of containing a random observation from the population. Related to the generalized sample variance of $\mathbf{S}$, we have, on the volume of ellipsoid,

$$\mathrm{Vol}\left(\left\{\mathbf{X} : (\mathbf{X}-\bar{\mathbf{X}})^t\mathbf{S}^{-1}(\mathbf{X}-\bar{\mathbf{X}}) \leq c^2\right\}\right) = k_p c^p \sqrt{|\mathbf{S}|},$$
$$k_p = \frac{2\pi^{p/2}}{p\Gamma(p/2)}. \qquad (7.32)$$

7.13 Ellipsoids and Their Projections into Lower Dimension Spaces

Let us consider an ellipsoid in $\mathbb{R}^p, p \geq 3$, determined by the quadratic formula $\{\mathbf{z} : \mathbf{z}^t\mathbf{A}^{-1}\mathbf{z} < c^2\}$. We can prove that its projection, or shadow in $\mathbb{R}^{p-1}$, is again an ellipsoid with parameters determined by the first ellipsoid. Consecutive projections in successive subspaces lead to an ellipse in $\mathbb{R}^2$ and then finally to a line related to a confidence interval in one dimension.

7.14   Checking Bivariate Normality

(a) To verify that a bivariate distribution is normal, we consider the scatterplots that should be elliptical. More precisely, using the set,

$$\left\{ \mathbf{X} : (\mathbf{X} - \bar{\mathbf{X}})^t \mathbf{S}^{-1} (\mathbf{X} - \bar{\mathbf{X}}) \leq \chi^2_{2,0.50} \right\},$$

which is an approximation of (7.32), there should be roughly half of the data points lying outside the ellipse (see Fig. A.9).

(b) For dimensions larger than two, we can check that all marginals are normal, and for any couple of components, their joint distribution is binormal. Failures will make the multinormality hypothesis suspect.

(c) We consider the sample Mahalanobis distance square above:

$$d^2_{\mathbf{X}} : (\mathbf{X} - \bar{\mathbf{X}})^t \mathbf{S}^{-1} (\mathbf{X} - \bar{\mathbf{X}}). \tag{7.33}$$

For the sample $\{X_i\}$, $i = 1, \ldots, n$, we then have n ordered square general distances $\left\{ d^2_{(i)} \right\}$, $i = 1, n$. We now consider the χ^2_p distribution and let $q\left((j - 1/2)/n \right)$ be the $100(j - 1/2)/n$th quantile of this distribution, i.e.

$$q\left(\frac{(j - 1/2)}{n} \right) = \chi^2_{p, \frac{n-j+1/2}{n}}. \tag{7.34}$$

Plotting the pair $\left(q\left((j - 1/2)/n \right), d^2_{(j)} \right)$, we should find a straight line.

More methods to check multinormality can be found in Rencher and Christensen (2012).

7.15 Transformation to Reach Binormality and Trinormality

The current practice is to consider marginal distributions and apply transformations so that we approach normality for each of them, although this does not guarantee that the joint distribution is multivariate normal. A few new approaches are suggested in the literature, but research results still have to confirm their effectiveness (see Rencher and Christensen (2012)).

7.16 Computing Probabilities for the Binormal

We know that in the univariate case, the cdf values of $Z \sim N_1(0,1)$ are tabulated and available in most textbooks, since the related integral cannot be computed in closed form.

For the *standard binormal*, there is only parameter ρ that varies. We have seen that

$$\mathbb{P}(X \le 0, Y \le 0) = \mathbb{P}(X \ge 0, Y \ge 0) = \int_{-\infty}^{0} \int_{-\infty}^{0} f(x,y) dxy$$

$$= \frac{1}{4} + \frac{\sin^{-1}\rho}{2\pi}. \tag{7.35}$$

Let's consider

$$\Phi(h,k;\rho) = F(X_1 \le h, X_2 \le k)$$

$$= \int_{-\infty}^{k} \int_{-\infty}^{h} \frac{\exp\left(-\frac{1}{2(1-\rho^2)}\left(x_1^2 - 2\rho x_1 x_2 + x_2^2\right)\right)}{2\pi\sqrt{1-\rho^2}} dx_1 dx_2 \tag{7.36}$$

and its companion

$$L(h,k;\rho) = \int_{k}^{\infty} \int_{h}^{\infty} \frac{\exp\left(-\frac{1}{2(1-\rho^2)}\left(x_1^2 - 2\rho x_1 x_2 + x_2^2\right)\right)}{2\pi\sqrt{1-\rho^2}} dx_1 dx_2. \tag{7.37}$$

Computationwise, we have $L(h,k;\rho) = \frac{1}{2\pi}\int_h^\infty Z(x_1)\int_{\frac{k-\rho x_1}{\sqrt{1-\rho^2}}}^\infty Z(x_2)dx_2dx_1$.

Using the binormal table: Using the univariate normal table is fairly easy. Tables for $L(h,k;\rho)$ are available (Pearson, 1931). Other authors involved in the continuation, expansion, and improvement of these tables include Everitt, Elderton, Lee, among others.

(a) Going to standard variables Z_1, Z_2, we have

$$(Z_1, Z_2) \sim N_2\left(0,0;1,1;\rho\right) \Rightarrow \begin{cases} (-Z_1, Z_2) \sim N_2(0,0;1,1;-\rho), \\ (Z_1, -Z_2) \sim N_2(0,0;1,1;-\rho), \\ (-Z_1, -Z_2) \sim N_2(0,0;1,1;\rho). \end{cases}$$

Naturally,

$$\mathbb{P}\left(\bigcap\{Y_i > a_i\}\right) = L\left(\frac{a_1 - \mu_1}{\sigma_1}, \frac{a_2 - \mu_2}{\sigma_2};\rho\right),$$

where $\mathbb{E}(X_i) = \mu_i$, $\mathrm{var}(X_i) = \sigma_i^2$, $i = 1,2$, denoted by $L^*(h,k;\rho)$ and $\Phi^*(h,k;\rho)$.

(b) The use of these tables requires attention to some relations between events to be used, since only values in the upper quadrant for $L(h,k;\rho)$ are given, i.e. $\mathbb{P}(Z_1 > h, Z_2 > k) = L^*(h,k;\rho)$ for positive values of h, k only, where $Z_1, Z_2 \sim N(0,1)$.

(c) The tables also rely on the relations on probabilities between single and double events, once ρ is chosen. Hence, for non-positive values of h, k in $L(h,k;\rho)$, we can use

$$\mathbb{P}(X \le h, Y \le k) = \mathbb{P}(X \le h) + \mathbb{P}(Y \le k) + \mathbb{P}(X > h, Y > k) - 1,$$

where $\Phi(h,\infty;\rho) = \Phi(h)$ and $\Phi(\infty,k;\rho) = \Phi(k)$, coming from the relation

$$\mathbb{P}(A \cap B) = 1 - \mathbb{P}(A^C) - \mathbb{P}(B^C) + \mathbb{P}(A^C \cap B^C)$$

or

$$\Phi(h,k;\rho) = 1 - L(h,-\infty;\rho) - L(-\infty,k;\rho) + L(h,k;\rho).$$

(d) Table 7.2 considers all four cases of signs of h, k in relation to $h, k \ge 0$.

Table 7.2 Formulas to obtain $\Phi^*(h, k; \rho)$ and $L^*(h, k; \rho)$ from $L^*(h, k; \rho)$ with $h, k \geq 0$.

Case	$\Phi^*(h, k; \rho)$	$L^*(h, k; \rho)$
$h \geq 0, k \geq 0$	$\Phi^*(h) + \Phi^*(k) + L^*(h, k; \rho)$	$L^*(h, k; \rho)$ (given by tables)
$h \geq 0, k < 0$	$\Phi^*(-k) - L^*(h, -k; -\rho)$	$1 - L^*(h, -k; -\rho) - F^*(h)$
$h < 0, k \geq 0$	$\Phi^*(-k) - L^*(-h, k; -\rho)$	$1 - L^*(-h, k; \rho) - F(k)$
$h < 0, k < 0$	$L^*(-h, -k; -\rho)$	$F^*(-h) + F^*(-k)$ $+ F^*(-h, -k; -\rho) - 1$

7.17 Recent Approaches to the Computation of the cdf of the Binormal and Trinormal

7.17.1 *cdf of the binormal*

Since the density depends on five parameters, computation can be difficult. One of the earliest approaches proposed is Drezner's Gauss quadrature method (Drezner, 1978). For the standard binormal, it considers $\Phi(h, k, \rho)$, as defined above. By Gauss quadrature, Drezner's method uses

$$\Phi(h, k, \rho) = \frac{\sqrt{1 - \rho^2}}{\pi} \sum_{i,j=1}^{k} B_i B_j f(x_i, x_j),$$

where

$$f(x, y) = \exp[h_1(2x - h_1) + k_1(2y - k_1) + 2\rho(x - h_1)(y - k_1)].$$

The values of the couples {coefficient, point}, $\{B_i, x_i\}$, $1 \leq i \leq k$, $k = 2, \dots, 15$, used in quadrature, are given in Steen, Byrne and Gelbard (1969) for the integrals $\int_0^\infty \exp(-x^2) f(x) dx$ and $\int_0^b \exp(-x^2) f(x) dx$. The related errors are quite small, going from 1.1×10^{-4} to 1.1×10^{-12}.

7.17.2 *cdf of the standard trinormal*

Let's consider the correlation matrix

$$\mathbf{R} = \begin{bmatrix} 1 & \rho_{12} & \rho_{13} \\ \rho_{12} & 1 & \rho_{23} \\ \rho_{13} & \rho_{23} & 1 \end{bmatrix}.$$

Table 7.3   Equivalence between $L_3(.)$, $L_2(.)$, and $L_1(.)$.

ρ_{12}	ρ_{13}	ρ_{23}	$L_3(h_1, h_2, h_3; \{\rho_{ij}\}; \mathbf{R})$
0	0	0	$L_1(h_1)L_1(h_2)L_1(h_3)$
0	0	ρ	$L_1(h_1)L_2(h_2, h_3; \rho)$
ρ	ρ	1	$L_2(h_1, \max\{h_2, h_3\}; \rho)$

The trivariate probability is

$$L_3(h_1, h_2, h_3; \{\rho_{ij}\}; \mathbf{R})$$

$$= \frac{1}{\sqrt{8\pi^3 |\mathbf{R}|}} \int_{h_1}^{\infty} \int_{h_2}^{\infty} \int_{h_3}^{\infty} \exp(-x^t \mathbf{R}^{-1} x/2) dx_1 dx_2 dx_3,$$

while the bivariate and univariate probabilities are $L_2(h_1, h_2; \rho)$ and $L_1(h)$ We also have $|\mathbf{R}| = 1 - \rho_{12}^2 - \rho_{13}^2 - \rho_{23}^2 + 2\rho_{12}\rho_{13}\rho_{23}$. In several cases, L_3 can be expressed in terms of L_2 and L_1, defined similarly, as shown in Table 7.3, and results already obtained for these cases of lower dimensions can be applied.

7.18   A Result of Plackett

(a) Plackett (1954) showed that

$$\frac{\partial L_3}{\partial \rho_{12}} = Z_2(h_1, h_2; \rho_{12})L_1(h^*), \tag{7.38}$$

where

$$P = Z_2(h_1, h_2; \rho_{12})$$

$$= \int_{-\infty}^{h} \int_{-\infty}^{k} \frac{\exp\left(-\frac{1}{2(1-\rho^2)}(x_1{}^2 - 2\rho x_1 x_2 + x_2{}^2)\right)}{2\pi\sqrt{1-\rho^2}} dx_1 dx_2,$$

density of L_2, while

$$h^* = \frac{h_3(1-\rho_{12}^2) - h_1(\rho_{13}-\rho_{23}\rho_{12}) - h_2(\rho_{23}-\rho_{13}\rho_{12})}{\sqrt{(1-\rho_{12}^2)|\mathbf{R}|}}.$$

This relation permits us to compute the orthant probability, e.g.

$$L_3(0,0,0;\{\rho_{ij}\};\mathbf{R}) = \frac{1}{8} + \frac{1}{4\pi}\sum_{i<j}\sin^{-1}\rho_{ij}$$

(see Drezner (1994)).

(b) Other interesting approaches to compute binormal probability include Owen (1956), Divgi (1979), and Drezner and Wesolowsky (1990), who promoted different approaches and expressions to compute $L(h,k;\rho)$. For the trivariate normal, we have Gupta (1963), who established

$$\Phi_3(-\infty,\mathbf{b};\mathbf{R}) = \int_{-\infty}^{b_1} \Phi_2\left(\frac{b_2-\rho_{21}y}{\sqrt{1-\rho_{21}^2}}, \frac{b_3-\rho_{31}y}{\sqrt{1-\rho_{31}^2}};\right.$$
$$\left.\frac{\rho_{32}-\rho_{21}\rho_{31}}{\sqrt{(1-\rho_{21}^2)(1-\rho_{31}^2)}}\right)\phi(y)dy,$$

where, in general, for $k \geq 3$,

$$\Phi_k(\mathbf{a},\mathbf{b};\boldsymbol{\Sigma}) = \frac{1}{\sqrt{(2\pi)^k|\boldsymbol{\Sigma}|}}\int_{a_1}^{b_1}\ldots\int_{a_k}^{b_k}\exp(-x^t\boldsymbol{\Sigma}^{-1}x)dx_k\ldots dx_1,$$

$$(7.39)$$

and ϕ is the pdf of the univariate standard normal.

For the evaluation of multivariate normal integrals in higher dimensions, see Chapter 9.

7.19   Independence of Sample Mean and Sample Variance

In one variable, we have seen the proof that the sample mean and sample variance of a sample of normal observations are independent of each other by using Helmert's change of variables method in Chapter 4. How about in $\mathbb{R}^2$? An extension of Helmert's approach is possible, but is very complicated. Section 7.20 proposes another approach using sample decomposition.

The sample mean $(\bar{X}_n, \bar{Y}_n)$ can be shown to have distribution

$$(\overline{X}, \overline{Y}) \sim N_2(\mu_1, \mu_2; \sigma_1^2/n, \sigma_2^2/n; \rho),$$

with

$$\mathbb{E}(\bar{X}) = \mu_1, \ \mathrm{Var}(\bar{X}) = \frac{\sigma_1^2}{n}, \ Cov(\bar{X}, \bar{Y}) = \frac{\sigma_1\sigma_2}{n}\rho,$$

and similarly for $\bar{Y}$.

7.20   Decomposition of an n-Sample of Binormal Observations

In the five-parameter model, Eq. (7.9), let's write the joint probability of the n-sample as

$$dF = \frac{1}{(2\pi\sigma_1\sigma_2)^n(1-\rho^2)^{n/2}}$$

$$\times \exp\left\{-\frac{n}{2(1-\rho^2)}\left[\sum\left(\frac{\bar{x}-\mu_X}{\sigma_1}\right)^2 + \sum\left(\frac{\bar{y}-\mu_Y}{\sigma_2}\right)^2\right.\right.$$

$$\left.\left.-2\rho\sum\left(\frac{\bar{x}-\mu_X}{\sigma_1}\right)\left(\frac{\bar{y}-\mu_Y}{\sigma_2}\right)\right]\right\}dv.$$

With the volume element, $dv = dx_1dy_1\ldots dx_ndy_n$. Let's replace the five parameters by their sample equivalents $\{\bar{X}, \bar{Y}, s_1^2, s_2^2, r\}$. It can be shown that $dv = s_1^{n-2}s_2^{n-2}ds_1ds_2(1-r^2)^{(n-4)/2}drd\bar{x}d\bar{y}$ so that dF can be factorized into two parts:

$$dF = \alpha \exp\left\{-\frac{n}{2(1-\rho^2)}\left[\sum\left(\frac{\bar{x}-\mu_X}{\sigma_1}\right)^2 + \sum\left(\frac{\bar{y}-\mu_Y}{\sigma_2}\right)^2\right.\right.$$

$$\left.\left. - 2\rho\sum\left(\frac{\bar{x}-\mu_X}{\sigma_1}\right)\left(\frac{\bar{y}-\mu_Y}{\sigma_2}\right)\right]\right\}d\bar{x}d\bar{y}$$

$$\times K\left\{-\frac{n}{2(1-\rho^2)}\left(\left(\frac{s_1}{\sigma_1}\right)^2 + \left(\frac{s_2}{\sigma_2}\right)^2 - \frac{2\rho r s_1 s_2}{\sigma_1\sigma_2}\right)\right\}$$

$$\times s_1^{n-2}s_2^{n-2}(1-r^2)^{(n-4)/2}ds_1ds_2dr, \tag{7.40}$$

with

$$K = \frac{1}{\pi}\left(\frac{n}{\sigma_1\sigma_2}\right)^{n-1}\frac{1}{(1-\rho^2)^{(n-1)/2}\Gamma(n-2)}.$$

So, the joint distribution of $(\bar{X}, \bar{Y})$ is independent of the joint distribution of (s_1^2, s_2^2, r). The sample can be decomposed into two parts. This is discovered by Fisher and characterizes the binormal distribution in $\mathbb{R}^2$.

7.21 A Distribution Directly Derived from the Binormal: The Bivariate Half-Normal Distribution

It has density

$$f(x_1, x_2) = 2\exp\left\{-\frac{\left(\frac{x_1}{\sigma_1}\right)^2 + \left(\frac{x_2}{\sigma_2}\right)^2}{2(1-\rho^2)}\right\}\frac{\cosh\left\{\frac{\rho x_1 x_2}{(1-\rho^2)\sigma_1\sigma_2}\right\}}{\pi\sigma_1\sigma_2\sqrt{1-\rho^2}},$$

$$x_1, x_2 > 0. \tag{7.41}$$

Each of the marginal variables X_1 and X_2 has a half-normal distribution and the conditional of $X_2|X_1$ and $X_1|X_2$ is a folded normal.

The regression function of X_2 on X_1 is

$$|\rho|\, X_1 + 2Z \left(\frac{\rho X_1}{\sqrt{1 - \rho^2}} \right).$$

7.22   Domains of Applications of the Binormal Distribution

There are numerous applications of the bivariate normal presented in Hutchinson and Lai (1990). Among these applications, we have applications in biology (size and shape of animals, dimensions of trees, home ranges of animals), medicine (epidemiology, blood pressure measurements), psychology (mental abilities, signal detection theory), air, earth, water, and climate (windspeed, meteorological conditions at two times, rainfall duration and amount, errors in tropical cyclones forecasts, river flows and rain, tides at two specific times), earthquake motion (trinormal model), physical and technological sciences (structural reliability, strength of lumber, thermodynamics, radar, target shooting, forces in wood grinding), sociology and economics (household size and income, commodities trading, blackjack gambling).

Naturally, we have to go to each cited article to really appreciate its content and its interesting approach.

7.23   Special Bivariate Normal Distributions Used in Applied Fields

In some applications, when we are not able to determine different values for the standard deviations, we often end up in giving them the same value (see also Computational Appendix).

7.23.1   *The circular normal distribution*

In case $\sigma_1 = \sigma_2 = \sigma$, $\rho = 0$, Eq. (7.9) is called circular and it can be expressed as (when means are taken at the origin)

$$f(x, y) = \frac{1}{2\pi\sigma^2} \exp\left(-\frac{x^2 + y^2}{2\sigma^2} \right).$$

Computations have resulted in tabulated values for convenient use.

Ellipses become circles and we have again

$$\mathbb{P}(M \in E_k) = 1 - \exp(-k^2/2).$$

The circle of radius 1.774σ is named 50% *probability circle*. The radius of this circle is called circular probable error (CPE). Hence, CPE $= 1.1774\sigma$. Some military organizations are interested in the mean diameter δ of a circle of about (μ_X, μ_Y) within which 75% of the observations would fall. Hence, we have $\delta = 3.330\sigma$. For convenient use, tables are constructed to have values of the radius $c\sigma$ in the function of p. For example, to have a probability of 75%, we need a radius of 1.665σ.

7.23.2 *Polar form*

A polar form is always convenient when there is symmetry between variables. Again with $\sigma_1 = \sigma_2 = \sigma$, $\rho = 0$ and the origin at the mean, the probability that a random point would fall within a region S is

$$p = \int \int\limits_{S} g(r) r \, dr \, d\theta,$$

where

$$g(r) = \frac{1}{2\pi\sigma} \exp\left(-\frac{r^2}{2\sigma^2}\right),$$

with

$$r^2 = \frac{x^2 + y^2}{\sigma^2}, \tan\theta = \frac{y}{x}.$$

We now consider the case where the mean (μ_X, μ_Y) is *not* at the origin, with d being the distance from $(0, 0)$ to (μ_X, μ_Y), and consider a disc C centered at $(0, 0)$ and having radius R. We wish to compute the probability contained in the disc C, i.e. compute $P = \int\int_C f(r) r \, dr \, d\theta$. Two parameters are considered d/σ and d/R, with $0 \le d/\sigma \le 6$, $0.1 \le R/\sigma \le 3$ and values of P have been tabulated in the function of the values of these two parameters for convenient use.

For example, for $d/\sigma = 0$, $R/\sigma = 3$, we have $P = 0.989$, the maximum value of the table. This is understandable since the disc is centered at the origin, with a radius three times the common standard deviation value. On the other hand, $P = 0.001$ for several couples of values of d/σ, d/R for $d/\sigma = 3$, $R/\sigma = 0.1$, for example, i.e. when the disc is fairly far away from the origin, with a small radius. These results are taken from the Table of Circular Normal Probabilities, Bell Aircraft Corporation, Buffalo, New York, 1956, as reported by Burington and May (1970, p. 135). Guenther and Terragno (1963) have published a good survey on coverage problems, related to this topic.

7.24   Trivariate Normal Distribution

The general trivariate normal has nine parameters, three of each: means, variances, and correlation coefficients. We have the following expression of the density, when using the matrix of correlations and its inverse:

$$f(x_1, x_2, x_3) = \frac{\exp(-\Sigma/2)}{(2\pi)^{3/2}\sqrt{|R|}\sigma_1\sigma_2\sigma_3},$$

$$\Sigma = \sum_{i,j=1}^{3} \frac{R_{ij}^*(x_i - \mu_i)(x_j - \mu_j)}{|R|\,\sigma_i\sigma_j},$$

where $[R_{ij}^*]$ is the transpose of the matrix of cofactors of the ith row and jth column of the matrix of correlations $\mathbf{R}$, with

$$\mathbf{R} = \begin{bmatrix} 1 & \rho_{12} & \rho_{13} \\ \rho_{12} & 1 & \rho_{23} \\ \rho_{13} & \rho_{23} & 1 \end{bmatrix},$$

and $|\mathbf{R}|$ is its determinant.

7.24.1   *The standard trivariate normal*

Let the standard trivariate normal distribution, with means zeros and variances one, be defined by the symmetric correlation matrix. Explicitly, we have here a normal distribution in three variables, and only three parameters, which has probability density

$$f(x_1, x_2, x_3) = \frac{\exp(-W/(2\Delta))}{2\sqrt{2}\pi^{3/2}\sqrt{\Delta}}, \tag{7.42}$$

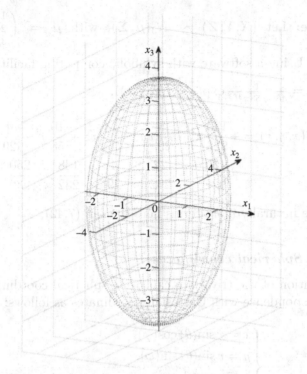

Fig. 7.20 Ellipsoid for standard trinormal distribution.

where $\Delta = 1 + 2\rho_{12}\rho_{13}\rho_{23} - (\rho_{12}^2 + \rho_{13}^2 + \rho_{23}^2)$ and

$$W = x_1^2(1 - \rho_{23}^2) + x_2^2(1 - \rho_{13}^2) + x_3^2(1 - \rho_{12}^2)$$
$$+ 2x_1x_2(\rho_{13}\rho_{23} - \rho_{12}) + 2x_1x_3(\rho_{12}\rho_{23} - \rho_{13})$$
$$+ 2x_2x_3(\rho_{12}\rho_{13} - \rho_{23}).$$

This convenient standard form can serve in many, but not all, cases. Figure 7.20 shows the graph of the ellipsoid determined by $f(x_1, x_2, x_3) = 0.05$, with $\rho_{12} = 0.20$, $\rho_{23} = 0.30$, $\rho_{13} = 0.40$.

7.24.2 *The general trivariate normal*

The precise expression of the density of a trinormal vector can be complicated.

Example: Let $(X, Y, Z) \sim N_3(\boldsymbol{\mu}, \boldsymbol{\Sigma})$, with $\boldsymbol{\mu} = \begin{pmatrix} 1 \\ 2 \\ 3 \end{pmatrix}$, $\boldsymbol{\Sigma} = \begin{pmatrix} 9 & 4 & 1 \\ 4 & 9 & 4 \\ 1 & 4 & 9 \end{pmatrix}$. Using a software with symbolic computing facility, we find $f(x, y, z) = \frac{1}{464\sqrt{58}\pi^{3/2}} \exp(-\frac{1}{2}G(x, y, z))$, with

$$
\begin{aligned}
G(x, y, z) = &+\frac{130}{928}x^2 - \frac{4}{29}xy - \frac{14}{464}xz + \frac{10}{58}y^2 - \frac{4}{29}yz \\
&+\frac{130}{928}z^2 - \frac{22}{232}x - \frac{4}{29}y - \frac{138}{232}z + \frac{250}{232}.
\end{aligned}
$$

There are naturally more terms here than in (7.42).

7.24.3 *Spherical coordinates*

The equation of the trivariate can be in spherical coordinates, with the correspondence with Cartesian coordinates as follows:

$$
\begin{cases}
x = r\sin(\theta)\cos(\varphi), \\
y = r\sin(\theta)\sin(\varphi), \\
z = r\cos(\theta), \\
0 \le \theta \le \pi, \quad 0 \le \varphi \le 2\pi, \quad r \ge 0.
\end{cases}
$$

Let $(X, Y, Z) \sim N_3(0, 0, 0; \sigma, \sigma, \sigma; 0, 0, 0)$, with density

$$
f(x, y, z) = \frac{1}{(2\pi\sigma^2)^{3/2}} \exp\left(-\frac{x^2 + y^2 + z^2}{2\sigma^2}\right). \tag{7.43}
$$

We can see that (X, Y, Z) is a vector with independent components. Then the joint pdf of (r, θ, φ) is

$$
\begin{aligned}
f(r, \theta, \varphi) &= f(x, y, z) \left| \det\left[J\left(\begin{matrix} x & y & z \\ r & \theta & \varphi \end{matrix}\right)\right]\right| \\
&= \frac{r^2\sin(\theta)}{(2\pi\sigma^2)^{3/2}} \exp\left(-\frac{r^2}{2\sigma^2}\right) \\
&= f(r).f(\theta).f(\varphi),
\end{aligned}
$$

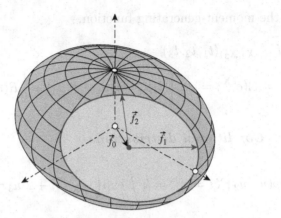

Fig. 7.21   Intersection of an ellipsoid with a plane.

with

$$f(r) = \left(\frac{2}{\pi}\right)^{1/2} \left(\frac{r^2}{\sigma^3}\right) \exp\left(-\frac{r^2}{2\sigma^2}\right), \quad f(\theta) = \frac{\sin(\theta)}{2}, f(\varphi) = \frac{1}{2\pi}.$$

$$(7.44)$$

So, these three spherical variables, defined on disjoint domains, are also independent. We have volume $V = \frac{4}{3}abc$, where a, b, and c are measures of halves of the principal axes. Also, $V(\text{inscribed box}) = \frac{8}{3\sqrt{3}}abc$ while $V(\text{surcumscribed box}) = 8abc$. The intersections of planes with ellipsoids, like with normal surfaces, are ellipses too (see Fig. 7.21).

7.25   Basic Properties of the Standard Trivariate Normal

7.25.1   *The characteristic function*

Let $R(t_1, t_2, t_3) = t_1^2 + 2\rho_{12}t_1t_2 + t_2^2 + 2\rho_{23}t_2t_3 + t_3^2 + 2\rho_{13}t_1t_3$ for the standard case. We obtain

$$\varphi_{X_1,X_2,X_3}(t_1, t_2, t_3) = \exp\left[i(t_1\mu_1 + t_2\mu_2 + t_3\mu_3) - \frac{1}{2}R(t_1, t_2, t_3)\right],$$

where the imaginary part involves only the mean. X_1 has chf $\varphi_{X_1,X_2,X_3}(t_1, 0, 0)$ and (X_1, X_2) has chf $\varphi_{X_1,X_2,X_3}(t_1, t_2, 0)$.

For the moment-generating function,

$$M_{(X_1, X_2, X_3)}(t_1, t_2, t_3)$$

$$= \mathbb{E}(e^{t\mathbf{X}}) = \exp\left[(t_1\mu_1 + t_2\mu_2 + t_3\mu_3) + \frac{1}{2}R(t_1, t_2, t_3)\right].$$

7.25.2   *Conditional distributions*

$$\varphi(u_1, u_2 \,|\, X_3 = x_3^0) = k \int \exp\left(-\frac{1}{2}\left\{R + 2iu_3 x_3^0\right\}\right) du_3.$$

The conditional distribution of $\left((X_1, X_2)\,|\,X_3 = x_3^0\right)$ is binormal, with

$$\left((X_1, X_2)\,|\,X_3 = x_3^0\right) \sim N_2\left\{\mathbb{E}(X_1)\,|\,x_3^0, \mathbb{E}(X_2)\,|\,x_3^0;\right.$$

$$\left.\sigma_{X_1}^2\,|\,x_3^0, \sigma_{X_{12}}^2\,|\,x_3^0; \rho_{X_1 X_2}\,|\,x_3^0\right\},$$

where

$$\mathbb{E}(X_1)\,|\,x_3^0 = \rho_{13}x_3^0,$$

$$\mathbb{E}(X_2)\,|\,x_3^0 = \rho_{23}x_3^0,$$

$$\sigma_{X_1}^2\,|\,x_3^0 = 1 - \rho_{13}^2,$$

$$\sigma_{X_{12}}^2\,|\,x_3^0 = 1 - \rho_{23}^2,$$

$$\rho_{X_1 X_2}\,|\,x_3^0 = \frac{\rho_{12} - \rho_{13}\rho_{23}}{\sqrt{1 - \rho_{12}{}^2}\sqrt{1 - \rho_{23}{}^2}}.$$

We can obtain, similarly, the distribution of $(X_1\,|\,X_2 = x_2^0, X_3 = x_3^0)$.

7.26   Orthant Probability for the Trinormal

For standard multivariate normal vectors, there is the probability that all components $X_1, \ldots, X_p$ are positive (or negative). Orthant probabilities have numerous applications.

(a) For the standardized trivariate normal distribution, with zero means and unit variances, the quadrant probability is then given analytically by

$$\mathbb{P}\left(\bigcap_{i=1}^{3}(X_i \leq 0)\right) = \int_{-\infty}^{0}\int_{-\infty}^{0}\int_{-\infty}^{0} f(x_1, x_2, x_3)dx_1dx_2dx_3$$

$$= \frac{1}{8} + \frac{1}{4\pi}(\sin^{-1}\rho_{12} + \sin^{-1}\rho_{13} + \sin^{-1}\rho_{23}).$$
(7.45)

(b) For any integer p, let $A_i = \{X_i > 0\}, i = 1,\ldots,p$. We have $\mathbb{P}(A_i) = 1/2$. With some conditions on the correlation coefficients, we have the following:

$$\rho_{ij} = 0, \forall i, j, \Rightarrow \mathbb{P}(\bigcap A_j) = 2^{-p},$$
$$\rho_{ij} = 1/2, \forall i, j, \Rightarrow \mathbb{P}(\bigcap A_j) = \frac{1}{p+1}.$$

7.27 Transformation of a Standard Trinormal Vector into a Trinormal Vector with Independent Components

Let $\{X_1, X_2, X_3\} \sim N_3(0, \boldsymbol{\Sigma})$, with unit variances.

Let's consider the change to independent normal variables $\{Z_1, Z_2, Z_3\}$

$$\begin{cases} X_1 = Z_1, \\ X_2 = \rho_{12}Z_1 + \sqrt{1 - \rho_{12}^2}Z_2, \\ X_3 = \rho_{13}Z_1 + aZ_2 + bZ_3, \end{cases}$$
(7.46)

with

$$a = \frac{\rho_{23} - \rho_{12}\rho_{13}}{\sqrt{(1 - \rho_{12}^2)}}, \quad b = \left\{1 - \rho_{13}^2 - \frac{(\rho_{23} - \rho_{12}\rho_{13})^2}{1 - \rho_{12}^2}\right\}^{1/2}.$$

This is a transformation to independent trinormal variates $\{Z_1, Z_2, Z_3\}$. Integrate over an appropriate region Ω of $\{Z_1, Z_2, Z_3\}$

by taking

$$\Omega = \left\{ Z_1 > 0, \rho_{12}Z_1 + \sqrt{1 - \rho_{12}^2}\,Z_2 > 0, \rho_{13}Z_1 + aZ_2 + bZ_3 > 0 \right\}.$$

This is a spherical triangle on a unit sphere and using appropriate considerations, we can then establish (Moran, 1968, p. 468)

$$\int \int_{\Omega} \int f(z_1, z_2, z_3)dz_1 dz_2 dz_3$$

$$= \frac{2\pi - \cos^{-1}\rho_{12} - \cos^{-1}\rho_{13} - \cos^{-1}\rho_{23}}{4\pi}, \tag{7.47}$$

as shown earlier.

7.28   cdf of the Standard Trinormal

Let

$$\Phi(h_1, h_2, h_3; \rho_{23}, \rho_{13}, \rho_{12}) = \mathbb{P}[(X_1 < h_1) \cap (X_2 < h_2) \cap (X_3 < h_3)]$$

and

$$L(h_1, h_2, h_3; \rho_{23}, \rho_{13}, \rho_{12}) = \mathbb{P}[(X_1 > h_1) \cap (X_2 > h_2) \cap (X_3 > h_3)].$$

Several relations exist between these two quantities. For example,

$$\Phi(0, 0, 0; \rho_{12}, \rho_{13}, \rho_{23}) = L(0, 0, 0; \rho_{12}, \rho_{13}, \rho_{23})$$

and

$$\begin{aligned}\Phi(h_1, h_2, h_3; \rho_{12}, \rho_{13}, \rho_{23}) &= L(h_1, h_2; \rho_{12}) + \Phi(h_1) + \Phi(h_2) \\ &\quad - \{1 + \Phi(h_1, h_2, -h_3; -\rho_{23}, \rho_{13}, \rho_{12})\}.\end{aligned}$$

Note: An interesting approach was suggested by Steck (1958), who has expressed Φ in terms of his function $S(h, a, b)$ defined by

$$\begin{aligned}S(h, a, b) &= \frac{1}{4\pi}\tan^{-1}\frac{b}{\sqrt{1 + a^2 + a^2 b^2}} \\ &\quad + \mathbb{P}([0 \le U_1 \le U_2 + bU_3] \cap [0 \le U_2 \le h] \cap [U_3 \ge aU_2]),\end{aligned}$$

with $U_i \overset{\text{iid}}{\sim} N_1(0, 1)$ and $h, a, b > 0$.

Steck (1958) provided tables of $S(h, a, b)$ to seven decimal places for different values of h, a, b.

7.29 Iso-probability Surfaces for a Standard Trinormal

Let's consider the general case

$$f(x, y, z) = \frac{\exp(-\Sigma/2)}{(2\pi)^{3/2}\sigma_X\sigma_Y\sigma_Z},$$
(7.48)

where

$$\Sigma = \frac{(x - \mu_X)^2}{\sigma_X^2} + \frac{(y - \mu_Y)^2}{\sigma_Y^2} + \frac{(z - \mu_Z)^2}{\sigma_Z^2}.$$

Like the bivariate case, for a constant k, the section of the normal surface by a plane at height h gives the ellipsoid $\Phi = k^2$, and the probability that a random point (X, Y, Z) will be inside the ellipsoid is

$$\beta = \sqrt{\frac{2}{\pi}} \int_0^k t^2 \exp\left(-\frac{t^2}{2}\right) dt,$$

using, as before, the chi distribution with 3 dof. For $k = 1.5382$, we have $\beta = 0.50$ and this ellipsoid is called the 50% probability ellipsoid. Similarly, for $k = 2.795$, we have the 95% probability ellipsoid.

7.30 Spherical Case

In some applications, we are unable to determine the values of the three standard deviations and hence give them the same value $\sigma_X = \sigma_Y = \sigma_Z = \sigma$, and we have a spherical trivariate normal. From (7.44), its density is

$$f(r) = \left(\frac{2}{\pi}\right)^{1/2} \left(\frac{r^2}{\sigma^3}\right) \exp\left(-\frac{r^2}{2\sigma^2}\right),$$
(7.49)

with $r^2 = (x - \mu_X)^2 + (y - \mu_Y)^2 + (z - \mu_Z)^2$, and ellipsoids become spheres. Now, for $k = 1.5382$, we have $\beta = 0.50$ and this sphere of radius 1.5382σ is the 50% probability sphere. Similarly, the probabilities p that a point (X, Y, Z) taken at random will fall within the sphere $r^2 = (C\sigma)^2$ for particular values of C are given by tables. For instance, for $C = 1.5832$, the sphere of radius 1.5832σ is the 50% probability sphere.

7.31   Specialized Tools for Computation of the Bivariate and Trivariate Normal Distributions

There are tables to provide the values of these probabilities in voluminous tables. These tables cover several particular cases. Some examples include Tables of the Bivariate Normal Distribution and Related Functions by National Bureau of Standards (1959) and Table of Circular Normal Probabilities, Report No. 02-919-106, by Bell Aircraft Corporation (1956).

For the univariate normal, normal probability graph paper was used manually in the past and is still used in some software to have a quick check on normality. Normal probability ruler was another convenient device to compute the normal probability value between two real values.

For the bivariate normal, there was a rectangular normal probability integral chart drawn for one quadrant. We can then compute approximately the probability for the normal in any area of the plane. Naturally, these tools were useful prior to personal computer days. For four quadrants, there is a polar form of normal probability integral chart. Details on these devices can be obtained in Burington and May (1970).

7.32   Generation of Observations from a Bivariate and a Trivariate Normal Distribution

7.32.1   *Bivariate case*

We have the transformation from

$$(X_1, X_2) \sim N_2(\mu_1, \mu_2; \sigma_1, \sigma_2, \rho) \to (Y_1, Y_2) \sim N_2(0, 0; 1, 1, 0),$$

with

$$Y_1 = \frac{X_1 - \mu_1}{\sigma_1}, \quad Y_2 = \frac{1}{\sigma_2\sqrt{1 - \rho^2}}(X_2 - \mu_2) - \rho\frac{\sigma_2}{\sigma_1}(X_1 - \mu_1),$$

which gives, conversely,

$$X_1 = \mu_1 + Y_1\sigma_1, \quad X_2 = \mu_2 + \rho\sigma_2 Y_1 + \sigma_2\sqrt{1 - \rho^2}Y_2.$$

We, hence, obtain an observation from $(X_1, X_2) \sim N_2(\boldsymbol{\mu}, \boldsymbol{\Sigma})$ by performing the last transformation on $(Y_1, Y_2) \sim N_2(0, 0; 1, 1, 0)$, which can be obtained using the uniform random generator at the start.

7.32.2 *Trivariate case*

Let

$$\mathbf{X} = (X_1, X_2, X_3)^t \sim N_3\left(\mu_1, \mu_2, \mu_3; \{\sigma_i\}, \{\rho_{ij}\}\right).$$

For the trivariate normal, we have an explicit formula for $\mathbf{L}$ from the symmetric

$$\boldsymbol{\Sigma} = \begin{pmatrix} \sigma_1^2 & \rho_{12}\sigma_1\sigma_2 & \rho_{13}\sigma_1\sigma_3 \\ & \sigma_2^2 & \rho_{23}\sigma_2\sigma_3 \\ sym. & & \sigma_3^2 \end{pmatrix}.$$

The lower triangular matrix is

$$\mathbf{L} = \begin{pmatrix} \sigma_1 & 0 & 0 \\ \sigma_2\rho_{12} & \sigma_2\sqrt{1 - \rho_{12}^2} & 0 \\ \sigma_3\rho_{13} & \dfrac{\sigma_3(\rho_{23} - \rho_{12}\rho_{13})}{\sqrt{1 - \rho_{12}^2}} & \sigma_3\sqrt{(1 - \rho_{12}^2)(1 - \rho_{13}^2) - (\rho_{23} - \rho_{12}\rho_{13})^2} \end{pmatrix}. \quad (7.50)$$

We conclude this chapter by presenting the interesting notion of copula.

7.32.3 *General case*

For the general multivariate normal $\mathbf{X} = (X_1, \ldots, X_p) \sim N_p(\boldsymbol{\mu}, \boldsymbol{\Sigma})$, let $\mathbf{L}$ be the lower triangular matrix of the Cholesky decomposition of $\boldsymbol{\Sigma}$, i.e. $\boldsymbol{\Sigma} = \mathbf{L}\mathbf{L}^t$. Then, for p independent univariate standard normal variates $\mathbf{Y}^t = (Y_1, \ldots, Y_p)^t$, which we can obtain by repeated application of the uniform random generator, an observation from $\mathbf{X}$ is obtained by performing the transformation $\mathbf{X} = \mathbf{L}\mathbf{Y} + \boldsymbol{\mu}$.

7.32.4   *Elicitation of a binormal distribution*

This problem occurs when we wish to obtain a binormal prior (see Chapter 6). Using data and sampling values of the parameters,

$$\bar{x} = \frac{1}{n} \sum_{i=1}^{n} x_i, \quad \bar{y} = \frac{1}{n} y_i,$$

$$s_x^2 = \frac{1}{n-1} \sum_{i=1}^{n} (x_i - \bar{x})^2, \quad s_y^2 = \frac{1}{n-1} \sum_{i=1}^{n} (y_i - \bar{y})^2,$$

$$r = \frac{\sum_{i=1}^{n} (x_i - \bar{x})(y_i - \bar{y})}{\left\{ \sum_{i=1}^{n} (x_i - \bar{x})^2 \cdot \sum_{i=1}^{n} (y_i - \bar{y})^2 \right\}^{1/2}},$$

the first approximation to the binormal distribution is hence $N_2(\bar{x}, \bar{y}; s_1, s_2; r)$, and it can be refined until we have satisfaction.

For the standard trinormal case, a similar operation can be done using Eq. (7.42).

We conclude this chapter by presenting the interesting notion of copula.

7.33   Normal Copula

The notion of copula is very useful in multivariate analysis to study the relationship between marginal distributions. Sklar's theorem states the following: Any joint multivariate distribution can be decomposed into univariate marginal distributions which are uniform and a copula which gives the dependence structure between the variables.

Hence, $C : [0,1]^p \to [0,1]$ is a p-dim copula if C is a joint cdf of a p-dim random vector on the unit cube $[0,1]^p$ with uniform marginals.

Hence, for cdf's, we have the copula $C(.)$ such as

$$H(x_1, \ldots, x_p) = C(F_1(x_1), \ldots, F_p(x_p)). \tag{7.51}$$

There are several well-known copulas in the literature.

The binormal copula is established as

$$C_\rho(u, v, \rho) = \Phi_2(\Phi^{-1}(u), \Phi^{-1}(v), \rho)$$

$$= \int_{-\infty}^{\Phi^{-1}(u_1)} \int_{-\infty}^{\Phi^{-1}(u_2)} \frac{1}{2\pi\sqrt{1-\rho^2}} \exp\left(-\frac{x^2 - 2\rho xy + y^2}{2(1-\rho^2)}\right) dxdy,$$

where $\phi(.)$ and $\Phi(.)$ are the density and cdf of the univariate standard normal, respectively, and $\phi_2(.)$ and $\Phi_2(.)$ are the density and cdf of the bivariate standard normal, respectively, i.e.

$$\phi_2(.) = \frac{1}{2\pi\sqrt{1-\rho^2}} \exp\left(\frac{x^2 - 2\rho xy + y^2}{2(1-\rho^2)}\right),$$

$$\Phi_2(h, k; \rho) = \int_{-\infty}^{h} \int_{-\infty}^{k} \phi_2(x, y)dxdy. \tag{7.52}$$

Here, we suppose that $\rho \neq -1$ or 1. But the normal copula can be extended continuously by defining

$$C(u, v, -1) = \lim_{\rho \to -1} C(u, v, \rho) = \max(u = v - 1, 0),$$

or Fréchet lower bound, and

$$C(u, v, +1) = \lim_{\rho \to 1} C(u, v, \rho) = \min(u, v),$$

or Fréchet upper bound.

The numerical evaluation of the bivariate normal copula relies on that of the cdf of the bivariate normal. Figures 7.22 and 7.23 show the general binormal density and the binormal copula (Wikipedia).

7.33.1 *MATLAB example*

$y = $ copulapdf('Gaussian',u,rho) returns the probability density of the Gaussian copula with linear correlation parameters, rho, evaluated at the points in u.

Let $h(x, y) = w_1\phi_1(x, y) + w_2\phi_2(x, y)$ with ϕ_i being std bivariate normal density, with correlation coefficient ρ_i and w_i as the two weights.

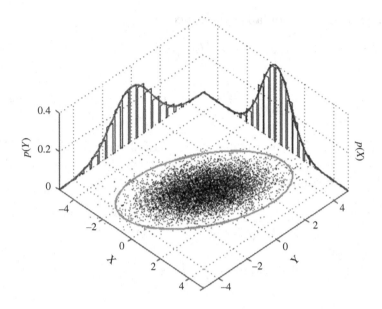

Fig. 7.22   The binormal and its marginal densities.

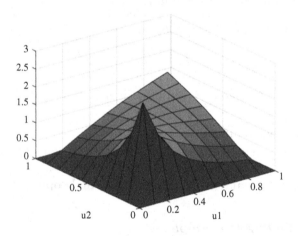

Fig. 7.23   Normal copula.

7.33.2   *Marginal distributions*

Let F and G be two univariate distribution functions. We define

$$H_0(x, y) = \max\{0, F(x) + G(y) - 1\}$$

and

$$H_1(x, y) = \min\{F(x), G(y)\}.$$

Then both H_0, H_1 have F and G as marginals. Let

$$H(x, y) = w_1 H_0(x, y) + w_2 H_1(x, y)$$

be a mixture of H_0 and H_1.

We consider the case where $F = G = \Phi$, where Φ is the cdf of $Z \sim N(0, 1)$. We can prove that H, H_0, H_1 have normal marginals, but the joint densities do not exist. Furthermore, for any bivariate distribution $L(.)$ with marginals F and G, we have

$$H_0(x, y) \leq L(x, y) \leq H_1(x, y),$$

which gives

$$\max\{0, \Phi(x) + \Phi(y) - 1\} \leq L(x, y) \leq \min\{\Phi(x), \Phi(y)\}$$

in the normal case.

7.34   Conclusion

In this chapter, we have presented some basic results related to the most important cases of the multinormal distribution. Graphs and tables, which can be drawn in the binormal case, will help the reader understand some more abstract concepts in the latter chapters. The following Computational Appendix should also be of interest.

Computational Appendix

A.1   Introduction

This appendix addresses the following issues:

(a) some practical volumes to use in computing and programming with the normal;
(b) the critical and prediction ellipses;
(c) the horizontal sectional cuts of the binormal surfaces and their relations to (b).

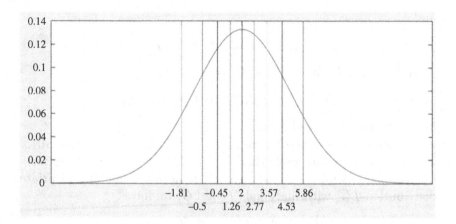

Fig. A.1   Division of the standard normal into 10 parts.

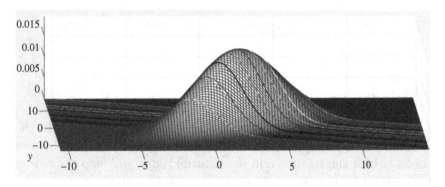

Fig. A.2   PDF of bivariate normal distribution $N_2(\boldsymbol{\mu}, \boldsymbol{\Sigma})$ (with the volume between vertical plans Si and $S(i+1)$ (perpendicular to the principal axis) equal to 10%).

Several questions, raised and solved in each issue, will help to better understand the bivariate and trivariate distributions. This appendix naturally complements this chapter. The above three parts concern various problems encountered, when working and programming with the normal distribution.

First, certain well-known results in univariate normal theory can readily be extended to the binormal distribution. For example, the division of the univariate normal into 10 parts (Fig. A.1) is extended here, into the binormal case, where, e.g. $\mathbf{X} \sim N_2(\boldsymbol{\mu}, \boldsymbol{\Sigma})$, with $\boldsymbol{\mu} = \begin{bmatrix} 2 \\ 2 \end{bmatrix}$, $\boldsymbol{\Sigma} = \begin{bmatrix} 9 & 8 \\ 8 & 16 \end{bmatrix}$ (see Figs. A.2 and A.3).

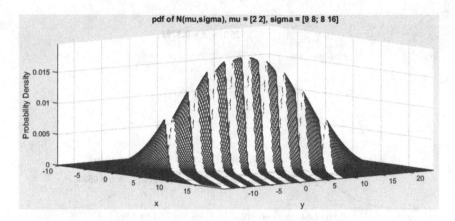

Fig. A.3   Separated (exploded) parts of the 10-division of the binormal.

A.2   Practical Volumes for Computing and Programming with the Multinormal

In dealing with the univariate normal, we have to face the fact that the variable goes from $-\infty$ to ∞.

Since most probability of the univariate normal distribution is between $(\mu - 3\sigma, \mu + 3\sigma)$, this interval can be taken as a basis for computation and programming in most *practical* problems.

We give here similar surfaces and volumes for other multinormal distributions, in a progressive way, so that distributions can be seen as successive generalizations of the univariate standard model. We will see that going from $N_1(0,1)$ to the general $N_p(\boldsymbol{\mu}, \boldsymbol{\Sigma})$ can be accomplished by making successive elementary steps. In the process, we will establish the relation between the height and width of a normal surface (see Chapter 9) whenever possible.

A.2.1   *Univariate normal*

A.2.1.1   *Standard univariate normal*

$$f(z) = \frac{1}{\sqrt{2\pi}} e^{-\frac{z^2}{2}}, \quad -\infty < z < \infty.$$

Practical interval $-3 < z < 3$ contains 0.9974 probability.
The horizontal axis is an asymptote for $z \to \pm\infty$.

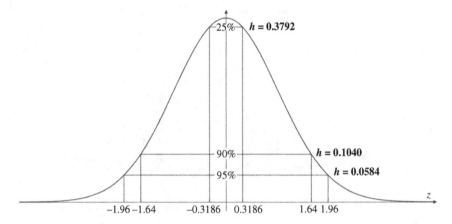

Fig. A.4   The standard normal and its characteristics.

Maximum density height, $H = \frac{1}{\sqrt{2\pi}} = 0.3989$.

95% probability interval $[-1.96, 1.96]$ $(\alpha = 0.05)$.

Relation between height h and width w (interval $[0, w]$ by reason of symmetry):

$$h = f(w) = \frac{1}{\sqrt{2\pi}} e^{\frac{-w^2}{2}}, \Rightarrow w^2 = 2\ln\left(\frac{1}{h\sqrt{2\pi}}\right).$$

We can limit w to $[0, 3.5]$. We have 95% probability for $z \in [-1.96, 1.96]$, at height $h_1 = 0.0584$ (approximately), and 90% probability for $z \in [-1.64, 1.64]$ at height $h_2 = 0.1040$ (Fig. A.4). Hence,

$$\int_0^{1.96} f(z)dz = \int_0^{w} f(z)dz = \frac{0.95}{2} = 0.475,$$

or $\int_0^{w(\alpha)} f(z)dz = (1 - \alpha)/2$, or using the Φ^{-1} function, $w(\alpha) = \Phi^{-1}(0.975)$. In general, we have $w(\alpha) = \Phi^{-1}(1 - (\alpha/2))$, when we work with the horizontal axis.

(a) **Height of an intersection line**

Conversely, there is approximately 90.38%, 76.01%, 54.98% probability for the intervals $\{A_j\}$, $j = 1, 2, 3$, at height $h_j = 0.1, 0.2, 0.3$, and we can determine A_j from these heights (Fig. A.5).

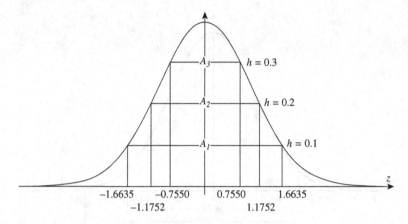

Fig. A.5   Probabilities at $\mathbf{A}_j, j = 1, 3$.

(b) In general, if we cut the standard normal density with a horizontal line, at a height h, between 0 and 0.3989, we have an intersection segment, with projection on the horizontal axis as $[-a, a]$, and $\frac{1}{\sqrt{2\pi}} \int_{-a}^{a} \exp\left(-\frac{x^2}{2}\right) dx$ is the probability content of $[-a, a]$, denoted $(1 - \alpha)$. This probability goes from 1 to 0 when h goes from 0 to 0.3989. The determination of $[-a, a]$ and the related probability is quite easy to do in the univariate case, and here, width $w = [0, a]$ has $(1 - \alpha)/2$ probability.

A.2.1.2   *General univariate normal*

Practical interval $[\mu - 3\sigma < x < \mu + 3\sigma]$ contains 0.9974 probability (Fig. A.6).

95% probability interval: $[\mu - 1.96\sigma, \mu + 1.96\sigma]$.
Maximum density height, $H = \frac{1}{\sigma\sqrt{2\pi}}$.

Note: We can bring this case to the previous one by the change of variable $z = \frac{x-\mu}{\sigma}$. See also Fig. A.7 and Table A.1.

A.2.1.3   *Generalization to the binormal*

These intervals will generalize to ellipses, in the binormal case, when the normal curve becomes the bivariate normal surface. Although the correspondence between height h of the intersecting plane, or

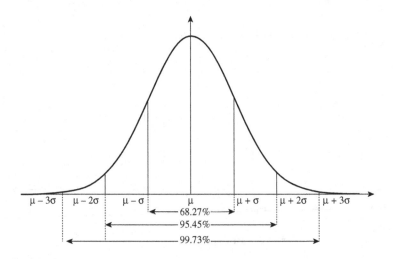

Fig. A.6   The general univariate normal density.

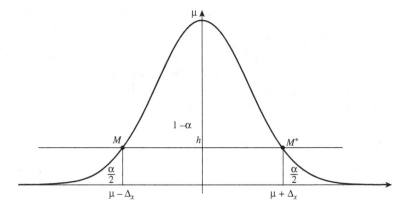

Fig. A.7   Intersection of the general normal density with the horizontal line at height h.

hyperplane, and the widths w of the normal surface remains similar, integration difficulties make this correspondence less obvious and we have to limit ourselves to considering ellipses that have some determined probability content. Further generalizations to dimensions higher than 2 can be similarly made. We then have intersections of hyperplanes and ellipsoids.

Table A.1 Chi-square and chi critical values.

$k^2 = \chi_1^2$	$k = \chi_1$	Probability
0.064	0.253	0.20
0.455	0.675	0.50
1.074	1.036	0.70
2.706	1.645	0.90
3.841	1.960	0.95
6.635	2.576	0.99

A.2.2 *Bivariate normal*

A.2.2.1 *Product binormal standard or circular normal probability density*

$$f(x,y) = \frac{\exp\left(-\frac{x^2+y^2}{2}\right)}{2\pi}, \quad -\infty < x, y < \infty.$$

Practical square $[-3,3]^2$ contains 0.9946 probability.

Maximum density height, $H = \frac{1}{2\pi} = 0.159$.

90.25% probability square $[-1.96, 1.96] \times [-1.96, 1.96]$.

The horizontal plane is an asymptote for $(x,y) \to \pm\infty$. So, the base of the surface drawn in Fig. A.8 is just the intersection of the normal surface with a horizontal plane at a very small height.

Relation between height h and width w ($w^2 = u^2 + v^2$):

$$h = \frac{\exp\left(-\frac{u^2+v^2}{2}\right)}{2\pi} \Rightarrow w^2 = 2\ln\left(\frac{1}{2\pi h}\right).$$

A.2.2.2 *General binormal with independent components*

$$f(u,v;\sigma_1^2,\sigma_2^2) = \frac{\exp\left\{-\frac{1}{2}\left(\frac{u^2}{\sigma_1^2} + \frac{v^2}{\sigma_2^2}\right)\right\}}{2\pi\sigma_1\sigma_2}, \quad -\infty < u, v < \infty, \quad \text{(A.1)}$$

with σ_1^2, σ_2^2 given, $\Delta^2 = \frac{u^2}{\sigma_1^2} + \frac{v^2}{\sigma_2^2}$ is the Mahalanobis square distance of a point to the mean as origin.

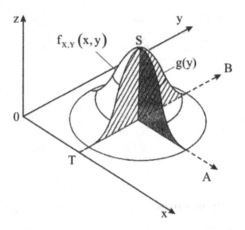

Fig. A.8   Circular normal density.

A.2.2.3   *Prediction ellipse $\frac{u^2}{\sigma_1^2} + \frac{v^2}{\sigma_2^2} \leq k^2$, and probability β for a point to be inside the ellipse, in function of k*

We can compute the probability for a point to belong to the ellipse E_k determined by the Mahalanobis square distance:

$$\Delta^2 = \frac{(x - \mu_X)^2}{(k\sigma_X)^2} + \frac{(y - \mu_Y)^2}{(k\sigma_Y)^2} = 1,$$

or $\Delta^2 = k^2, 0 \leq k \leq 3.5$. This sum follows a central chi-square with 2 dof, χ_2^2, which is represented by the ellipse, $E_k, 0 \leq k \leq 3.5$. Using the chi distribution with 2 dof for $\Delta = k$, the probability that a point M lies inside E_k is

$$\beta = \mathbb{P}(M \in E_k) = \mathbb{P}(\chi_2 < k) = 1 - e^{-k^2/2},$$

which gives a relation

$$k = \sqrt{-2\ln(1 - \beta)}. \tag{A.2}$$

This is the only case where the relation between k and β is algebraic. Alternately, we can relate β to k by using the chi cdf.

$$\beta = \sqrt{\frac{2}{\pi}} \int_0^k t \exp(-t^2/2)dt,$$

with 2 dof. This approach applies to all ellipsoids.

Table A.2 Probability in function of k and h.

β	w width(k)	Height h (approx.) ($h = \frac{1}{2\pi\sigma_1\sigma_2}\exp(-k^2/2)$)
0.25	1.0785	0.0890
0.50	1.4050	0.0593
0.75	1.8331	0.0297
0.90	2.2788	0.0119
0.95	2.5650	0.0059
0.99	3.1312	0.0012

We have Table A.2. We can see that on the vertical axis of maximum height H of a binormal surface, we can have a horizontal cut at height h, which gives a width w of the ellipse and by the cdf of the corresponding chi distribution, we have the probability β, as shown in Table A.2.

Notes:

(1) Using the chi-square distribution, we have $k^2 = -2\ln(1 - \beta)$. Both relations have been used by various authors. We will use the chi distribution in this book.

(2) As $k \uparrow$, for $0 \leq k \leq 3.5$, the ellipse gets larger and there is more probability coverage, and $p \uparrow$ while $h \downarrow$, for $0 \leq h \leq H$ ($h =$ distance from the horizontal plane to the intersection ellipse). We see that, for $E3$, we obtain about 99% of the observations.

(3) The general intersection ellipse has general equation:

$$\frac{(x - \mu_X)^2}{(k\sigma_X)^2} + \frac{(y - \mu_Y)^2}{(k\sigma_Y)^2} - 2\rho\frac{(x - \mu_X)(y - \mu_Y)}{k\sigma_X k\sigma_Y} = 1,$$

or $\Delta_x^2 = k^2$. \hfill (A.3)

This expression is a non-central chi-square with 2 dof and is represented by an ellipse with slanted axes. As seen in the section on conics, the rectangular term could be eliminated by a rotation of axes to make the transformed variables independent.

A.2.2.4   *Case of equal variances and $\rho = 0$*

If $\sigma_1 = \sigma_2 = \sigma$, we have

$$f(u, v; \sigma) = \frac{\exp\left(-\frac{u^2+v^2}{2\sigma^2}\right)}{2\pi\sigma^2},$$

called circular normal, with sd σ, and radial density

$$f(r) = \frac{1}{\sigma^2} r \exp\left(-\frac{r^2}{2\sigma^2}\right), \quad r^2 = (x - \mu_X)^2 + (y - \mu_Y)^2$$

(see A.3).

A.2.2.5   *General binormal with dependent components (with origin at (μ_X, μ_Y)) and max height H*

$$H = \frac{1}{2\pi\sigma_X\sigma_Y\sqrt{1-\rho^2}}.$$

This is a function of ρ, which is minimum at 0, and $\to \infty$ for $|\rho| \to 1$.

$$f(x, y; 0, 0; \sigma_X, \sigma_Y; \rho) = \frac{1}{2\pi\sigma_X\sigma_Y\sqrt{1-\rho^2}}$$

$$\times \exp\left[-\frac{1}{2(1-\rho^2)}\left\{\frac{x^2}{\sigma_X^2} - 2\rho\frac{xy}{\sigma_X\sigma_Y} + \frac{y^2}{\sigma_Y^2}\right\}\right].$$

The general ellipse with probability content α has equation:

$$\frac{(x-\mu_X)^2}{\sigma_X^2} - 2\rho\frac{(x-\mu_X)(y-\mu_Y)}{\sigma_X\sigma_Y} + \frac{(y-\mu_Y)^2}{\sigma_Y^2}$$

$$= -2(1-\rho^2)\log(1-\alpha). \tag{A.4}$$

For $\mu_x = \mu_Y = 0$, it becomes $\frac{x^2}{\sigma_X^2} - 2\rho\frac{xy}{\sigma_X\sigma_Y} + \frac{y^2}{\sigma_Y^2} = -2(1-\rho^2)\log(1-\alpha)$.

Example A.1: We have here a typical case:

$$f(x, y; 0, 0; \sigma_X, \sigma_Y; \rho) = \frac{1}{2\pi\sigma_X\sigma_Y\sqrt{1-\rho^2}} \exp\left[-\frac{Q_2(x,y)}{2}\right],$$

$$Q_2(x,y) = \frac{1}{(1-\rho^2)}\left\{\frac{x^2}{\sigma_X^2} - 2\rho\frac{xy}{\sigma_X\sigma_Y} + \frac{y^2}{\sigma_Y^2}\right\}.$$

(A.5)

Let, for example, $\sigma_X = \sqrt{1.8}$, $\sigma_Y = 1$, $\rho = 0.0$ and $\rho = -0.45$.

We can see that the introduction of ρ into Eq. (A.1) has two effects:

(a) Heightening of the maximum point at the origin by the factor $\frac{1}{\sqrt{1-\rho^2}}$.

(b) Rotation of the surface by an angle

$$\theta^* = \frac{1}{2}\tan^{-1}\left(\frac{2\rho\sigma_1\sigma_2}{\sigma_1^2 - \sigma_2^2}\right).$$

(A.6)

We have the following values for these maxima (Fig. A.9):

$$h_1 = \frac{1}{2\pi\sigma_X\sigma_Y} = 0.1186,$$

$$h_2 = \frac{1}{2\pi\sigma_X\sigma_Y\sqrt{1-(-.45)^2}} = 0.1328.$$

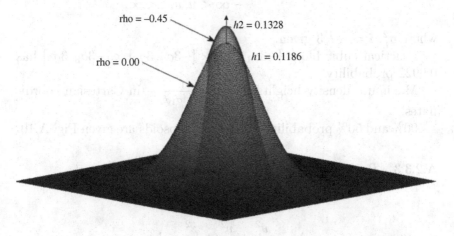

Fig. A.9 Two normal surfaces, with $\rho = 0.00$ and $\rho = -0.45$.

A.2.3   *Trivariate normal*

A.2.3.1   *Product trinormal standard (circular trinormal)*

$$f(u,v,z) = \frac{\exp\left(-\frac{u^2+v^2+z^2}{2}\right)}{(2\pi)^{3/2}}, \quad -\infty < u,v,z < \infty. \qquad (A.7)$$

Practical cubic block $[-3,3]^3$, contains 0.9922 probability.
Maximum density height, $H = \frac{1}{(2\pi)^{3/2}}$.

Confidence cube: $[-1.96, 1.96] \times [-1.96, 1.96] \times [-1.96, 1.96]$ has 0.857 probability.

Relation between height h and width w ($w^2 = u^2 + v^2 + z^2$):

$$w^2 = 2\ln\left(\frac{1}{(2\pi)^{3/2}h}\right) \Rightarrow h = \frac{\exp\left(-\frac{w^2}{2}\right)}{(2\pi)^{3/2}}. \qquad (A.8)$$

A.2.3.2   *General trinormal with independent components*

$$f(u,v,z;\sigma_1^2,\sigma_2^2,\sigma_3^2) = \frac{\exp\left(-\frac{1}{2}\left(\frac{u^2}{\sigma_1^2} + \frac{v^2}{\sigma_2^2} + \frac{z^2}{\sigma_3^2}\right)\right)}{(2\pi)^{3/2}\sigma_1\sigma_2\sigma_2},$$

$$-\infty < u,v,z < \infty,$$

where σ_i^2, $i = 1,2,3$ given.

Practical cubic block $[-3\sigma_1, 3\sigma_1] \times [-3\sigma_2, 3\sigma_2] \times [-3\sigma_3, 3\sigma_3]$ has 0.9922 probability.

Maximum density height, $H = \frac{1}{(2\pi)^{3/2}\sigma_1\sigma_2\sigma_3}$ in Cartesian coordinates.

90% and 50% probability prediction ellipsoids are given Fig. A.10.

A.2.3.3   *Prediction ellipsoid, and probability β for a point to be inside the ellipsoid, in function of k*

$$\Delta_x^2 = \frac{u^2}{\sigma_1^2} + \frac{v^2}{\sigma_2^2} + \frac{z^2}{\sigma_3^2} \leq k^2.$$

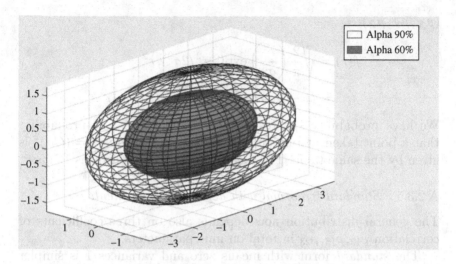

Fig. A.10   Prediction ellipsoid E_k.

Table A.3   Probability in function of k.

β	k
0.25	1.109
0.50	1.538
0.75	2.027
0.90	2.506
0.95	2.795
0.99	3.368

Using chi distribution with 3 dof, we have the relation

$$\beta = \sqrt{\frac{2}{\pi}} \int_0^k t^2 \exp(-t^2/2)dt. \qquad (A.9)$$

Hence, we have Table A.3 relating k to β (Burington and May, 1970, p. 140).

A.2.3.4   *Case of equal variances*

The normal spherical distribution: This is obtained when $\sigma_i = \sigma, \forall i$, and the density has the form (in terms of r), with center at

(μ_X, μ_Y, μ_Z):

$$f(r) = \sqrt{\frac{2}{\pi}} \left(\frac{r^2}{\sigma^3}\right) \exp(-r^2/2\sigma^2), r^2$$

$$= (x - \mu_X)^2 + (y - \mu_Y)^2 + (z - \mu_Z)^2.$$

We have probability spheres instead of ellipsoids. The probability that a point taken at random falls within the sphere $r^2 = (C\sigma)^2$ is given by the same table. The mean spherical radial error is 1.5958σ.

A.2.3.5 *Standard trinormal with dependent components*

The general distribution now depends also on three coefficients of correlation $\rho_{12}, \rho_{13}, \rho_{23}$ in total on nine parameters.

The *standard* form with means zero and variances 1 is simpler and has three parameters only.

$$f(x_1, x_2, x_3) = \frac{\exp(-W/(2\Delta))}{2\sqrt{2}\pi^{3/2}\sqrt{\Delta}}, \tag{A.10}$$

where $\Delta = 1 + 2\rho_{12}\rho_{13}\rho_{23} - (\rho_{12}^2 + \rho_{13}^2 + \rho_{23}^2)$ and

$$W = x_1^2(1 - \rho_{23}^2) + x_2^2(1 - \rho_{13}^2) + x_3^2(1 - \rho_{12}^2) + 2x_1x_2(\rho_{13}\rho_{23} - \rho_{12})$$
$$+ 2x_1x_3(\rho_{12}\rho_{23} - \rho_{13}) + 2x_2x_3(\rho_{12}\rho_{13} - \rho_{23}).$$

This convenient standard form can serve in many, but not all, cases.

Maximum density height, $H = \frac{1}{(2\pi)^{3/2}\sqrt{\Delta}}$ in Cartesian coordinates.

A.2.4 *n-variate normal ($n \geq 4$)*

A.2.4.1 *n-product normal standard (circular k-normal)*

$$f(u_1, \ldots, u_n) = \frac{\exp\left(\frac{-\sum_{i=1}^{n} u_i^2}{2}\right)}{(2\pi)^{n/2}}, \quad -\infty < u_1, \ldots, u_n < \infty.$$

Practical n-dim block $[-3, 3]^n$ has $(0.9974)^n$ probability.
Max density height, $H = \frac{1}{(2\pi)^{n/2}}$.

k-cube: $[-1.96, 1.96] \times [-1.96, 1.96] \cdots \times [-1.96, 1.96]$ has $(0.95)^n$ probability.

Relation between height h and square width $w^2 = \sum_{i=1}^n u_i^2$ is $w^2 = 2 \ln \left(\frac{1}{h(2\pi)^{n/2}} \right)$, and conversely, $h = \frac{1}{(2\pi)^{n/2}} \exp \left(-\frac{1}{2} w^2 \right)$.

A.2.4.2 *General n-normal with n independent components*

$$f(u_1, \ldots, u_n; \sigma_1^2, \ldots, \sigma_n^2) = \frac{\exp \left(-\frac{1}{2} \sum_{i=1}^n \frac{u_i^2}{\sigma_i^2} \right)}{(2\pi)^{n/2} \sigma_1 \ldots \sigma_n},$$

$$-\infty < u_1, \ldots, u_n < \infty.$$

Practical cubic block $[-3\sigma_1, 3\sigma_1] \times [-3\sigma_2, 3\sigma_2] \ldots [-3\sigma_n, 3\sigma_n]$ has probability $(0.9974)^n$.

Maximum density height, $H = \frac{1}{(2\pi)^{n/2} \sigma_1 \ldots \sigma_n}$ in Cartesian coordinates.

A.2.4.3 *Prediction ellipsoid and probability β that a point belongs to the ellipsoid: $\sum_{j=1}^n \left(\frac{u_j}{\sigma_j} \right)^2 \leq k^2$*

A table, relating β to k, can be set, like in the three-dimensional case, using the chi distribution with n dof,

$$\beta = \frac{1}{2^{\frac{n}{2}-1} \Gamma(n/2)} \int_0^k t^{n-1} \exp(-t^2/2) dt. \tag{A.11}$$

A.2.4.4 *Case of equal variances*

The n-normal hyperspherical distribution is obtained when all variances are equal, and the density has form, in terms of r, with center at $(\mu_1, \ldots, \mu_n)$:

$$f(r) = \sqrt{\frac{2}{\pi}} \left(\frac{r^2}{\sigma^n} \right) \exp(-r^2/2\sigma^2), \quad r^2 = \sum_{i=1}^n (x_i - \mu_i)^2,$$

and we have probability hyperspheres instead of ellipsoids. The probability that a point taken at random falls within the sphere $r^2 = (C\sigma)^2$ is given by the same table. The mean spherical radial error is 1.5958σ.

A.2.4.5   *General n-normal with dependent components $(n > 3)$*

$$f(x_1, \ldots, x_n) = C \exp\left(-\frac{1}{2} Q_n(x_1, \ldots, x_n)\right),$$

$Q_n(x_1, \ldots, x_n) = (\mathbf{x} - \boldsymbol{\mu})^t \boldsymbol{\Sigma}^{-1}(\mathbf{x} - \boldsymbol{\mu})$, where $\boldsymbol{\Sigma}, \boldsymbol{\mu}$ given, with $\sigma_{ij} \neq 0$ for at least one couple (i, j).

$$h = f(\mathbf{X}) = \frac{1}{(2\pi)^{n/2}\sqrt{|\boldsymbol{\Sigma}|}} \exp\left\{-\frac{(\mathbf{X} - \boldsymbol{\mu})^t \boldsymbol{\Sigma}^{-1}(\mathbf{X} - \boldsymbol{\mu})}{2}\right\}$$

$$= \frac{1}{(2\pi)^{n/2}\sqrt{|\boldsymbol{\Sigma}|}} \exp\left\{-\frac{\Delta_X^2}{2}\right\},$$

and conversely, $w^2 = \Delta_x^2 = 2\ln\left(\frac{1}{h(2\pi)^{n/2}\sqrt{|\boldsymbol{\Sigma}|}}\right)$.

For the ellipsoid $(\mathbf{x} - \boldsymbol{\mu})^t \boldsymbol{\Sigma}^{-1}(\mathbf{x} - \boldsymbol{\mu}) = w^2$, its principal axes can be computed, associated with eigenvalues $\{\lambda_1, \ldots, \lambda_n\}$ of $\boldsymbol{\Sigma}$, with half-lengths $\sqrt{w^2\lambda_j}$. If $\mathbf{x}$ are normal rv's and w^2 is a percentile value of the chi-square, with n dof, i.e. $w^2 = \chi_{n,\alpha}^2$, then the ellipsoid has probability α.

Maximum density height, $H = \frac{1}{(2\pi)^{n/2}\sqrt{|\boldsymbol{\Sigma}|}}$ in Cartesian coordinates.

A.3   Critical and Prediction Ellipses and Ellipsoids

A.3.1   *Prediction ellipses (also called dispersion or prevision, or error ellipses), and critical ellipses and ellipsoids, in function of space dimensions*

These are geometrical figures, with probability of containing a percentage of randomly generated observations from the normal distribution. They can be computed, and graphically drawn, in the bivariate and trivariate cases, as presented by Theorem 7.1 (Chapter 7). They are, in fact, transversal cuts of the multivariate density of the random vector $\mathbf{X}$.

A.3.2 *Some notations used in this book*

Percentiles of the chi-square distribution with p dof: By $\chi^2_{p,\delta}$ we mean the $100 \times \delta\%$ percentile of $Y \sim \chi^2_p$. Hence, $\mathbb{P}(Y \leq \chi^2_{p,\delta}) = \delta$, and $\mathbb{P}(Y > \chi^2_{p,\delta}) = 1 - \delta$. For example, $\mathbb{P}(Y \leq \chi^2_{2,0.20} = 3.219) = \delta = 0.20$, and $\mathbb{P}(Y > 3.219) = 1 - \delta = 0.80$, as given in Table A.4.

Hence, in this book, α is the probability between the origin and the numerical value of $\chi^2_{2,\alpha}$ (Fig. A.11).

We use the following basic result: For $\mathbf{X} \sim N_p(\boldsymbol{\mu}, \boldsymbol{\Sigma})$, we have $(\mathbf{X} - \boldsymbol{\mu})^t \boldsymbol{\Sigma}^{-1}(\mathbf{X} - \boldsymbol{\mu}) \sim \chi^2_p$, a random variable, and the ellipsoid defined by $(\mathbf{X} - \boldsymbol{\mu})^t \boldsymbol{\Sigma}^{-1}(\mathbf{X} - \boldsymbol{\mu}) \leq \chi^2_{p,\alpha}$ has probability α.

Another formulation: If K is the upper β-percentile of χ^2_p, then

$$\mathbb{P}\left\{(\mathbf{X} - \boldsymbol{\mu})^t \boldsymbol{\Sigma}^{-1}(\mathbf{X} - \boldsymbol{\mu}) \leq K\right\} = 1 - \beta. \qquad (A.12)$$

Hence, the following points are to be noted:

Table A.4   Some special values of the chi-square with 2 and 3 dof.

α	0.01	0.02	0.05	0.10	0.20	0.50	0.75	0.80	0.90	0.95	0.99	0.999
$1-\alpha$	0.99	0.98	0.95	0.90	0.80	0.50	0.25	0.20	0.10	0.05	0.01	0.001
$\chi^2_{2,\alpha}$	0.020	0.040	0.103	0.211	0.446	1.386	2.776	3.219	4.605	5.091	9.20	13.815
$\chi^2_{3,\alpha}$	0.115	0.185	0.352	0.584	1.005	2.366	4.115	4.642	6.251	7.815	11.345	16.268

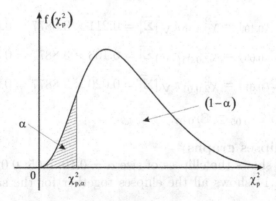

Fig. A.11   Chi-square distribution.

(a) For prediction ellipses, we take large values of α, e.g. $\alpha = 0.90$, 0.95, 0.99 and write $\wp_{0.90}$, $\wp_{0.95}$, $\wp_{0.99}$ as prediction ellipses, and their areas, with probability 0.90, 0.95, 0.99, respectively.

(b) For critical ellipses, denoted $\mathbb{C}r$, we take small values of α. The word *critical* is defined only in this context, associated with a covariance matrix.

Example A.2: Case $p = 2$

Given $\boldsymbol{\mu} = \begin{pmatrix} 1.5 \\ 1.0 \end{pmatrix}$, $\boldsymbol{\Sigma} = \begin{pmatrix} 1.8 & 0.75 \\ 0.75 & 1.0 \end{pmatrix}$, we have $\pi|\boldsymbol{\Sigma}| = 3.14159 \times 1.2375 = 3.8877$.

For prediction ellipses, we have the following:

$$\mathrm{mes}(\wp_{0.90}) = \chi^2_{2,0.90}\pi\sqrt{|\boldsymbol{\Sigma}|} = 4.605 \times 3.8877 = 17.89,$$

$$\mathrm{mes}(\wp_{0.95}) = \chi^2_{2,0.95}\pi\sqrt{|\boldsymbol{\Sigma}|} = 5.091 \times 3.8877 = 19.79,$$

$$\mathrm{mes}(\wp_{0.99}) = \chi^2_{2,0.99}\pi\sqrt{|\boldsymbol{\Sigma}|} = 9.20 \times 3.8877 = 35.76,$$

and $\wp_{0.90} \subset \wp_{0.95} \subset \wp_{0.99}$.

Moreover, the ellipse for $\alpha = 0.50$, denoted by $\wp_{0.50}$, lies inside the above three ellipses. Generating 200 random points from the above binormal density $N_2(\boldsymbol{\mu}, \boldsymbol{\Sigma})$, we can see that roughly 50% of them lie inside $\wp_{0.50}$ (Fig. A.12).

For the same binormal vector, we have similar ellipses.

For critical ellipses, we have the following:

$$\mathrm{mes}(\mathbb{C}r_{0.10}) = \chi^2_{2,0.10}\pi\sqrt{|\boldsymbol{\Sigma}|} = 0.211 \times 3.8877 = 0.820,$$

$$\mathrm{mes}(\mathbb{C}r_{0.05}) = \chi^2_{2,0.05}\pi\sqrt{|\boldsymbol{\Sigma}|} = 0.103 \times 3.8877 = 0.3993,$$

$$\mathrm{mes}(\mathbb{C}r_{0.01}) = \chi^2_{2,0.01}\pi\sqrt{|\boldsymbol{\Sigma}|} = 0.020 \times 3.8877 = 0.0774,$$

and $\mathbb{C}r_{0.10} \supset \mathbb{C}r_{0.05} \supset \mathbb{C}r_{0.01}$.

Critical ellipses graphs

Figure A.13 shows the ellipses of size $\alpha = 0.10;\ 0.05;\ 0.01$.

Figure A.14 shows all the ellipses together for the same normal distribution.

See also Abbreviations at the beginning of the book.

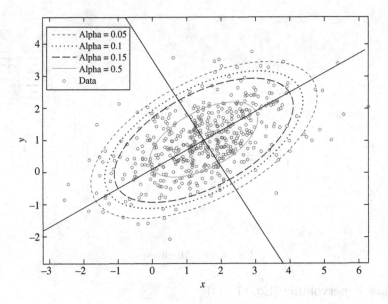

Fig. A.12   50%, 90%, 95%, and 99% prediction ellipses.

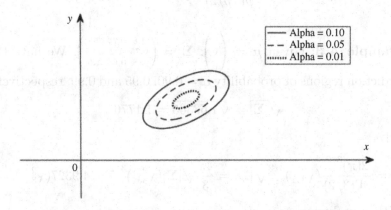

Fig. A.13   Critical ellipses of size $\alpha = 0.10$; 0.05 and 0.01.

A.3.3   *Prediction ellipsoids in $\mathbb{R}^p$ of size α, for $p \geq 3$*

The inequation to use is

$$(\mathbf{x} - \boldsymbol{\mu})^t \boldsymbol{\Sigma}^{-1} (\mathbf{x} - \boldsymbol{\mu}) \leq \chi^2_{p,\alpha}.$$

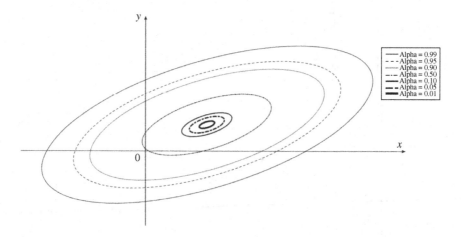

Fig. A.14   All ellipses.

It has hypervolume (Eq. (7.31)):

$$V = \frac{2(\pi)^{p/2}}{p\Gamma(p/2)}(\chi^2_{p,\alpha})^{p/2}\sqrt{|\Sigma|}. \qquad (A.13)$$

Example A.3: Given $\mu = \begin{pmatrix} 0 \\ 0 \\ 0 \end{pmatrix}$, $\Sigma = \begin{pmatrix} 1.8 & 0.75 & 0.8 \\ 0.75 & 1.2 & 0.65 \\ 0.8 & 0.65 & 1.5 \end{pmatrix}$. We have the prediction regions of probability $\alpha = 0.90, 0.95$ and 0.99, respectively.

$$\sqrt{|\Sigma|} = \sqrt{1.63305} = 1.1776$$

and

$$V = \frac{2(\pi)^{3/2}}{3\Gamma(3/2)}(\chi^2_{3,\alpha})^{3/2}\sqrt{|\Sigma|} = \frac{4\pi}{3}\sqrt{|\Sigma|}(\chi^2_{3,\alpha})^{3/2} = 4.9327(\chi^2_{3,\alpha})^{3/2}.$$

Hypervolumes:

$$\mathrm{mes}(\wp_{0.90}) = 4.9327 \times (\chi^2_{3,0.90})^{3/2} = 4.9327 \times (6.251)^{3/2} = 77.092,$$

$$\mathrm{mes}(\wp_{0.95}) = 4.9327 \times (\chi^2_{3,0.95})^{3/2} = 4.9327 \times (7.815)^{3/2} = 107.765,$$

$$\mathrm{mes}(\wp_{0.99}) = 4.9327 \times (\chi^2_{3,0.99})^{3/2} = 4.9327 \times (11.345)^{3/2} = 188.491,$$

and $\wp_{0.90} \subset \wp_{0.95} \subset \wp_{0.99}$.

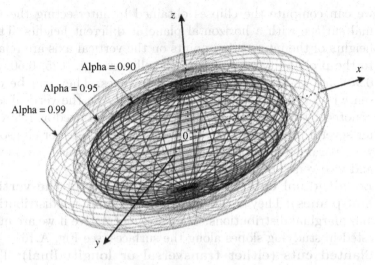

Fig. A.15   Prediction ellipsoids.

The prediction ellipsoids, for 99%, 95%, and 90%, are drawn in Fig. A.15.

For critical ellipsoids, we have their hypervolumes:

$$\mathrm{mes}(\mathbb{C}r_{0.01}) = 4.9327 \times \left(\chi^2_{3,0.01}\right)^{3/2} = 4.9327 \times (0.115)^{3/2} = 0.1923,$$

$$\mathrm{mes}(\mathbb{C}r_{0.05}) = 4.9327 \times \left(\chi^2_{3,0.05}\right)^{3/2} = 4.9327 \times (0.352)^{3/2} = 1.031,$$

$$\mathrm{mes}(\mathbb{C}r_{0.10}) = 4.9327 \times \left(\chi^2_{3,0.10}\right)^{3/2} = 4.9327 \times (0.584)^{3/2} = 2.201,$$

and $\mathbb{C}r_{0.10} \supset \mathbb{C}r_{0.05} \supset \mathbb{C}r_{0.01}$.

A.4   Study of Cuts of the Normal Surface in $\mathbb{R}^2$ and $\mathbb{R}^3$

The binormal distribution provides a variety of numerical applications, with clear equations and figures.

A.4.1   *Different kinds of cuts*

(a) **Transversal cuts (made at different heights, parallel to the horizontal plane):** We note that, for the bivariate normal, we can compute the ellipses obtained by intersecting the normal surface with a horizontal plane at different heights. These heights of the intersection points on the vertical axis are related to the probability coverages of the ellipses (e.g. 0.25, 0.50, and 0.95) and the measures of their main axes. They can be chosen wrt the maximum value of the density on the vertical axis, denoted by H, given here in Section A.2, and used as reference for several problems. These are, generally, ellipses, or ellipsoids, with parameters obtained from the normal surface parameters and vice versa. Details are in Section A.4.2.

(b) **Longitudinal cuts (made parallel to one of the vertical half-planes):** They are associated with conditional distributions and marginal distributions. They are informative if we are interested in studying slopes along the surfaces, see Fig. A.16.

(c) **Slanted cuts (either transversal or longitudinal):** They can be performed and the related equations determined, but expressions are complicated, mostly in parametric forms. They are useful when we wish to travel across a normal surface, following a certain direction.

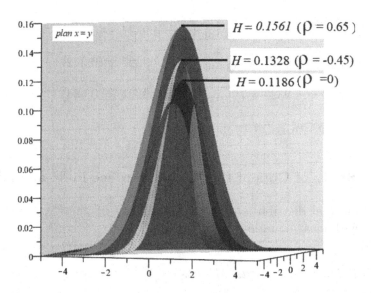

Fig. A.16   Front view (cut by vertical plane $x = y$) of the three normal surfaces, with different values of ρ, considered in Example A.1.

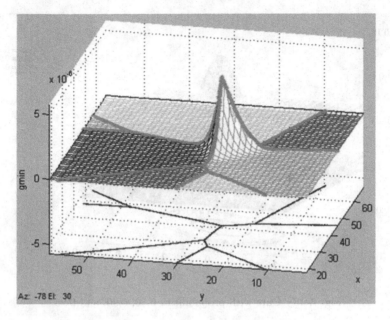

Fig. A.17   Minimum of several densities.

(d) **Cuts for two or several normal surfaces at the same time:** This problem is very complex and is present in mixtures of several multinormals. In this book, we are mostly interested in their intersections in $\mathbb{R}^3$ and their projections on the plane $\mathbb{R}^2$, or their maximum or minimum, (see Fig. A.17 and Chapter 10).

In this appendix, we only treat transversal cuts, but information on other cuts can be obtained from AMSARG.

A.4.2   *Transversal cuts (made at different heights, parallel to the horizontal plane)*

In what follows, we informally set, and solve, various problems related to these ellipses, using either the general, or the standard, bivariate normal density. Simple solutions and computations are available here

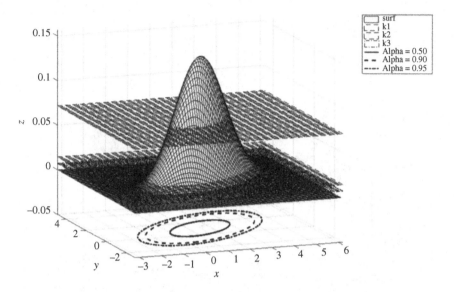

Fig. A.18   Cuts with probabilities 0.50, 0.90, and 0.95.

by using the chi-square distribution, and we do not have to deal with elaborate mathematical equations (Fig. A.18).

We first note the following:

(1) The ellipse has probability close to 1 for a height close to zero.
(2) The larger the height gets, the smaller the ellipse gets, and is nil when $h = H$ (tip of the surface).
(3) The axes of the ellipse are parallel to those of the system, when no rectangular term is present or $\rho = 0$.
(4) A rotation of axes will make components independent. But then, we have to work in the new system of axes.

Note: Ellipses obtained here are the same as those obtained in Section A.3, where analytic geometry is used. Hence, an extension of the univariate operation of the intersection of a line with the univariate normal can be made here to the intersection of a plane and a bivariate normal surface.

A.4.3 Practical problems related to cuts

Problem A.1: Prescribed probability for ellipse given $\Rightarrow$ equation of ellipse = ?
 Solution: General binormal with dependent components (origin at (μ_X, μ_Y)).

$$f(x, y; 0, 0; \sigma_X, \sigma_Y; \rho) = \frac{1}{2\pi\sigma_X\sigma_Y\sqrt{1-\rho^2}}$$

$$\times \exp\left[-\frac{1}{2(1-\rho^2)}\left\{\frac{x^2}{\sigma_X^2} - \frac{2\rho xy}{\sigma_X\sigma_Y} + \frac{y^2}{\sigma_Y^2}\right\}\right].$$
(A.14)

Maximum density height, $H = \frac{1}{2\pi\sigma_X\sigma_Y\sqrt{1-\rho^2}}$ (see Section A.2) is a function of ρ, which is minimum at 0, and $\to \infty$ for $|\rho| \to 1$. This fact can be used to study the general normal surface wrt the independent components surface.
 Hence, using $\mathbb{P}(\chi_2^2 \leq A) = 1 - \exp(-\frac{A}{2})$, the ellipse with probability α has equation:

$$\frac{(x-\mu_X)^2}{\sigma_X^2} - \frac{2\rho(x-\mu_X)(y-\mu_Y)}{\sigma_X\sigma_Y} + \frac{(y-\mu_Y)^2}{\sigma_Y^2} = -2(1-\rho^2)\log(1-\alpha).$$
(A.15)

For $\mu_X = \mu_Y = 0$, it becomes

$$\frac{x^2}{\sigma_X^2} - \frac{2\rho xy}{\sigma_X\sigma_Y} + \frac{y^2}{\sigma_Y^2} = -2(1-\rho^2)\log(1-\alpha).$$
(A.16)

Problem A.2: Setting $\alpha = 0.25, 0.50, 0.90$, we have three ellipses, with known probability content, within each other. The three ellipses are drawn here (Fig. A.19), for $\mathbf{X} \sim N_2(\boldsymbol{\mu}, \boldsymbol{\Sigma})$, with $\boldsymbol{\mu} = \begin{pmatrix} 0 \\ 0 \end{pmatrix}$, $\boldsymbol{\Sigma} = \begin{pmatrix} 1.8 & -0.75 \\ -0.75 & 1.0 \end{pmatrix}$.

Problem A.3: Height of cut given $\Rightarrow$ Equation of ellipse = ?

Solution: We can determine, alternately, the ellipse at height H/k, with H being the maximum height given in tables of the previous section and $k_1 = 20$, $k_2 = 2$, $k_3 = 4$, $k_4 = 8$, $k_5 = 10$, for example.

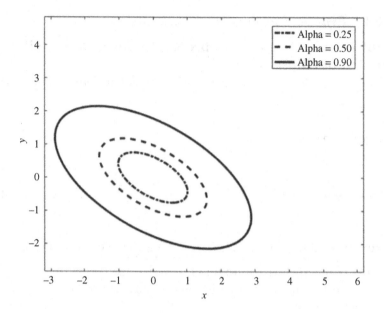

Fig. A.19   Ellipse with given probability content.

Using Eq. (A.14),

$$f(x, y; 0, 0; \sigma_X, \sigma_Y; \rho) = \frac{1}{2\pi\sigma_X\sigma_Y\sqrt{1-\rho^2}}$$

$$\times \exp\left[-\frac{1}{2(1-\rho^2)}\left\{\frac{x^2}{\sigma_X^2} - \frac{2\rho xy}{\sigma_X\sigma_Y} + \frac{y^2}{\sigma_Y^2}\right\}\right]$$

$$= H/k.$$

We draw two ellipses with k_1 and k_2. The equations of two ellipses can be determined and computable ellipse areas can be given:

$$\frac{1}{2\pi\sigma_X\sigma_Y\sqrt{1-\rho^2}}\exp\left[-\frac{1}{2(1-\rho^2)}\left\{\frac{x^2}{\sigma_X^2} - \frac{2\rho xy}{\sigma_X\sigma_Y} + \frac{y^2}{\sigma_Y^2}\right\}\right] = H/k,$$

or

$$\exp\left[-\frac{1}{2(1-\rho^2)}\left\{\frac{x^2}{\sigma_X^2} - \frac{2\rho xy}{\sigma_X\sigma_Y} + \frac{y^2}{\sigma_Y^2}\right\}\right] = \frac{1}{k},$$

or

$$\left[-\frac{1}{2(1-\rho^2)}\left\{\frac{x^2}{\sigma_X^2} - \frac{2\rho xy}{\sigma_X \sigma_Y} + \frac{y^2}{\sigma_Y^2}\right\}\right] = \ln\left(\frac{1}{k}\right),$$

or

$$\frac{1}{2(1-\rho^2)}\left\{\frac{x^2}{\sigma_X^2} - \frac{2\rho xy}{\sigma_X \sigma_Y} + \frac{y^2}{\sigma_Y^2}\right\} = \ln(k),$$

or

$$\frac{x^2}{\sigma_X^2} - 2\rho\frac{xy}{\sigma_X \sigma_Y} + \frac{y^2}{\sigma_Y^2} = 2(1-\rho^2)\ln(k). \qquad (A.17)$$

We know σ_X, σ_Y, ρ and can determine the above ellipse in $\mathbb{R}^2$ and $\mathbb{R}^3$.

Problem A.4: Volumes above the intersection ellipses for the standard binormal density = ?

Solution: Using the standard density, the relation

$$I = \mathbb{P}(\chi_2^2 \le k(1-\rho^2)) = 1 - \exp\left(-k\frac{1-\rho^2}{2}\right),$$

based on the value of the integral of

$$\frac{1}{2\pi\sqrt{1-\rho^2}}\exp\left[-\frac{1}{2(1-\rho^2)}(x^2 - 2\rho xy + y^2)\right]$$

over the interior of the ellipse $x^2 - 2\rho xy + y^2 = k$, we have $I_1 > I_2$ and the volume between two ellipses determined by $k_2 = 2$ and $k_1 = 20$ is

$$I_1 - I_2 = 1 - \exp\left(-k_1\frac{1-\rho^2}{2}\right) - \left[1 - \exp\left(-k_2\frac{1-\rho^2}{2}\right)\right]$$

$$= 0.9990 - 0.4972 = 0.5018 \text{ (Fig. A.20)}.$$

Remarks: We can see the following:

(a) I_1 is almost unity and hence the ellipse determined by the section height $H/20$ could be used as a computing base for this normal surface.

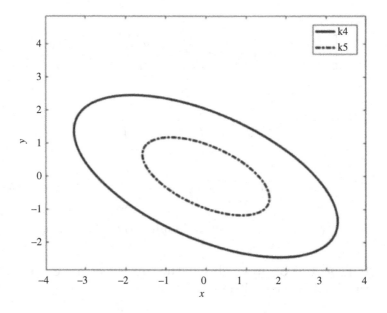

Fig. A.20   Ellipses obtained by cutting normal surface at $H/2$ and $H/20$.

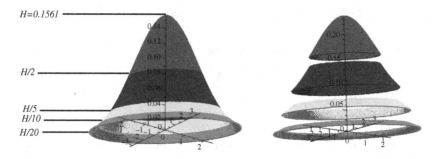

Fig. A.21   Intersection ellipses at different heights and volumes between them.

(b) As shown in Fig. A.20, this ellipse is well inside the practical rectangle $[-4, 4] \times [-3, 3]$ and could serve as base computing area as well.

(c) The volume determined by the normal surface between heights $H/20$ and $H/2$ is 0.5018.

(d) Intermediary cuts are at $H/10$ and $H/5$, and volumes between successive cuts can be computed (Fig. A.21).

Problem A.5: Determine the heights of 10 ellipses s.t. the volume between two consecutive ellipses is 0.10. The first ellipse at the bottom will be at a very low altitude while the last ellipse has the volume between it and the top of the normal surface equal to 0.10. This exercise should produce a graph similar to Fig. A.2, but with transversal cuts.

Problem A.6: For cuts made at different heights, areas of ellipses = ?

Solution: They can be computed (see Section 7.12) using area $(E_i) = \pi a_i b_i$, $i = 1, \ldots, 5$, where $\{a_i, b_i\}$ are the principal half-axes of the ellipses. Their ratios

$$\xi_{ik} = \frac{\text{area}(E_i)}{\text{area}(E_k)} = \frac{a_i b_i}{a_k b_k}$$

will provide information on the shape of the normal surface. Moreover, in the general p-dimensional case, the length of the jth longest axis is proportional to $\sqrt{\lambda_j}$ the square root of the jth eigenvalue.

Problem A.7: Let α probability be desired within the intersection ellipse. At what height h should we cut the general normal surface?

Solution: Given the density function of bivariate normal distribution,

$$f(x, y) = \frac{1}{2\pi \sigma_X \sigma_Y \sqrt{1 - \rho^2}} \times \exp\left\{ -\frac{1}{2(1 - \rho^2)} \left[\frac{(x - \mu_X)^2}{\sigma_X^2} \right.\right.$$

$$\left.\left. -\frac{2\rho(x - \mu_X)(y - \mu_Y)}{\sigma_X} \frac{}{\sigma_Y} + \frac{(x - \mu_Y)^2}{\sigma_Y^2} \right] \right\}.$$

Let E_a be the interior of projection of the intersection curve between $f(x, y)$ and the horizontal plane $z = a$ (Fig. A.22):

$$E_a = \left\{ (x, y) : \frac{(x - \mu_X)^2}{\sigma_X^2} - \frac{2\rho(x - \mu_X)(y - \mu_Y)}{\sigma_X} \frac{}{\sigma_Y} \right.$$

$$\left. + \frac{(x - \mu_Y)^2}{\sigma_Y^2} \leq C_a \right\}, \qquad (A.18)$$

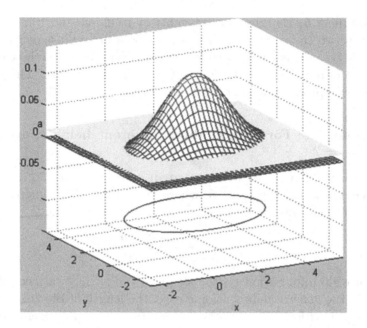

Fig. A.22   Ellipse with predetermined probability content.

where

$$C_\alpha = -2(1 - \rho^2) \ln \left(a2\pi\sigma_X\sigma_Y \sqrt{1 - \rho^2} \right). \qquad (A.19)$$

Although the analytic solution to this problem is available, we will use Eqs. (A.16) and (A.17) to provide practical solutions for the standard bivariate normal.

By Eq. (A.16), ellipse

$$\frac{x^2}{\sigma_X^2} - \frac{2\rho xy}{\sigma_X\sigma_Y} + \frac{y^2}{\sigma_Y^2} = -2(1 - \rho^2) \log(1 - \alpha)$$

has probability α.

By Eq. (A.17), ellipse

$$\frac{x^2}{\sigma_X^2} - \frac{2\rho xy}{\sigma_X\sigma_Y} + \frac{y^2}{\sigma_Y^2} = 2(1 - \rho^2) \ln(k)$$

is the ellipse at height H/k, where H is the maximum height and k need not be an integer. Equalling the two rhs, we have $k = \frac{1}{1-\alpha}$. So, we need to determine k in order to make the cut at $h = H/k$.

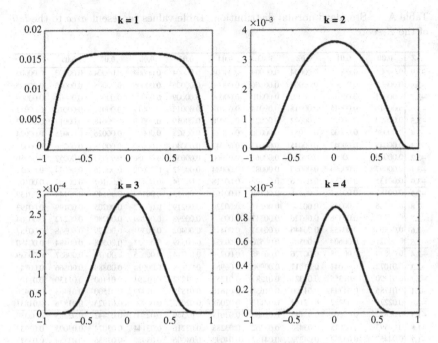

Fig. A.23 Binormal density as function of ρ.

For $\alpha = 0.10$, for example, we have $k = 1.11$, and the cut should be made at $h = H/1.11$.

Problem A.8: How does the binormal standard density behave within a iso-level ellipse $x^2 - 2\rho\,xy + y^2 = k^2$?

Solution: $f(x,y) = \dfrac{1}{2\pi\sigma_X\sigma_Y\sqrt{1-\rho^2}} \exp\left(\dfrac{-k^2}{2(1-\rho^2)}\right) = h(\rho)$ is a function of ρ alone (see Fig. A.23). $h(\rho)$ is a curve, symmetrical wrt the vertical axis. Hence, by limiting the variations of (x,y) to the interior and on an iso-level curve, $f(x,y)$ can be studied as a function of ρ alone.

A.5   Conclusion

In this appendix, we have presented a few real issues on computation with the bivariate normal distribution under either its standard form or its general form. Some original answers are suggested and some of them could be worked out into more developed ideas. We hope that readers will find them of interest (see Table A.5).

Table A.5   Standard normal distribution: Table values represent *area* to the *left* of the Z score.

Z	0.00	0.01	0.02	0.03	0.04	0.05	0.06	0.07	0.08	0.09
-3.9	0.00005	0.00005	0.00004	0.00004	0.00004	0.00004	0.00004	0.00004	0.00003	0.00003
-3.8	0.00007	0.00007	0.00007	0.00006	0.00006	0.00006	0.00006	0.00005	0.00005	0.00005
-3.7	0.00011	0.00010	0.00010	0.00010	0.00009	0.00009	0.00008	0.00008	0.00008	0.00008
-3.6	0.00016	0.00015	0.00015	0.00014	0.00014	0.00013	0.00013	0.00012	0.00012	0.00011
-3.5	0.00023	0.00022	0.00022	0.00021	0.00020	0.00019	0.00019	0.00018	0.00017	0.00017
-3.4	0.00034	0.00032	0.00031	0.00030	0.00029	0.00028	0.00027	0.00026	0.00025	0.00024
-3.3	0.00048	0.00047	0.00045	0.00043	0.00042	0.00040	0.00039	0.00038	0.00036	0.00035
-3.2	0.00069	0.00066	0.00064	0.00062	0.00060	0.00058	0.00056	0.00054	0.00052	0.00050
-3.1	0.00097	0.00094	0.00090	0.00087	0.00084	0.00082	0.00079	0.00076	0.00074	0.00071
-3.0	0.00135	0.00131	0.00126	0.00122	0.00118	0.00114	0.00111	0.00107	0.00104	0.00100
-2.9	0.00187	0.00181	0.00175	0.00169	0.00164	0.00159	0.00154	0.00149	0.00144	0.00139
-2.8	0.00256	0.00248	0.00240	0.00233	0.00226	0.00219	0.00212	0.00205	0.00199	0.00193
-2.7	0.00347	0.00336	0.00326	0.00317	0.00307	0.00298	0.00289	0.00280	0.00272	0.00264
-2.6	0.00466	0.00453	0.00440	0.00427	0.00415	0.00402	0.00391	0.00379	0.00368	0.00357
-2.5	0.00621	0.00604	0.00587	0.00570	0.00554	0.00539	0.00523	0.00508	0.00494	0.00480
-2.4	0.00820	0.00798	0.00776	0.00755	0.00734	0.00714	0.00695	0.00676	0.00657	0.00639
-2.3	0.01072	0.01044	0.01017	0.00990	0.00964	0.00939	0.00914	0.00889	0.00866	0.00842
-2.2	0.01390	0.01355	0.01321	0.01287	0.01255	0.01222	0.01191	0.01160	0.01130	0.01101
-2.1	0.01786	0.01743	0.01700	0.01659	0.01618	0.01578	0.01539	0.01500	0.01463	0.01426
-2.0	0.02275	0.02222	0.02169	0.02118	0.02068	0.02018	0.01970	0.01923	0.01876	0.01831
-1.9	0.02872	0.02807	0.02743	0.02680	0.02619	0.02559	0.02500	0.02442	0.02385	0.02330
-1.8	0.03593	0.03515	0.03438	0.03362	0.03288	0.03216	0.03144	0.03074	0.03005	0.02938
-1.7	0.04457	0.04363	0.04272	0.04182	0.04093	0.04006	0.03920	0.03836	0.03754	0.03673
-1.6	0.05480	0.05370	0.05262	0.05155	0.05050	0.04947	0.04846	0.04746	0.04648	0.04551
-1.5	0.06681	0.06552	0.06426	0.06301	0.06178	0.06057	0.05938	0.05821	0.05705	0.05592
-1.4	0.08076	0.07927	0.07780	0.07636	0.07493	0.07353	0.07215	0.07078	0.06944	0.06811
-1.3	0.09680	0.09510	0.09342	0.09176	0.09012	0.08851	0.08691	0.08534	0.08379	0.08226
-1.2	0.11507	0.11314	0.11123	0.10935	0.10749	0.10565	0.10383	0.10204	0.10027	0.09853
-1.1	0.13567	0.13350	0.13136	0.12924	0.12714	0.12507	0.12302	0.12100	0.11900	0.11702
-1.0	0.15866	0.15625	0.15386	0.15151	0.14917	0.14686	0.14457	0.14231	0.14007	0.13786
-0.9	0.18406	0.18141	0.17879	0.17619	0.17361	0.17106	0.16853	0.16602	0.16354	0.16109
-0.8	0.21186	0.20897	0.20611	0.20327	0.20045	0.19766	0.19489	0.19215	0.18943	0.18673
-0.7	0.24196	0.23885	0.23576	0.23270	0.22965	0.22663	0.22363	0.22065	0.21770	0.21476
-0.6	0.27425	0.27093	0.26763	0.26435	0.26109	0.25785	0.25463	0.25143	0.24825	0.24510
-0.5	0.30854	0.30503	0.30153	0.29806	0.29460	0.29116	0.28774	0.28434	0.28096	0.27760
-0.4	0.34458	0.34090	0.33724	0.33360	0.32997	0.32636	0.32276	0.31918	0.31561	0.31207
-0.3	0.38209	0.37828	0.37448	0.37070	0.36693	0.36317	0.35942	0.35569	0.35197	0.34827
-0.2	0.42074	0.41683	0.41294	0.40905	0.40517	0.40129	0.39743	0.39358	0.38974	0.38591
-0.1	0.46017	0.45620	0.45224	0.44828	0.44433	0.44038	0.43644	0.43251	0.42858	0.42465
-0.0	0.50000	0.49601	0.49202	0.48803	0.48405	0.48006	0.47608	0.47210	0.46812	0.46414

Table A.5 (*Continued*)

Z	0.00	0.01	0.02	0.03	0.04	0.05	0.06	0.07	0.08	0.09
0.0	0.50000	0.50399	0.50798	0.51197	0.51595	0.51994	0.52392	0.52790	0.53188	0.53586
0.1	0.53983	0.54380	0.54776	0.55172	0.55567	0.55962	0.56356	0.56749	0.57142	0.57535
0.2	0.57926	0.58317	0.58706	0.59095	0.59483	0.59871	0.60257	0.60642	0.61026	0.61409
0.3	0.61791	0.62172	0.62552	0.62930	0.63307	0.63683	0.64058	0.64431	0.64803	0.65173
0.4	0.65542	0.65910	0.66276	0.66640	0.67003	0.67364	0.67724	0.68082	0.68439	0.68793
0.5	0.69146	0.69497	0.69847	0.70194	0.70540	0.70884	0.71226	0.71566	0.71904	0.72240
0.6	0.72575	0.72907	0.73237	0.73565	0.73891	0.74215	0.74537	0.74857	0.75175	0.75490
0.7	0.75804	0.76115	0.76424	0.76730	0.77035	0.77337	0.77637	0.77935	0.78230	0.78524
0.8	0.78814	0.79103	0.79389	0.79673	0.79955	0.80234	0.80511	0.80785	0.81057	0.81327
0.9	0.81594	0.81859	0.82121	0.82381	0.82639	0.82894	0.83147	0.83398	0.83646	0.83891
1.0	0.84134	0.84375	0.84614	0.84849	0.85083	0.85314	0.85543	0.85769	0.85993	0.86214
1.1	0.86433	0.86650	0.86864	0.87076	0.87286	0.87493	0.87698	0.87900	0.88100	0.88298
1.2	0.88493	0.88686	0.88877	0.89065	0.89251	0.89435	0.89617	0.89796	0.89973	0.90147
1.3	0.90320	0.90490	0.90658	0.90824	0.90988	0.91149	0.91309	0.91466	0.91621	0.91774
1.4	0.91924	0.92073	0.92220	0.92364	0.92507	0.92647	0.92785	0.92922	0.93056	0.93189
1.5	0.93319	0.93448	0.93574	0.93699	0.93822	0.93943	0.94062	0.94179	0.94295	0.94408
1.6	0.94520	0.94630	0.94738	0.94845	0.94950	0.95053	0.95154	0.95254	0.95352	0.95449
1.7	0.95543	0.95637	0.95728	0.95818	0.95907	0.95994	0.96080	0.96164	0.96246	0.96327
1.8	0.96407	0.96485	0.96562	0.96638	0.96712	0.96784	0.96856	0.96926	0.96995	0.97062
1.9	0.97128	0.97193	0.97257	0.97320	0.97381	0.97441	0.97500	0.97558	0.97615	0.97670
2.0	0.97725	0.97778	0.97831	0.97882	0.97932	0.97982	0.98030	0.98077	0.98124	0.98169
2.1	0.98214	0.98257	0.98300	0.98341	0.98382	0.98422	0.98461	0.98500	0.98537	0.98574
2.2	0.98610	0.98645	0.98679	0.98713	0.98745	0.98778	0.98809	0.98840	0.98870	0.98899
2.3	0.98928	0.98956	0.98983	0.99010	0.99036	0.99061	0.99086	0.99111	0.99134	0.99158
2.4	0.99180	0.99202	0.99224	0.99245	0.99266	0.99286	0.99305	0.99324	0.99343	0.99361
2.5	0.99379	0.99396	0.99413	0.99430	0.99446	0.99461	0.99477	0.99492	0.99506	0.99520
2.6	0.99534	0.99547	0.99560	0.99573	0.99585	0.99598	0.99609	0.99621	0.99632	0.99643
2.7	0.99653	0.99664	0.99674	0.99683	0.99693	0.99702	0.99711	0.99720	0.99728	0.99736
2.8	0.99744	0.99752	0.99760	0.99767	0.99774	0.99781	0.99788	0.99795	0.99801	0.99807
2.9	0.99813	0.99819	0.99825	0.99831	0.99836	0.99841	0.99846	0.99851	0.99856	0.99861
3.0	0.99865	0.99869	0.99874	0.99878	0.99882	0.99886	0.99889	0.99893	0.99896	0.99900
3.1	0.99903	0.99906	0.99910	0.99913	0.99916	0.99918	0.99921	0.99924	0.99926	0.99929
3.2	0.99931	0.99934	0.99936	0.99938	0.99940	0.99942	0.99944	0.99946	0.99948	0.99950
3.3	0.99952	0.99953	0.99955	0.99957	0.99958	0.99960	0.99961	0.99962	0.99964	0.99965
3.4	0.99966	0.99968	0.99969	0.99970	0.99971	0.99972	0.99973	0.99974	0.99975	0.99976
3.5	0.99977	0.99978	0.99978	0.99979	0.99980	0.99981	0.99981	0.99982	0.99983	0.99983
3.6	0.99984	0.99985	0.99985	0.99986	0.99986	0.99987	0.99987	0.99988	0.99988	0.99989
3.7	0.99989	0.99990	0.99990	0.99990	0.99991	0.99991	0.99992	0.99992	0.99992	0.99992
3.8	0.99993	0.99993	0.99993	0.99994	0.99994	0.99994	0.99994	0.99995	0.99995	0.99995
3.9	0.99995	0.99995	0.99996	0.99996	0.99996	0.99996	0.99996	0.99996	0.99997	0.99997

Bibliography

Burington, R.S. and May, D.C. (1970). *Handbook of Probability and Statistics, with Tables* (McGraw-Hill, New York).

Divgi, D.R. (1979). Calculation of univariate and bivariate normal probability functions, *Annals of Statistics*, **7**, 903–910.

Drezner, Z. (1978). Computation of the bivariate normal integral, *Mathematics of Computation*, **32**(141), 277–279.

Drezner, Z. (1994). Computation of the trivariate normal integral, *Mathematics of Computation*, **62**(203), 289–294.

Drezner, Z. and Wesolowsky, G.O. (1990). On the computation of the bivariate normal integral, *Journal of Statistical Computation and Simulation*, **3**, 101–107.

Genz, A. and Bretz, F. (2009). *Computation of Multivariate Normal and t Probabilities*, *Lecture Notes in Statistics*, 195 (Springer-Verlag, Berlin Heidenberg).

Guenther, W.C. and Terragno, P.J. (1963). A review of the literature on a class of coverage problems, *Annals of Mathematical Statistics*, **35**(1), 232–260.

Gupta, S.S. (1963). Probability integrals of multivariate normal and multivariate t, *Annals of Mathematical Statistics*, **34**, 792–828.

Hutchinson, T. and Lai, C.D. (1990). *Continuous Bivariate Distributions, Emphasizing Applications* (Rumsby Scientific Publishing, Adelaide, Australia).

The Maplesoft Company, *Maple 2014* (Waterloo, Canada).

The Wolfram Company, *Mathematica 2017* (New York).

National Bureau of Standards (1959). *Tables of the Bivariate Normal Distribution Function and Related Functions*, *Applied Mathematics Series*, 50 (US Government Printing Office, Washington, DC).

Moran, P.A.P. (1968). *Introduction to Probability Theory* (Clarendon Press, Oxford, London).

Owen, D.B. (1956). Tables for computing bivariate normal probabilities, *Annals of Mathematical Statistics*, **27**, 1075–1090.

Pearson, K. (1931). *Tables for Statisticians and Biometricians* (Cambridge University Press, London).

Plackett, R.L. (1954). A reduction formula for normal multivariate integrals, *Biometrika*, **41**, 341–360.

Rencher, A.C. and Christensen, W.F. (2012). *Methods of Multivariate Analysis*, 3rd ed. (John Wiley and Sons, New Jersey).

Sheppard, W.F. (1898). On the Application of the theory of error to cases of normal distributions and normal correlations, *Philosophical Transactions of the Royal Society*, **A192**, 101–167.

Steck, G.P. (1958). A table for computing trivariate normal probabilities, *Annals of Mathematical Statistics*, **29**, 780–800.

Steen, N.M., Byrne, G.O. and Gelbard, E.M. (1969). Gaussian quadratures, *Mathematics of Computation*, **23**, 661–671.

Stuart, A. and Ord, J.K. (1987). *Kendall's Advanced Theory of Statistics*, Vol. I, 5th ed. (Oxford University Press, New York).

Thomas, J.P. (1971). *An Introduction to Probability and Random Processes* (John Wiley and Sons, New York).

Thomasian, A. (1969). *The Structure of Probability Theory with Applications* (McGraw-Hill, New York).

Tong, Y.L. (1989). *The Multivariate Normal Distribution* (Springer-Verlag, New York).

Wold, H. (1948). *Random Normal Deviates, Tracts for Computers*, 35 (Cambridge University Press, London).

Chapter 8

Probability, Historical Survey, and Discussion

8.1 Introduction

In this chapter, we discuss the concepts of probability and its development through the ages. The normal distribution, with its numerous applications, is highly dependent on how we interpret the probability of an event.

In the statistical literature, we frequently encounter different terms, such as game theory, gambling theory, simulation, and operations research, all of which seem related to probability. It is important to distinguish between them, although the boundaries are sometimes blurry. We start with a brief history of probability through the ages. We also discuss two special topics: gambling theory and theories of probability, using two well-known reference books. We conclude the chapter with a famous result, the de Finetti theorem on exchangeable sequences of rv's and its application to normal vectors. It is shown that concrete applications can be found with quite abstract probability concepts.

8.2 How Did Probability Develop with Time?

8.2.1 *Antiquity*

The Greek philosophers made their marks on ancient western civilizations, in philosophy, navigation, astronomy, physics, mathematics, chemistry, but few of them cared about probability, chance, and random phenomena. Aristotle (384–322 B.C.) would identify chance as indefinite, inscrutable to the human intellect. Whether events happened or not, those which did not happen do not exist yet and should not be discussed. The two main schools of thought did not differ much on the subject of chance. The Stoics (Democritus (460–379 B.C.), Plato (427–347 B.C.), and Aristotle (384–322 B.C.)) claimed that all events happen according to fate and that only two concepts exist: impossible and necessary. Chance and possible are only due to our incomplete knowledge.

The Epicureans (Epicurus (341–270 B.C.) and Lucretius (99–55 B.C.)) thought that the perfect functioning of celestial bodies leaves no room for chance. Hence, probability was never considered. Furthermore, since classical Greeks thought of experimentation as a lowly undertaking, statistics, under any form, was never considered as of interest.

8.2.2   *The middle ages*

For Aquinas (1225–1274), chance is the coincidence of two causes. Spinoza (1632–1677) and, to some extent, Aquinas, attributed chance to the deficiency of our knowledge. It can be said that, in general, this negative attitude toward all empirical methods, and complete devotion to speculative knowledge, infused a retarding influence on the development of empirical sciences. Only the great discoveries made in applied domains, such as astronomy and medicine, forced these thinkers to accept experimentation and experiential data as alternate ways to acquire new knowledge. They had to accept that experiments conducted under identical conditions could lead to different results, i.e. chance does interfere with outcomes. The whole issue of randomness had to be recognized.

A scientist that stood tall in this debate was Francis Bacon (1561–1626), who could be considered as the father of new empiricism. In his writings, he exhorted his contemporaries not to rely blindly on speculative intellect and to accept applied domains, so as to develop new and sound methods to process experimental data, which was, in short, the beginning of a new science, much later called *statistics*.

8.2.3 The beginning

It is often stated that the probability theory started with gambling, a universal pastime throughout the world. It had a mathematical beginning when, in medieval France, the Chevalier de Méré, his real name being Antoine Gombaud (1607–1684), wrote to Blaise Pascal (1623–1662), asking for some advice. De Méré was not only an astute gambler but also a well-rounded *honnête homme*, with several serious writings to his credit. Here, we have at least two problems that he submitted to Pascal on 29 July 1654:

(1) If one wishes to get a 6 in 4 throws of a die, the advantage is 671 to 625. With 2 dice and 24 throws, the advantage of obtaining 2 sixes should be the same, since 24 is to 36 as 4 is to 6. He wondered why it was not so.

Without the present day's Boolean algebra knowledge to work with, computations done abstractly were indeed difficult. Nowadays, using complementary events, we can easily determine the two probabilities:

- One die: $\mathbb{P}(\text{at least one 6 in 4 throws}) = 1 - \left(\frac{5}{6}\right)^4 = \frac{671}{1296} = 0.5177$.
- Two dice: $\mathbb{P}(\text{at least 2 sixes in 24 throws}) = 1 - \left(\frac{35}{36}\right)^{24} = \frac{636}{1296} = 0.4914$.

So, it is slightly less than the first probability.

Pascal, who solved the problem, proved de Méré correct. In fact, de Méré's observation remains true even if two dice are thrown 25 times, since the probability of getting at least 1 double-6 is then $1 - \left(\frac{35}{36}\right)^{25} = 0.5055$.

(2) Another question posed by de Méré was the number of rolls required to have a probability greater than $1/2$ of having a 6 or a double-6 for two rolls. Solving $n > \frac{\ln(1-P_n)}{\ln(1-p)}$ with $P_n = 1/2$, $p = 1/6$, we have $n > 3.81$ or four rolls. For two dice, $n > \frac{\ln(1-P_n)}{\ln(1-p)} = 24.61$, or 25 rolls.

De Méré again asked Pascal for advice on the following more important problem.

(3) **Split the pot (or the problem of points or division of stakes):** Two players put a certain amount of money in a pot, and a series of m games is played ($m = $ odd number). The pot will

be given to the player winning the first $(m - 1)/2 + 1$ games. (In the well-known NHL Hockey yearly tournament, for the two finalist teams, we have $m = 7$ and the team winning four games first wins the Stanley cup).

Pascal enlisted the help of his friend Fermat (1601–1665) and together, through an exchange of letters, they came up with the same final answer, using two different approaches. Pascal used mathematical expectation while Fermat listed all possible future outcomes.

This problem of splitting the common pot has come up previously and several thinkers have given their own solutions. Paciolli (1494), for example, thought that the pot should be split proportionately to the number of points already obtained. Cardano and Tartaglia suggested other solutions.

Fermat reasoned that if one player needs r more rounds to win and the other needs s, the game will certainly end by one of the players, after $r + s - 1$ additional rounds, which in total have 2^{r+s-1} different possible outcomes that are equally likely. Fermat then computed the odds for each player to win by writing down a table of all 2^{r+s-1} continuations. He then counted how many of them would lead to each player winning and divide the stakes in proportion to these odds.

Pascal, on the other hand, used expectations and showed that the correct division of the stakes is in the ratio of $\sum_{k=0}^{s-1} \binom{r+s-1}{k}$ to $\sum_{k=s}^{r+s-1} \binom{r+s-1}{k}$. Pascal and Huygens (1629–1695) were the early users of expectations in probability. In fact, Huygens published the first treatise on probability in 1657 entitled *De Ratiociniis in Aleae Ludo* (*Calculations in Games of Chance*), which had a great impact on scholars at that time.

In general, let A have the probability p to win a single point but needs n points to win while for B, it is q and m. Then de Montmort (1678–1719) has found that A's probability of winning is $P_A(n, m, p) = \sum_{k=n}^{n+m-1} \binom{k-1}{n-1} p^n (1-p)^{k-n}$. A fair division of the stakes should be in the ratio of $P_A(n, m, p)$ for A and $1 - P_A(n, m, p)$ for B. If $p = q = 1/2$ (fair coin flipping) and the first player to gain 10 points wins the game, if A already has 7 points and B, 5 points, then $P_A(3, 5, 1/2) = \sum_{k=3}^{7} \binom{k-1}{2} (1/2)^k = 99/128$. Hence, the pot should be divided in the ratio 99 to 29. If $p + q < 1$, then $1 - (p + q)$ represents the probability of a draw, but the same ratio applies.

Prior to the Chevalier de Méré's initiative, some isolated rough work on probability and gambling has been done by Gerolamo Cardano (1501–1576), Professor of Mathematics at the University of Bologna, in Italy, who was a genius, a fool, a charlatan, a scoundrel, who committed suicide on the day of his self-predicted death to redeem his reputation.

Gambling theory is closely associated with probability, especially the law of large numbers. An authoritative source of reference is the book by Epstein (2009), which we will discuss later.

8.2.4 *The 17th and 18th centuries*

Some researchers have questioned why the subject of probability arose as late as the 16th and 17th centuries in Europe, even though the notions of *probable* and *likely* were familiar to ancient civilizations in the West, and in the East, since antiquity. Without a solid scientific basis, the notion of probability progressed through the ages, getting more adepts, and attracting the attention of the best minds, without being precise. Although the domain of *numerical probability*, as described above, is just a very narrow sub-domain of probability, it contains, however, some very interesting results, such as the occupancy problem, which led later to the well-known results in quantic physics.

The works of Fermat (1601–1665), Pascal (1623–1662), Huygens (1629–1695), Laplace (1749–1827), and Bernoulli (1654–1705) marked the beginning of probability theory for the 17th and 18th centuries. An ecclesiastic, Rev. Thomas Bayes (1702–1761), left his mark on probability. In his book, *An Essay Towards Solving a Problem in the Doctrine of Chance*, published in 1763, after his death, he presented his formula of posterior probability. He also proved that the probability of success being between a and b is $\int \frac{1}{B(p,q)} x^p (1 - x)^q dx$, the integral that he calculated by approximation.

It is a bit surprising to learn that the Bayesian approach was used in the middle ages in Europe prior to any formal consideration of probability. However, it was a combination of data, prior knowledge, and formal deduction that could be intuitive, but confusing at the same time. But since the publication of Bayes's work, the Bayesian approach has become influential for almost a century and a half until another point of view took over at the beginning of the 20th century. More precisely, it was developed by three statisticians: Fisher (1890–1962), Neyman (1894–1981), and Pearson (1857–1936). It is called the *frequentist approach*, where the classical definition of probability of an event is adopted, and distribution parameters are considered as constants.

Bernoulli (1654–1705) published posthumously in 1713, *Arts Conjectandi*, which was a significant contribution, where he developed the theory of permutations and combinations, and applied them to games of chance (dice, cards, etc). Some years before, de Montmort (1678–1719) also published some work on combinatorial analysis related to cards and dice. De Moivre (1677–1754) followed suit with his work *Doctrine of Chances* and, on the basis of his approximation of the binomial distribution, is commonly credited as the initiator of the normal distribution. Lagrange solved several problems posed by other authors, among them Buffon (1707–1788), who discussed a number of gambling problems and believed that a better understanding of probability would provide "a powerful antidote against the evil of the gambler's passion".

8.2.5　*The 19th century*

In the early 19th century, the understanding of chance events has progressed much from its naive level. Laplace (1749–1827) declared in his *Essai philosophique sur les probabilités*: "the theory of probabilities is nothing more than common sense reduced to calculations". His influential book *Théorie analytique des probabilités*, in 1812, summarized all knowledge on probability at that time. He also elaborated on the idea of inverse probability considered by Bayes and can claim credit for providing the greatest advance in probabilistic methodology.

Poisson (1781–1840) was probably the first to define probability as a limit frequency in his book in 1837, which also mentioned the law of large numbers. Gauss (1777–1858) was the dominant figure in mathematics and used probability for astronomical phenomena. Gauss invented several statistical techniques to better analyze astronomy data and championed the use of a bell-shaped distribution. Poincaré

(1854–1912) wrote an essay on chance and random events. Quetelet (1796–1874), a Belgian academic, has reported on surprising statistical regularities in social and biological phenomena. He championed the use of probability models throughout Europe.

In Russia, probability has attracted the interest of several competent mathematicians, among them Chebychev (1821–1894), who generalized Bernoulli's work on the law of large numbers. His influential school of mathematical probability had two famous students: Liapounov (1859–1918) and Markov (1856–1922).

Two philosophers worked to establish probability as an independent and deep scientific discipline, with practical importance: Russell (1872–1970), with his *Principia Mathematica* published in 1903, and Keynes (1883–1946), who extended the procedures of inductive inference. About 1927, Russell made two famous statements on probability:

> *How dare we speak of the laws of chance? Is not chance the antithesis of all law?*

> *Probability is amongst the most important science, not least because no one understands it.*

We can confidently say that the 19th century was a time of outstanding evolution of empirical science, and reshaping of several branches of mathematics, where the discovery of paradoxes and antinomies has led to an appeal for their abstract treatments and axiomatizations. Set theory was clearly named, and a new field, probability, was also aimed at.

8.2.6 *The 20th century*

At the beginning of the 20th century, two notions were still missing: a solid mathematical basis for probability as a science and the meaning of the probability of an event. On the first topic, at the *International Congress of Mathematicians* in Paris in 1900, Hilbert posed 23 problems (later, problem number 24 was added) to be solved for the century, among which was problem number 6.

Although it seemed to concern physics, Hilbert clarified later the title, and the first part of the problem concerned the foundations of the probability theory.

In 1929, Kolmogorov came up with a solid theory, published later in 1933, in German, which is now considered modern mathematical probability theory. Previously, Bernstein (1880–1968) had published in 1927 some work on probability theory, using only three axioms.

With the development of measure theory pioneered by Lebesgue (1875–1941) and Borel (1871–1956), Lévy (1886–1971) produced important works based on that theory. Borel's memoir in 1909 marked the beginning of modern probability theory. His important ideas were then expanded by Khintchine (1894–1981), Kolmogorov, and Lévy, who in the 1930s took credit in the development of stochastic processes. Von Mises (1883–1953) developed the idea of sample space and furthered the idea of statistical probability.

Because of its close connection with abstract mathematical analysis, theoretical probability has been, for a long time, unjustly *considered as just a chapter of measure theory*, and hence of secondary importance by mathematical analysts. Fortunately, the works of Feller (1906–1970) and other applied probabilists have shown, convincingly, the different nature of probability, in view of its numerous applications in different domains. It must be mentioned that game theory, as a mathematical discipline, was developed by von Neumann (1903–1957) in 1928, who, later in 1944, expanded these ideas with Morgenstern (1902–1977) in the *Theory of Games and Economic Behavior*.

We should also mention the works of Einstein (1879–1955), Maxwell (1831–1879), Wiener (1894–1964), Cramer (1893–1985), and Lindenberg (1876–1932), related to probability. Finally, a number of very competent statisticians worked in domains adjacent to applied probability, and we wish to name a few of them here: Pearson (1857–1936), Fisher (1890–1962), Mahalanobis (1893–1972), Hotelling (1895–1973), Wald (1902–1950), Caratheodory (1873–1950) and Tukey (1915–2000).

In relation to other sciences, probability and statistics have become companions and tools for their development. For example, it has been reported that 20% of the Nobel prizes awarded in economics during the last 40 years were given to scholars working on probability/statistics (see Roussas (2009)) or on an associated field.

Probabilistic models abound in the study of diverse natural and economic phenomena, such as turbulent flow, population growth, and fluctuations in stock prices. Results of practical interest are those related to the mass behavior of a large population in the long run.

Mention should be made of the rapid and impressive development of the internet at the end of the last century, and the beginning of this century, which has helped develop several applied probability concepts associated with signal processing, Bayesian networks, pattern recognition, and data mining. Among the recent advancements in theoretical probability, we should mention the work of Voiculescu, who received the 2004 National Academy of Sciences (NAS) Award in Mathematics for *the theory of free probability*, in particular, using random matrices, and a new concept of entropy to solve several intractable problems in von Neumann algebras.

8.2.7 *Probability in the orient*

Outside the western world, modern probability theory made some inroads into China and India. It must be mentioned that *I Ching*, the oldest Chinese mathematical text which dates to 1150 B.C., presented basic notions on permutations and the Pascal triangle. In their article, "Transmission of Probability Theory into China at the End of the Nineteenth Century", Siu Man Keung and Lih Ko Wei (2015) mentioned the first book on this discipline, *Jueyi Shuxue*. It was a Chinese translation of a work by Thomas Galloway, about 1880, written for *Encyclopedia Britannica* back in 1839 and published only in 1896. But this article was published separately *as A Treatise on Probability*. Galloway was a Scottish mathematician, teaching at the Royal Military College at Sandhurst, beginning in 1823. The book seemed to have exerted little influence, however, on the subsequent development of probability and statistics, which took place in the Chinese Republic. It was in the 1930s only that serious study of this domain, in the modern sense of the subject, got under way in China.

In his article, "Probability in Ancient India", Raju (2011) relates that the theory of permutations and combinations was basic to the Indian understanding of meter and music. The Vedic, and post-Vedic composers, depended on combinations of two syllables called *guru* (deep, long) and *laghu* (short). On the computation side, binomial expansion was used in the orient. The pyramidal expansion is identical to the Pascal triangle, and the Chinese mathematician Shih-Chieh had some Pascal triangle results, up to $n = 8$, as early as 1303. In Persia (present-day Iran), it was known as Khayyam's triangle, after the work of an even earlier mathematician Omar Khayyám (1048–1131).

An excellent document on the history of probability is Sheynin's *Theory of Probability. A Historical Essay* (see Sheynin (2016)).

8.2.8   *The history of normal distribution*

The history of the normal distribution is naturally tied up with the history of probability.

Back in the 17th century, Galileo has already revealed several of the characteristics of the normal distribution when he expressed his opinion on the measurements of distances to the stars made by astronomers. But serious mathematical expression of this law began with de Moivre.

Religious persecution in France became severe when King Louis XIV issued the Edict of Fontainebleau in 1685, which revoked the Edict of Nantes, which had given substantial rights to French Protestants. It forbade Protestants to worship and required that all children be baptized by Catholic priests. De Moivre (1667–1754) was sent to the Prieure de Saint-Martin, a school where the authorities sent Protestant children for indoctrination into Catholicism.

Records of the Prieure de Saint-Martin indicate that he left the school in 1688, but de Moivre and his brother presented themselves as Huguenots admitted to the Savoy Church in London on 28 August 1687.

The normal distribution was first introduced by de Moivre in an article in 1734 (reprinted in the second edition of his *The Doctrine of Chances* (1738)) in the context of approximating certain binomial distributions for large n, although it is doubtful that de Moivre had any knowledge of a probability density function. His result was extended by Laplace in his book *Analytical Theory of Probabilities* (1812) and is now called the de Moivre–Laplace theorem (see Chapter 3). Laplace used the normal distribution in the analysis of errors of experiments and made some important contributions to this topic by computing the integral $\int_{-\infty}^{\infty} \exp(-t^2)dt = \sqrt{\pi}$ in 1782 and, in 1810, by proving the fundamental version of the central limit theorem.

The important method of least squares was introduced by Legendre in 1805. In 1809, Gauss (1777–1855), who claimed to have used the method since 1794, justified it rigorously by assuming a normal distribution of the errors. He published an important work, where he demonstrated that the law that rationalizes the arithmetic mean as an estimator of the location parameter is the normal law of errors. This law is denoted by $\varphi\Delta$, with expression $\varphi\Delta = \frac{h}{\sqrt{\pi}}\exp(-hh\Delta\Delta)$, where h is the "measure of the precision of the observations". Prior to Gauss, however, Adrain (1775–1843), an American mathematician, had already published a general treatise on the theory of errors, which, unfortunately, was not much read. Adrian adopted the expression of errors as $\phi(x) = Ce^{-k^2x^2}$. The name "bell curve" goes back to Jouffret (1837–1904), who first used the term "bell surface" in 1872 for a bivariate normal with independent components. The name "normal distribution" was coined independently by Peirce (1839–1914), Galton (1822–1911) and Lexis (1837–1914) around 1875. This terminology is unfortunate, since it reflects and encourages the fallacy that many probability distributions are "normal", while those which are not are abnormal. Maxwell showed that "the number of particles with velocity in a certain direction, which lie in $(x, x + dx)$, is $\frac{N}{\alpha\sqrt{\pi}}\exp(-\frac{x^2}{\sigma^2})dx$", reinforcing the role of this distribution. In 1915, Fisher (1890–1962) added the location parameter to the formula of the standard normal, giving it the present-day form, which textbooks in the 1950s helped popularize. See also Section 5.3.1.

This distribution is called the normal, or Gaussian distribution is an instance of Stigler's law of eponymy: "No scientific discovery is named after its original discoverer".

Concerning the normal, also known as the "law of frequency of error" Sir Francis Galton wrote as follows in 1889: "The law would have been personified by the Greek and deified, if they had known of it. It reigns with serenity and in complete self-effacement amidst the wildest confusion". However, Galton eventually questioned the universality of the normal when he observed that the log-normal could be a more suited model in several instances. In the English-speaking world, normal distribution and Gaussian distribution are in common use, and are considered equivalent, although Gaussian refers to a much more larger approach promoted by Gauss.

The binormal distribution (see Chapter 7) has a separate history of its own, with the pioneering work of Francis Galton (1822–1911), a cousin of Charles Darwin. His work was associated with genetics, with the goal of explaining variations between generations raised by Darwin in his *On the Origin of Species*. In 1885, appearing before *the British Association for the Advancement of Sciences*, Galton used the bivariate normal distribution to explain the relationship between two heights: the father's and his eldest son's. Galton found that fathers' and sons' heights, X and Y, respectively, are normally distributed, with equal means of 68 inches and variances of 3 inches. He also found that if he considers fathers of a fixed height x_0, then the heights of the sons are normally distributed with the average height being a linear function of x_0. In other terms, the conditional distribution of $Y|x_0$ is normal, with parameters functions of x_0. Galton found that if the father is taller, or shorter, than the average, so is the son, but, in general, sons regress more to the average than fathers, hence coining the term *regression* line in statistical analysis.

Correlation is credited to Bravais (1811–1863). Statistics had important promoters, among them Pearson (1857–1936) who introduced the method of moments and the normal curve in 1893. Later, he devised the chi-square test for goodness of fit and authored an important treatise on the Monte Carlo roulette.

8.2.9   *The occupancy problem*

In numerical probability, there are many interesting applications to real-life problems and to more abstract physical phenomena.

It is posed as follows: There are r balls to be put into n cases. How many ways can we do it? We have two situations:

(a) **Distinguishable balls:** This model gives rise to many interesting applications when we assign different meanings to balls and cases: birthday problem, coupon problem, elevator discharge of passengers, etc.

The problem of birthdays is based on the following facts: Although there are 365 days in a year, except for leap years, we need a group of 23 individuals only (chosen at random) to have a probability larger than 1/2 that "at least two of them share the same birthday". This number goes up to 254, for "at least one shares the same birthday as a fixed individual". But this solution is based on the hypothesis that birthdates are random throughout the year, although data have shown a higher frequency for summer months. Several variants of this problem exist:

$$\text{Balls} = \text{elementary particles,}$$
$$\text{Cases} = \text{known levels of energy,}$$

and look for the repartition of elementary particles among energy levels. We come up with three models.

First, the probability that energy levels numbered $1, 2, \ldots, n$, contains $\{r_1, \ldots, r_n\}$ particles with $\sum_{j=1}^{n} r_j = r$ is as follows:

(1) Under the Maxwell–Boltzmann statistics, $\frac{r!}{r_1! \ldots r_n!} n^{-r}$ (but no known particle follows this distribution). They apply to "classical" particles with non-quantized energy levels.

In quantum physics, we take the following.

(b) **Indistinguishable balls:**

(2) Under Bose–Einstein, $\binom{n+r-1}{r}^{-1}$ (photons, nuclei, atoms with even number of elementary particles follow this distribution). They apply to quantum particles with the property that any number of particles can occupy any level. It turns out these are the particles with integral "spin" such as photons and He 4 (but not He 3) atoms. *Bosons* is the related name.

(3) Under Fermi–Dirac statistics, $\binom{n}{r}^{-1}$ (electrons, neutrons, protons). Apply to quantum particles with the property that only one particle can occupy any particular level. It turns out these are the particles with half-integral "spin" such as electrons and neutrons. *Fermions* is the related name.

Example of deriving Bose–Einstein statistic (this expression is peculiar to physics): Let us consider the distribution of three distinguishable balls a, b, c in three cells. There are $3^3 = 27$ possible outcomes that can be seen in Table 8.1.

But when the balls become indistinguishable, several of these outcomes come together, and we have only 10 distinct outcomes, as shown in Table 8.2.

But Bose–Einstein statistics gives the same probability to each outcome, or 1/10, instead of the above probabilities (Table 8.3).

Table 8.1   27 outcomes of putting three distinguishable balls into three cells.

1	abc(A)	—	—	10	a	bc	—(F)	19	—	bc(I)
2	—	abc(B)	—	11	b	ac	—(F)	20	—	ac(I)
3	—	—	abc(C)	12	c	ab	—(F)	21	—	ab(I)
4	ab	c	—(D)	13	a	—	bc(G)	22	a	c(J)
5	ac	b	—(D)	14	b	—	ac(G)	23	a	b(J)
6	bc	a	—(D)	15	c	—	ab(G)	24	b	c(J)
7	ab	—	c(E)	16	—	ab	c(H)	25	b	a(J)
8	ac	—	b(E)	17	—	ac	b(H)	26	c	b(J)
9	bc	—	a(E)	18	—	bc	a(H)	27	c	a(J)

Table 8.2   Outcomes when balls are indistinguishable.

1	***	—	—	6	*	**	—
2	—	***	—	7	*	—	**
3	—	—	***	8	—	**	*
4	**	*	—	9	—	*	**
5	**	—	*	10	*	*	*

Table 8.3   Probabilities suggested.

	A	B	C	D	E	F	G	H	I	J	
Outcome and related probability	1	1	1	3	3	3	3	3	3	6	
	1	1	1	3	3	3	3	3	3	6	/27

The formula for Bose–Einstein gives the same answer $1\big/\binom{6-1}{3} = 1\big/\binom{5}{3} = 1/10$ to each of these outcomes.

8.3   Gambling and Probability

8.3.1   *What is gambling?*

Although gambling gave rise to probability, as seen in Section 8.2.3, it is now a distinct domain of its own. Man has invented an activity to occupy his free time, which can be considered as a vice, a pleasure, a disease, a weakness, a business, an occupation. It is *gambling for profit*.

Gambling is the wagering of money, or something of value, on an event, with an uncertain outcome, with the primary intent of winning money, or material goods. Usually, there are three elements: consideration (amount wagered), risk (chance of winning), and prize (reward). This activity is often permitted by law.

Acceptance of gambling in a society depends on its culture and religion. Christians deem gambling a vice. But games of chance have entered Europe from the Arab world through the crusades. Cards appeared in the 14th century and were soon imported to Europe, where they enjoyed great popularity.

Gambling can be guided by numerical probability to achieve success. But probability is only one factor, among many others, affecting the gambler when playing a game.

The term "odds" is frequently used in gambling, and, usually, we define odds 5 to 1 for an event E to happen to mean that there are five chances against that event happening opposed to one chance that it happens, giving $\mathbb{P}(E) = 1/6$. This is the case of $E =$ having a two (snake's eyes), when rolling a die.

Hence, in relation to probability, the *odds* (in favor) of an event, or a proposition, is the ratio of the probability that the event will happen (or $1/6$) to the probability that the event will not happen (or $5/6$) $= \frac{5}{6}/\frac{1}{6} = \frac{5}{1}$, or 5 to 1.

We can see that if $\mathbb{P}(E) = m/n$, with m, n positive integers and $m \leq n$, then the odds that E happens is $n - m$ to m. The odds that E does not happen is then m to $n - m$.

Modern gambling theory takes its main sources from the fields of probability, statistics, and game theory, besides psychology and other behavioral sciences.

8.3.2 *Skill and luck*

Some games, such as chess, depend more on player skill, while many children's games require no decisions by the players and are decided purely by luck. Slot machines, too, require no skill. Many games require some level of both skill and luck.

Ancient China was the cradle of gambling, and several, if not most, forms of gambling practised today originated in China, hundreds, or even thousands, of years ago. These include cards, lotteries, roulettes, but, as in ancient Greece, probability, as a study and research discipline, was completely ignored. The profound reason was, probably, that intellectuals and scholars frowned upon the idea of undertaking a study of gambling, a pastime of dubious repute, often associated with illegal activities.

Two games of chance, however, are respected unanimously by intellectuals and laborers alike: the Chinese chess (*Xiangqi*), with its origins lost in time (in fact, as far back as the 11th century B.C. in southern China, a game which is thought to be an early form of *Xiangqi* was played) and *Mahjong*, which conquered the world since the 1920s. Both require more skill than luck.

Western chess is a two-player strategy board game, played on a checkered board with 64 squares arranged in an 8×8 grid. Chess is believed to be derived from the Indian game *chaturanga* sometime before the 7th century. *Chaturanga* is also the likely ancestor of the Chinese strategy games *xiangqi*, *janggi* (Korean chess), and *shogi* (Japanese chess). Chess reached Europe by the 9th century due to the conquest of Hispania. The pieces assumed their current powers in Spain, in the late 15th century; the modern rules were standardized in the 19th century.

Each player begins with 16 pieces: one king, one queen, two rooks, two knights, two bishops, and eight pawns. Each piece moves differently, with the most powerful being the queen and the least powerful the pawn.

8.3.3   *A book on gambling*

In this book authored by Epstein (2009), gambling is considered from antiquity to the present time. The mathematics go from moderate to advanced. Gambling is taken here in a very large sense, going from very old simple games: coin games, oddman out, paper, scissors, rocks, etc. to present-day extrasensory perception (ESP) and other paranormal phenomena. The book is full of interesting facts, and results, on various games which can be antique (Le Her game), classical (cards, dice, coins, blackjack, tic-tac-toe, and various casino games), as well as new (Parrondo's strategy of combining two losing games into a winning one, computer chess, psychological betting systems), or games of pure chance, where little experience is required, or of pure skill, where serious strategies have to be worked out. Also, there are games of non-stationary statistics, such as horse racing and stock market speculation.

A more practical reference is the book by Gros (1996), where most casino games are discussed.

An interesting detail is that the Reddere integral (p. 70), used in Baccarat, Craps, and Roulette, is, in fact, our unit normal loss function: $L(z) = \varphi(z) - z(1 - \Phi(z))$ encountered in the Bayesian decision theory, when the mean of the normal prior is itself normal and the loss is linear. Here, $\varphi(.)$ and $\Phi(.)$ are the pdf and cdf of the standard normal, respectively (see Chapter 6).

Usually, when variance(X_k) also exist, they permit the use of the central limit theorem (CLT) instead of only the law of large numbers (LLN) (see Chapter 3).

8.3.3.1 *Some interesting facts*

(a) For games considered of mixed chance and skill, there is *blackjack*, or the game of 21, to which a whole section is devoted. A well-known author in this domain, Thorp (1962), has mentioned a few times, since he has written in 1962 a comprehensive mathematical treatment of all blackjack strategies used at that time and provided the first quantitative synthesis of card-counting systems. By reducing the house edge in this game to a minimum, he has provoked several changes in the game rules. On the other hand, in the 18th century in France, Jean le Rond d' Alembert has devised a strategy based on martingales (known as progression betting system) and the equilibrium of happenings for two complementary events (e.g. tails are more probable after a run of heads, according to him), which proved to be completely non-effective.

(b) Gambling theory is concerned mainly of making decisions under risks and *utility functions* are fundamental in decision-making. But in spite of clear axioms adopted for the construction of individual utility functions, it is never certain that they are consistent and logical for effective use in gambling.

On the other hand, institutions like casinos and other large betting organizations like horse racing and lotteries are now, in most developed countries, major business enterprises with

important economic impact, in spite of the negative social problems they can create. In the USA, hundreds of millions of gam-
blers wager more than USD 100 billion each year. Institutions
can rely on the law of large numbers (LLN), which apply to
them, and not to individual players, to have their prosperity
guaranteed.

(c) **Fallacies**: In general, the *gambler's fallacy*, also known as the
Monte Carlo fallacy or the *fallacy of the maturity of chances*, is
the mistaken belief that if something happens more frequently
than normal during a given period, it will happen less frequently
in the future (or vice versa). In situations where the outcome
being observed is truly random and consists of independent trials
of a stochastic process, this belief is false.

In the book, there are 13 main fallacies reported by the author,
which might affect a gambler judgment. One, for example, is a tendency to overvalue the significance of a small sample taken from a
large population. These fallacies are prevalent in gambling as a result
of psychological influences.

8.3.3.2   *When is a game "fair" to the player?*

From the player's point of view, a game of chance is only of interest
if in the long run he will win some money. Let X_k be the positive or
negative gain at the kth trial, when the player keeps playing the same
game. We also suppose that the player has unlimited resources so that
he can keep on playing and that the games are independent of each
other, with a fee ξ to be paid before each game. Let $S_n = X_1 + \cdots + X_n$
be the accumulated gain and suppose that $\mathbb{E}[X_j] = \mu < \infty$, with
$\mathbb{E}[X]$ being the expectation of X. The variable $S_n - n\mu$ is likely to
be small wrt n. Hence, if $\xi < \mu$, the game is favorable to the player,
and for $\xi > \mu$, it is unfavorable. The case $\xi = \mu$ is a *fair game*.
But $S_n - n\mu$ can be positive or negative. So, *fair* does not mean
necessarily *favorable*.

8.3.3.3   *Fair pay-off and house edge*

In the card game above, suppose that you pay 1\$ and pick a card.
You lose the dollar if it is not a queen, but win 11\$ if it is a queen.

The final expected gain X for each game is

$$
\begin{array}{c|cc}
X & -1 & 10, \\
\hline
\mathbb{P}(X) & \dfrac{12}{13} & \dfrac{1}{13},
\end{array}
$$

$$
\mathbb{E}(X) = \frac{12 \times (-1) + 10 \times 1}{13} = -\frac{2}{13}.
$$

The pay-off of that game, if it is fair, should be 13\$, to make $\mathbb{E}(X) = 0$. So, your average loss per game is 2/13\$, and the edge of the house can be simply taken as $\frac{13}{13} - \frac{11}{13} = \frac{2}{13}$. It is this edge that makes you a loser in the long run. Following are the loss expectations of some casino games (these are unofficial information only)

- Baccarat: -1.2%
- Bingo: -1.5% to 0%
- Blackjack: -6.0% to $+1.5\%$
- Craps: -1.3% to -0.65%
- Lottery: -50% to 0%
- Roulette: -5.3%
- Slot machine: -20% to -2.7%.

We can see that rare are games with positive expectation for the player, and those with a small positive one require a lot of strategy and skill.

8.4 What is Probability?

8.4.1 *Meanings of probability*

Although the mathematical side of probability is on a good foundation now, thanks to Kolmogoroff's work, there is still no consensus on the meaning of probability. We find that most definitions are caught up in a vicious circle. For example, we define probability as likelihood, chance, etc., which, in turn, needs to be defined too.

Nevertheless, we can usually find two main ideas:

(a) **Frequentist probability or objective probability:** It is related to the notions of *relative frequency*; when dealing with

well-defined experiments where the number of favorable outcomes can be clearly identified, the probability of the event can
be computed. *Limit relative frequency*: It is related to the above
concept together with the limit notion. The weakness in this last
definition is that we cannot create exact conditions for an event
to happen except in the lab.

(b) **Subjective probability:** This is based on degrees of belief and
is mostly personal. If we are very convinced of the event happening, its probability is close to one. Conversely, if we believe that it
will not happen, its probability is close to zero. However, individuals need to be consistent, with themselves first, then with each
other, which could be a big problem. A system of bettings can
be set up to refine the probability assigned to an event. Statistical decision theory uses this kind of probability. It is reported
that Poincaré insisted on making the difference between these
two probabilities, but Keynes maintained that all probabilities
are subjective. There are many intermediary definitions. Some
authors think that it is non-productive to try to have different
meanings of probability. Instead, we should talk about *different
theories of probability* to acknowledge their fundamental differences. To extend the above list of two interpretations, we give
the following list of seven interpretations, which is satisfactory,
for most practical purposes.

8.4.2   *Main theories*

We give the following main theories:

(1) **Axiomatic interpretation:** The probability of an event is the
value that fulfills the various axioms of the theory of probability.
In 1933, Kolmogorov (1903–1987) formulated the first axiomatic
system for probability.

(2) **Classical interpretation:** The probability of an event is the
ratio of the favorable outcomes to the possible outcomes. Thus,
the probability of "2" or "snake's eyes" is $1/6$ since there are six
possible outcomes with only one favorable.

(3) **Limit frequency interpretation:** The probability of an event
is the limit of the relative frequency of that event when the
experience leading to it is repeated many times.This approach is

promoted by Richard Von Mises, and looks very attractive. However, it can only be applied to events that will happen repeatedly.

(4) **Subjective interpretation (Keynes and Jeffreys):** Probability is a measure of the degree of belief.

(5) **Probability as a degree of confirmation:** This is an approach supported by Rudolph Carnap and students of inductive logic. The probability of a statement is the degree of confirmation the empirical evidence gives to the statement. For example, the statement "the die shows 2" receives a partial confirmation by the evidence with degree of confirmation $1/6$.

(6) **Propensity interpretation:** The probability of an event is an objective property of the event. The physical properties of a die (the die is balanced, has six sides, and on every side, there is a different number between 1 and 6) explain the fact that the limit of the relative frequency of "the face shown is 2". This is proposed by Popper.

(7) **Just define probability as a number between 0 and 1**, attached to an event, with no further interpretation (Doob, Feller, Neyman, and Cramer).

These notions are presented in Fine's excellent book, *Theories of Probability* (see Fine (1973)). In mathematical terms, Fine introduces these theories in depth: their requirements, their weaknesses, their attractive features, etc. Furthermore, he often provides comparative studies between them.

Fine (1973) first identifies and defines the following criteria related to the theories of probability and then states that theories of probability can be classified into 11 categories:

(1) **Empirical:** based on experiments and observations.
(2) **Logical:** based on rational reasoning and deep thinking.
(3) **Objective:** independent of thought and feelings.
(4) **Pragmatic:** stress is made on the practical and good rather than the correct and complex.
(5) **Subjective:** utilizing individual opinions and judgments.

Fine's book is almost 50 years old now, but its criticisms and recommendations remain mostly valid. We have achieved huge progresses in mathematical probability definition, in Bayesian computing, in simulation, and in other areas, but as far as the meaning of the probability of an event is concerned, we haven't made much progress.

A review of the book was written in *IEEE Transactions on Information Theory* by Good (1974), who, in his own pamphlet (Good, 1950), thinks that although there are at least five kinds of probability, one can easily live with one, the subjective probability. Good (1974) raised several technical points that he wanted to contest, but overall, the book content was appreciated.

Naturally, subjective probability goes along with Bayesian statistics and decision theory, where a utility function, or a cost function, or a loss function, is often considered (see Chapter 6).

8.5   Exchangeable Variables

The notion of *exchangeable random variables* was introduced by de Finetti in 1931 and has proved to be valuable for the definition of probability. Basically, we now understand how the binomial distribution can be constructed. The following results show that statistical computations can be carried out with *exchangeable normal* variables.

Let $\{X_i\}_{i=1}^{\infty}$ be an infinite sequence of univariate rv's. It is called sequence of exchangeable rv's if for every finite n and every permutation $\{\pi_1, \ldots, \pi_n\}$ of $\{1, \ldots, n\}$, we have $\{X_1, \ldots, X_n\}^t$ and $\{X_{\pi_1}, \ldots, X_{\pi_n}\}^t$ having identical distributions. Equivalently, we have the following.

Definition 8.1. An infinite sequence of rv's $\{X_j\}_{j=1}^{\infty}$ is *exchangeable* if for all $n = 2, 3, \ldots$ $X_1, \ldots, X_n \overset{D}{=} X_{\pi(1)}, \ldots, X_{\pi(n)}$ for all permutation π of $\{1, \ldots, n\}$.

We can see that if $X_1, \ldots, X_n, \ldots$ are iid, they are exchangeable, but the converse is not true.

In 1931, de Finetti showed that all exchangeable binary sequences are mixtures of iid Bernoulli sequences. We have the following theorem.

Theorem 8.1. *A binary sequence $\{X_j\}_{j=1}^{\infty}$ is exchangeable iff there is a distribution function F on $[0, 1]$, s.t.*

$$\mathbb{P}(X_1 = x_1, \ldots, X_n = x_n) = \int_0^1 \theta^{t_n}(1 - \theta)^{n-t_n} dF(\theta),$$

with $t_n = \sum_{i=1}^{n} x_i$. *Furthermore, F is the cdf of the limiting frequency, i.e.*

$$F(y) = \mathbb{P}(Y \leq y), \quad with \ Y = \lim_{n \to \infty} \frac{1}{n} \sum_{i=1}^{n} X_i,$$

and the Bernoulli distribution can be obtained by conditioning on Y, i.e.

$$\mathbb{P}(X_1 = x_1, \ldots, X_n = x_n \,|\, Y = \theta) = \theta^{t_n} (1 - \theta)^{n - t_n}.$$

For a finite n, the sequence $\{X_i\}_{i=1}^{n}$ is exchangeable if there is an infinite sequence of exchangeable rv $\{X_i^*\}_{i=1}^{\infty}$ s.t. $\{X_1, \ldots, X_n\}^t$ and $\{X_1^*, \ldots, X_n^*\}^t$ are identically distributed.

Hence, *exchangeable observations are conditionally independent, given a latent variable to which a probability distribution is assigned.* For an exchangeable sequence of Bernoulli rv's, there are underlying variables, which are iid, and exchangeable sequences are mixtures of these iid sequences.

Another formulation of the above theorem is as follows.

Theorem 8.2. *To every infinite sequence of binary exchangeable variables $\{X_n\}$, there corresponds a probability distribution $F(.)$ on $[0, 1]$, such that*

$$\mathbb{P}(X_1 = 1, \ldots, X_k = 1, X_{k+1} = 0, \ldots, X_n = 0)$$

$$= \int_0^1 \theta^k (1 - \theta)^{n-k} dF(\theta)$$

and

$$\mathbb{P}(S_n = k) = \binom{n}{k} \int_0^1 \theta^k (1 - \theta)^{n-k} dF(\theta),$$

where $S_n = X_1 + \cdots + X_n$.

In what follows we deal with *exchangeable normal* variables, a subgroup of *exchangeable* variables.

Definition 8.2. $\{X_1, \ldots, X_n\}$ is called *exchangeable normal* variables if they are exchangeable and their joint distribution is $N_n(\boldsymbol{\mu}, \boldsymbol{\Sigma})$.

We have the following theorem.

Theorem 8.3. *We have the following equivalent statements:*

(1) $\{X_1, \ldots, X_n\}$ *are exchangeable normal variables.*
(2) $\mathbf{X} = (X_1, \ldots, X_n)^t \sim N_n(\boldsymbol{\mu}, \boldsymbol{\Sigma})$, *with X_i having same mean μ, same variance σ^2 and same correlation coefficient ρ, for any couple $\{X_i, X_j\}$, $i \neq j$.*
Furthermore, we have the following.
(3) *For $\rho \in [0, 1)$, $\mathbf{X}$ has density*

$$f(\mathbf{x}) = A \exp\left\{-\frac{1}{2}\left[\sum_{i=1}^{n}\frac{(x_i - \mu)^2}{(1-\rho)\sigma^2} - \frac{\rho\left(\sum_{i=1}^{n}(x_i - \mu)/\sigma\right)^2}{1 + (n-1)\rho}\right]\right\},$$

where

$$A = \frac{1}{(\sigma\sqrt{2\pi})^n(1-\rho)^{\frac{n-1}{2}}\sqrt{1 + (n-1)\rho}}.$$

(4) $\mathbf{X}$ *is identically distributed as*

$$\left(\sigma\left(\sqrt{1-\rho}Z_1 + \sqrt{\rho}Z_0\right) + \mu, \ldots, \sigma\left(\sqrt{1-\rho}Z_n + \sqrt{\rho}Z_0\right) + \mu\right)^t,$$

where $Z_0, \ldots, Z_n \overset{iid}{\sim} N(0, 1)$, $\rho \in [0, 1]$.
(5) *For $\rho = 1$, $\mathbf{X}$ and $(\sigma Z_0 + \mu, \ldots, \sigma Z_0 + \mu)^t$ are identically distributed,*

$$f(\mathbf{x}) = \int_{-\infty}^{\infty} \frac{\prod_{i=1}^{n}\phi(u_i + \lambda z)}{(\sigma\sqrt{1-\rho})^n}\phi(z)dz,$$

where

$$\lambda = \sqrt{\frac{\rho}{1-\rho}}, u_i = \frac{x_i - \mu}{\sigma\sqrt{1-\rho}}, \quad i = 1, \ldots, n.$$

Corollary 8.1 (consequences on the cdf and its computation).
Let $\{X_1, \ldots, X_n\}$ be exchangeable normal variables, with common $\mu, \sigma^2, \rho \neq 1, \rho \geq 0$. Then

(1) *The cdf of $(X_1, \ldots, X_n)^t$ is $F(\mathbf{x}) = \int_{-\infty}^{\infty}\prod_{i=1}^{n}\Phi\left(\frac{a_i + \sqrt{\rho}z}{\sqrt{1-\rho}}\right)\phi(z)dz$.*

(2) *The cdf of* $(|X_1|, \ldots, |X_n|)^t$ *is*

$$
F_{|\mathbf{X}|}(\mathbf{x}) = \int_{-\infty}^{\infty} \prod_{i=1}^{n} \left[\Phi \left(\frac{\frac{x_i}{\sigma} + \sqrt{\rho}z}{\sqrt{1-\rho}} \right) - \Phi \left(\frac{-\frac{x_i}{\sigma} + \sqrt{\rho}z}{\sqrt{1-\rho}} \right) \right] \phi(z)dz,
$$

with $\phi(.)$ *and* $\Phi(.)$ *being, respectively, the pdf and cdf of* $\mathbf{Z} \sim N_1(0,1)$ *variables.*

Proof. See Tong (1989, p. 116), where it is pointed out that the cdf of $|X|$ is obtained without deriving the density of $|X|$. □

8.6 Conclusion

This chapter presents a brief history of probability theory and of the normal distribution. It also gives the main focus points on two important topics related to probability: the theory of gambling and different theories of probability. Some discussions on these topics are made, and they are necessarily subjective. De Finetti's result on exchangeable sequences of random variables and results on exchangeable normal conclude the chapter and show that highly abstract concepts in probability can lead to concrete computations.

Bibliography

Epstein, R.A. (2009). *The Theory of Gambling and Statistical Logic*, 2nd ed. (Academic Press, New York).

Fine, T.L. (1973). *Theories of Probability* (Academic Press, New York and London).

Good, I.G. (1950). *Probability and the Weighing of Evidence* (Griffin, London).

Good, I.G. (1974). Review of theories of probability by Terrence L. Fine, *Transactions on Information Theory*, **20**(2), 298–300.

Gros, R. (1996). *How to Win at Casino Gambling* (Carlton Books, New Jersey).

Raju, C.K. (2011). Handbook of the Philosophy of Science, *Philosophy of Statistics*, **7**, 1175–1195.

Roussas, G. (2009). Probability and Statistics throughout the Centuries. Greek in the Proceedings of the Academy of Athens, **1**, 84.

Sheynin, O. (2016). *Theory of Probability. A Historical Essay, Revised and Enlarged Edition.*

Siu Man Keung and Lih Ko Wei (2015). Transmission of Probability Theory into China at the End of the Nineteenth Century. D.-E. Rowe and W. S. Horng (eds.). A Delicate Balance: Global Perspectives on Innovation and Tradition in History of Mathematics Trends in the history of Science. (Springer International Publishing, Switzerland) Do 1.10.1007/978-3-31912030-0_17.

Thorp, E. (1962). *Beat the Dealer* (Blaisdell Publishing, New York).

Tong, Y. (1989). *The Multivariate Normal Distribution* (Springer, New York).

Chapter 9

Normal Theory in $\mathbb{R}^p$

9.1 Introduction

This chapter treats the problem in a more general context of a distribution in $\mathbb{R}^p$, $p \geq 4$, the cases $p = 2$ and $p = 3$ have been treated in Chapter 7. There should be more general considerations here, but graphical illustrations become limited. We will also look at the Mahalanobis distance in conjunction with the normal. Applications concern the estimation of the mean vector of a multinormal distribution under various conditions. We will also discuss different forms of correlation coefficients encountered in applied research projects.

9.2 Basic Results in Quadratic Forms in p Variables

Since the density of p-normal vector is a quadratic form, we start by recalling the following basic result in linear algebra.

Let $\mathbf{C}$ be a $(p \times p)$ real symmetric matrix. It determines a quadratic form in p scalar variables $Q_C(x_1, \ldots, x_p) = \sum_{i=1}^{p} \sum_{j=1}^{p} c_{ij} x_i x_j = \mathbf{x}^t \mathbf{C} \mathbf{x}$. This form can be simplified by a homogeneous linear change of variables. Indeed, a theorem by Jacobi shows that it can be brought to a diagonal form:

$$Q_B(y_1, \ldots, y_p) = \lambda_1 y_1^2 + \cdots + \lambda_p y_p^2,$$

where λ_i are the eigenvalues of a matrix $\mathbf{B}$.

Proof. Let $\mathbf{A}$ be an orthogonal matrix, s.t. $\mathbf{A}^{-1}\mathbf{C}\mathbf{A} = B = diag(\lambda_1, \ldots \lambda_p)$. Writing $\mathbf{x} = \mathbf{A}\mathbf{y}$, we have $\mathbf{x}^t = \mathbf{y}^t\mathbf{A}^t = \mathbf{y}^t\mathbf{A}^{-1}$ and

$$Q_C(\mathbf{x}) = Q_B(\mathbf{y}) = \mathbf{y}^t\mathbf{A}^{-1}\mathbf{C}\mathbf{A}\mathbf{y} = \mathbf{y}^t\mathbf{B}\mathbf{y}$$

$$= \sum_{i=1}^{p}\sum_{j=1}^{p} b_{ij}y_iy_j = \lambda_1 y_1^2 + \cdots + \lambda_p y_p^2.$$

This quadratic form now contains only square terms with coefficients being the eigenvalues of $\mathbf{B}$, which are also eigenvalues of $\mathbf{C}$. It is hence not necessary to know $\mathbf{A}$ if we know these eigenvalues.

Computationwise, this approach has difficulties in some cases, however, as the determination of eigenvalues can lead to large errors in approximation (see Csaki (1970)).

When $\mathbf{X}_i$ is a random variable, with its own domain of definition, the above result can lead to the variables $\mathbf{Y}_i$ having null correlations, without being independent, however, because their domains of definition might not be disjoint (see Chapter 7 for results on the bivariate normal). □

9.3   Expression of the Density

We have similar explanations for the vector case. We start with the center, or mean $\boldsymbol{\mu} = (\mu_1, \ldots, \mu_p)^t$, and a covariance matrix $\boldsymbol{\Sigma}$, required to be symmetric and positive definite. The distance used now is the square of the statistical distance from $\mathbf{X}$ to $\boldsymbol{\mu}$, in $\mathbb{R}^p$, $p \geq 4$, but takes also into consideration the variance of each component and the covariances between different components, as given by the covariance matrix $\boldsymbol{\Sigma}$. The square of this statistical distance from $\mathbf{X}$ to $\boldsymbol{\mu}$ is $\Delta_X^2 = (\mathbf{X} - \boldsymbol{\mu})^t\boldsymbol{\Sigma}^{-1}(\mathbf{X} - \boldsymbol{\mu})$. This distance is often called the *Mahalanobis square distance*. Two different values of $\mathbf{X}$ give rise to two different Δ_X^2's, and the probability between them has to be computed using rectangles (see Section 7.9.1) unlike the unidimensional case.

We, finally, have the multivariate density defined as $f(\mathbf{X}) = C\exp(-\Delta_X^2/2)$, using Newton's principle on the square of the distance, together with the exponential function, or just by applying the formula. Integrating in $\mathbb{R}^p$, we find

$$\int\limits_{-\infty}^{\infty} ... \int\limits_{-\infty}^{\infty} \exp(-\Delta_X^2/2)dx_1...dx_p = (2\pi)^{p/2}\sqrt{|\boldsymbol{\Sigma}|}.$$

Hence, the density is

$$f(\mathbf{X}) = \frac{1}{(2\pi)^{p/2}\sqrt{|\boldsymbol{\Sigma}|}} \exp\left\{-\frac{(\mathbf{X} - \boldsymbol{\mu})^t\boldsymbol{\Sigma}^{-1}(\mathbf{X} - \boldsymbol{\mu})}{2}\right\}, \quad \mathbf{X} \in \mathbb{R}^p.$$

(9.1)

We see that (9.1) is also the density of the attraction of the point $\mathbf{X}$ toward the origin $\mathbf{0}$, using the Mahalanobis square distance and the inverse exponential. We write $\mathbf{X} \sim N_p(\boldsymbol{\mu}, \boldsymbol{\Sigma})$, where $|\boldsymbol{\Sigma}|$ is the determinant of $\boldsymbol{\Sigma}$.

We note that $(\mathbf{X} - \boldsymbol{\mu})^t\boldsymbol{\Sigma}^{-1}(\mathbf{X} - \boldsymbol{\mu})$ is a quadratic form in $\mathbf{X}$, denoted by $Q_p(\mathbf{X})$. Hence, we can also write

$$f(\mathbf{x}) = \frac{1}{(2\pi)^{p/2}\sqrt{|\boldsymbol{\Sigma}|}} \exp\left\{-\frac{Q_p(\mathbf{x})}{2}\right\}, \quad \mathbf{x} \in \mathbb{R}^p.$$

Notes:

(1) The matrix $\boldsymbol{\Sigma}$ is required to be *positive definite*, which is equivalent to be strictly positive for a scalar variable. It can be proven that a square matrix is positive definite if, for each principal minor matrix defined on its diagonal, its determinant is strictly positive.

(2) Another way of looking at the normal surface is as follows. The surface is built around the mean $\boldsymbol{\mu}$ in the following way: At each point $\mathbf{x}$ of $\mathbb{R}^p$, we compute its Mahalanobis square distance to the mean, Δ_x^2, and assign to this point the ordinate $C \exp\left(-\Delta_x^2/2\right)$, the normalizing constant being

$$\frac{1}{(2\pi)^{p/2}|\boldsymbol{\Sigma}|^{1/2}}.$$

This viewpoint explains the relationship between the height and width of a normal surface, as seen in the Computational Appendix in Chapter 7.

The density reduces to (2.2) for $p = 1$,

$$f(x) = \frac{1}{\sigma\sqrt{2\pi}} \exp\left(-\frac{(x-\mu)^2}{2\sigma^2}\right), \quad -\infty < x < \infty,$$

denoted $X \sim N_1\left(\mu, \sigma^2\right)$, and to (7.9) for $p = 2$, denoted $\mathbf{X} \sim N_2\left(\mu_X, \mu_Y; \sigma_X^2, \sigma_Y^2; \rho\right)$.

9.3.1 *Expression of the normal density*

We now deal with a p-dimensional vector $\mathbf{X}$, and vector and matrix notations are required now, for otherwise formulas become extremely complicated, as can be seen in the previous chapter on the trivariate normal. Hence, the *column vector* $\mathbf{X}$ is said to have a normal, or Gaussian, distribution if its density has the form

$$f(\mathbf{x}) = \frac{1}{(2\pi)^{p/2}|\mathbf{\Sigma}|^{1/2}} \exp\left(-\frac{1}{2}(\mathbf{x}-\mu)^t \mathbf{\Sigma}^{-1}(\mathbf{x}-\mu)\right),$$

$$\mathbf{x} = (x_1, \ldots, x_p)^t,$$

where μ is a p-vector in $\mathbb{R}^p$ called the *mean vector* and $\mathbf{\Sigma}$ is the $(p \times p)$ *covariance matrix*, which is symmetric wrt the first diagonal, where the entries are all positive. Also, it is supposed to be positive definite. We write $\mathbf{X} \sim N_p\left(\mu, \mathbf{\Sigma}\right)$.

9.3.2 *Examples*

We see that *in the simplest case* we have $\mu = \mathbf{0}$, $\mathbf{\Sigma} = \mathbf{I}_p$. Then $\mathbf{X}^t\mathbf{I}_p\mathbf{X} = \mathbf{Z}^t\mathbf{Z}$ and the density becomes $f(\mathbf{z}) = \exp(-\mathbf{z}^t\mathbf{z}/2)/(2\pi)^{p/2}$, a product of p independent univariate standard normal densities, with $\sigma_i^2 = 1$ and $\text{cov}(Z_i, Z_j) = 0$, $i \neq j$.

Example: Let $\mu = \begin{pmatrix} 1.5 \\ 1 \end{pmatrix}$, $\mathbf{\Sigma} = \begin{pmatrix} 1.8 & 0.2147 \\ 0.2147 & 1 \end{pmatrix}$. We have the graphs of $\mathbf{X} \sim N_2\left(\mu, \mathbf{\Sigma}\right)$ and of $\mathbf{X}^* \sim N_2\left(\mu, \mathbf{\Sigma}^{-1}\right)$ in Fig. 9.1, i.e. graphs

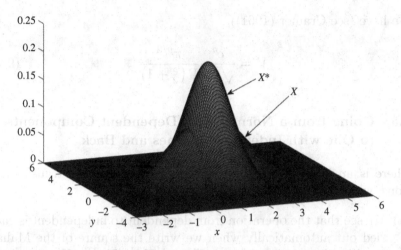

Fig. 9.1   Graphs of multinormal densities (quadratic form and reciprocal).

related to a quadratic form, and its reciprocal

$$f(\mathbf{X}) = \frac{1}{(2\pi)^{p/2}\sqrt{|\mathbf{\Sigma}|}} \exp\left\{-\frac{(\mathbf{X}-\boldsymbol{\mu})^t \mathbf{\Sigma}^{-1}(\mathbf{X}-\boldsymbol{\mu})}{2}\right\},$$

$$f(\mathbf{X}^*) = \frac{\sqrt{|\mathbf{\Sigma}|}}{(2\pi)^{p/2}} \exp\left\{-\frac{(\mathbf{X}^*-\boldsymbol{\mu})^t \mathbf{\Sigma}(\mathbf{X}^*-\boldsymbol{\mu})}{2}\right\}.$$

So, we can see here what happens when we forget to have the inverse sign on the covariance matrix in the formula of the multivariate normal.

9.4   Some Integrals Related to the Multivariate Normal

Let $Q_p(\mathbf{x}) = \mathbf{x}^t \mathbf{A} \mathbf{x}$ be a quadratic form in $\mathbf{x}$. For $p \geq 3$, the equation $Q_p(\mathbf{x}) = c^2$ defines an ellipsoid in p-dimension space and we are interested in the integral within that ellipsoid, which can be written as

$$V = \int\limits_{Q_p(x_1,\ldots,x_p)<c^2} \ldots \int dx_1 \ldots dx_p$$

$$= \frac{c^p}{\sqrt{|\mathbf{A}|}} \int\limits_{\sum\limits_{i=1}^{p} x_i^2 < 1} \ldots \int dx_1 \ldots dx_p.$$

We have (see Cramer (1961))

$$V = \frac{c^p}{\sqrt{|\mathbf{A}|}} \frac{\pi^{p/2}}{\Gamma\left(\frac{p}{2}+1\right)}. \qquad (9.2)$$

9.5   Going from a Normal with Dependent Components to One with Independent Ones and Back

There is an isomorphism between matrices and linear transformations, which can be used here:

(a) We see that the operation from dependent to independent is carried out automatically when we write the square of the Mahalanobis distance as $\Delta_X^2 = (\mathbf{X} - \boldsymbol{\mu})^t \boldsymbol{\Sigma}^{-1/2} \boldsymbol{\Sigma}^{-1/2} (\mathbf{X} - \boldsymbol{\mu})$ since $\boldsymbol{\Sigma}^{-1/2} (\mathbf{X} - \boldsymbol{\mu}) = \mathbf{Z}$, with $\mathbf{Z} \sim \mathbf{N}_p (\mathbf{0}, \mathbf{I}_p)$ (see Chapter 11 for the definition of $\boldsymbol{\Sigma}^{1/2}$). For $p \geq 2$, we have $\Delta_X \sim \chi_p$, the chi distribution with p dof, and $\Delta_X^2 \sim \chi_p^2$, the chi-square distribution with p dof.

(b) Conversely, let $\{U_i\}_{i=1}^p$ be p independent $N_1 (0,1)$ variables. Their joint density is $p_\mathbf{U} (\mathbf{u}) = \exp(-\mathbf{u}^t \mathbf{u}/2)/(2\pi)^{p/2}$. Let $\mathbf{Y}^t = (Y_1, \ldots, Y_p)$ and $\mathbf{H}$ be a matrix such that $|\mathbf{H}| \neq 0$. Let's perform the transformation: $\mathbf{U} = \mathbf{HY}$. Then $\mathbf{Y}$ has density

$$p_\mathbf{Y}(\mathbf{y}) = \frac{|\mathbf{H}| \exp(-\mathbf{y}^t \mathbf{H}^t \mathbf{H} \mathbf{y}/2)}{(2\pi)^{p/2}} = \frac{|\mathbf{A}|^{1/2} \exp(-\mathbf{y}^t \mathbf{A} \mathbf{y}/2)}{(2\pi)^{p/2}}$$

if we set $\mathbf{A} = \mathbf{H}^t \mathbf{H}$. We can easily show that $\mathrm{Var}\,(\mathbf{Y}) = \mathbf{A}^{-1}$. (A more general form can be obtained by setting $\mathbf{U}^t + \boldsymbol{\xi}^t = \mathbf{Z}^t \mathbf{H}^t$, and we have the density of $\mathbf{Z}$ as

$$p_Z(z) = \frac{|\mathbf{A}|^{1/2} \exp(-(\mathbf{z} - \boldsymbol{\xi})^t \mathbf{A} (\mathbf{z} - \boldsymbol{\xi})/2)}{(2\pi)^{p/2}},$$

with $\mathbb{E}\,(\mathbf{Z}) - \boldsymbol{\xi}$, $\mathrm{Var}\,(\mathbf{Z}) = \mathbf{A}^{-1}$. We have $\mathbf{Z} \sim N_p (\boldsymbol{\xi}, \mathbf{A}^{-1})$.) In general, if

$$f(x_1, \ldots, x_p) = C \cdot \exp\left(-\frac{Q_p(x_1, \ldots, x_p)}{2}\right), \qquad (9.3)$$

where $Q_p(x_1, \ldots, x_p)$ is a positive definite quadratic form, then we have a multinormal distribution $N_p(\boldsymbol{\mu}, \boldsymbol{\Sigma})$, where $\boldsymbol{\mu}$ and $\boldsymbol{\Sigma}$ can be determined from Q_p.

Notes:

(1) When $\boldsymbol{\Sigma}$ has rank $r < p$, the vector $\mathbf{X}$ has a singular normal distribution. It is then possible to transform $\mathbf{X}$ into $\mathbf{Y} = \mathbf{AX} + \mathbf{b}$, with $\mathbf{A}$ a $(r \times p)$ matrix of rank r. Then $\mathbf{Y}$ has a non-singular multivariate normal distribution in r-dimensions (Eq. (9.7)).
(2) Pierce and Dykstra (1969) showed that we can have a vector $(X_1, \ldots, X_n)$ s.t.:

   (i) Any proper subset of it are jointly normally distributed and mutually independent.
   (ii) Yet, the whole set together are not normally distributed and are dependent.
   (iii) Although p can take any positive integral value in theory, it is rarely larger than 8 in applications. Some statistical techniques, such as principal component analysis, propose to reduce the number of variables for a more meaningful analysis.

9.6 Properties

Let us consider $\mathbf{X} \sim N_p(\boldsymbol{\mu}, \boldsymbol{\Sigma})$. There are many properties of $\mathbf{X}$, like in the univariate case:

(1) As pointed out in the bivariate case, two jointly dependent normal variables can be transformed linearly into two non-correlated normal variables. This can be viewed as an axis rotation in $\mathbb{R}^2$. Matrices, as linear operations, perform similar transformations in $\mathbb{R}^p$, but it is difficult to visualize it. For higher dimensions, the multivariate density with covariance matrix $\boldsymbol{\Sigma}$ and mean column vector $\boldsymbol{\mu}$, $\mathbf{X} \sim N_p(\boldsymbol{\mu}, \boldsymbol{\Sigma})$, can be transformed into the multivariate normal density $\mathbf{Z}$ with covariance matrix $\mathbf{I}_p$ and mean column vector $\mathbf{0}$, i.e. $\mathbf{Z} \sim N_p(\mathbf{0}, \mathbf{I}_p)$, just as in the univariate case, $Z = \frac{X - \mu}{\sigma}$ for $X \sim N(\mu, \sigma^2)$.

   Here, as seen in the previous section, the transformation $\mathbf{Z} = \boldsymbol{\Sigma}^{-1/2}(\mathbf{X} - \boldsymbol{\mu})$ gives $\mathbf{Z} \sim N_p(\mathbf{0}, \mathbf{I}_p)$. The density of each component is

$$f(z) = \frac{1}{\sqrt{2\pi}} \exp(-z^2/2).$$

This operation is also called *whitening* of the vector $\mathbf{X}$, or of the matrix $\mathbf{\Sigma}$, since $\mathbf{Z}$ is associated with a *white noise* in the signal processing language. $\mathbf{X} \sim \mathbf{Z}^t\mathbf{Z}$ is hence a sum of p elementary independent Mahalanobis square distances $\Delta_z^2 \sim \chi_1^2$.

(2) A *whitening transformation* is a linear transformation of a vector of rv's, with a known covariance matrix, into a set of new variables whose covariance matrix is $\mathbf{I}_p$, the identity matrix. They are uncorrelated and all have variance 1.

(3) Several other transformations are closely related to whitening: (i) the decorrelation removes only the correlations, but leaves variances intact, (ii) the standardization transforms variances to 1, but leaves correlations intact, and (iii) coloring transform changes white vector into random vector with a specified covariance matrix. More precisely, we have the following:

(a) **Decorrelation:** Suppose that the mean of $\mathbf{X}$ is null, i.e. $\mathbf{X} \sim N_p(\mathbf{0}, \mathbf{\Sigma})$. Hence, $\mathbf{\Sigma}$ can be decomposed as $\mathbf{\Sigma} = \mathbf{\Phi}\mathbf{\Lambda}\mathbf{\Phi}^{-1}$ with $\mathbf{\Lambda}$ being the diagonal matrix of eigenvalues of $\mathbf{\Sigma}$ and $\mathbf{\Phi}$ is the matrix of eigenvectors and is orthonormal. We write $\mathbf{\Sigma} = \mathbf{\Phi}\mathbf{\Lambda}^{1/2}\mathbf{\Lambda}^{1/2}\mathbf{\Phi}^{-1}$. Now, set $\mathbf{Y} = \mathbf{\Phi}^t\mathbf{X}$. Then $\mathbf{Y}$ is decorrelated with variances λ_i on the diagonal of its covariance matrix. Finally, take $\mathbf{W} = \mathbf{\Lambda}^{-1/2}\mathbf{Y} = \mathbf{\Lambda}^{-1/2}\mathbf{\Phi}^t\mathbf{X}$. Then $\mathbf{W} \sim N_p(\mathbf{0}, \mathbf{I}_p)$.

(b) **Coloring:** This is the inverse of whitening and steps are taken in reverse order. Let $\mathbf{Z} \sim N_p(\mathbf{0}, \mathbf{I})$. Then $\mathbf{Y} = \mathbf{\Sigma}^{\frac{1}{2}}\mathbf{Z} + \boldsymbol{\mu} \sim N_p(\boldsymbol{\mu}, \mathbf{\Sigma})$. Conversely, $\mathbf{Y} \sim N_p(\boldsymbol{\mu}, \mathbf{\Sigma})$. Then $\mathbf{Z} = \mathbf{\Sigma}^{-\frac{1}{2}}(\mathbf{Y} - \boldsymbol{\mu}) \sim N_p(\mathbf{0}, \mathbf{I})$.

9.6.1   *Characteristic function*

For every real vector $\mathbf{t} \in \mathbb{R}^p$, we define the chf of a random vector $\mathbf{X}$ as $\varphi_X(\mathbf{u}) = \mathbb{E}\left(e^{i\mathbf{u}^t\mathbf{X}}\right)$. Hence, for $\mathbf{X} \sim N_p(\boldsymbol{\mu}, \mathbf{\Sigma})$, we have

$$\varphi_X(\mathbf{u}) = \mathbb{E}\left(e^{i\mathbf{u}^t\mathbf{X}}\right) = e^{i\mathbf{u}^t\boldsymbol{\mu}} \cdot e^{-\frac{\mathbf{u}^t\mathbf{\Sigma}\mathbf{u}}{2}}. \tag{9.4}$$

Conversely, the density of $\mathbf{X}$ can be obtained from the chf by integration, under some conditions,

$$f(\mathbf{x}) = \frac{1}{(2\pi)^p} \int\limits_{-\infty}^{\infty} \cdots \int\limits_{-\infty}^{\infty} e^{-iu^t\mathbf{x}} \varphi_X(u) du_1 \ldots du_p.$$

This shows that the characteristic function determines the density uniquely. Hence, the chf provides a powerful tool for the study of any density, and the exponential form of the chf in the case of the normal gives other advantages to this distribution.

The moment-generating function is $M_X(\mathbf{t}) = \mathbb{E}\left(e^{\mathbf{t}^t\mathbf{X}}\right)$.

Moments: We can also compute the moments of each component by partial differentiation or mixed moments of several components by mixed differentiation. For example, $\mathbb{E}(X_j) = \frac{1}{i}\frac{\partial\varphi}{\partial u_j}\Big|_{t=0} = \mu_j$ and

$$\mathbb{E}(X_h X_j) = \frac{1}{i^2}\frac{\partial^2\varphi}{\partial u_h \partial u_j}\Big|_{t=0} = \sigma_{hj} + \mu_h\mu_j.$$

9.6.2 *Contours of the density surface*

This topic has been treated in Chapter 7 and the related Computational Appendix. Some complementary information on cases $p \geq 3$ are given here.

Curves obtained as intersections of the density surface with a horizontal plan at height denoted by h are called contours at height $h > 0$. They are given by $f(\mathbf{x}) = h > 0$, which gives the truncation at $k^2 > 0$ of the related square Mahalanobis distance, i.e.

$$(\mathbf{x} - \boldsymbol{\mu})^t \boldsymbol{\Sigma}^{-1} (\mathbf{x} - \boldsymbol{\mu}) = k^2,$$

where k obtained from h is

$$k^2 = 2\log\left(\frac{1}{h(2\pi)^{p/2}|\boldsymbol{\Sigma}|}\right)$$

and, conversely,

$$h = \frac{1}{(2\pi)^{p/2}|\boldsymbol{\Sigma}|^{1/2}} \exp\left(-\frac{1}{2}k^2\right).$$

We can see that when $h \uparrow$ (intersection at a higher altitude), then $k \downarrow$ (less probability inside the contour). See Chapter 7 and its Computational Appendix.

9.7   Operations on the Distribution of $X \sim N_p(\mu, \Sigma)$

As we have noted, the exponential function lies at the heart of the multivariate normal density, which benefits from many of its properties. One such property is $\exp(c) = \exp(a) \exp(b)$ if $a + b = c$. In univariate distribution, it gives to the normal the property of infinite divisibility. Moreover, the chf and mgf of the multinormal are the only ones, among all chf's of distributions, that have expressions directly in terms of the exponential function alone. We can see that we can conveniently deal with sub-block matrices.

9.7.1   *Linear combination*

The linear combinations are on components of a normal vector or on several independent normal vectors with same, or different, covariance matrices. Any linear combination of *components* of $\mathbf{X}$ is univariate normal $N_1(\mu^*, \Sigma^*)$, where μ^*, Σ^* are corresponding linear combinations of means and covariances.

(a) More precisely, if $X^* = \mathbf{A}^t \mathbf{X}$, then $X^* \sim N_1(\mathbf{A}^t \mu, \mathbf{A}^t \Sigma \mathbf{A})$, where $\mathbf{A} = (a_1, \ldots, a_p)^t$ is a vector of constants.

Conversely, we have $\mathbf{X} \sim N_p(\mu, \Sigma)$ if, for any vector of constants $\mathbf{A} = (a_1, \ldots, a_p)^t$, we have $X^* = \mathbf{A}^t \mathbf{X} \sim N_1(\mu^*, \Sigma^*)$. This property has been taken as a definition of a multivariate normal vector.

(b) Any linear combination of independent normal vectors (of the same size) is normal, with mean and matrix of variances obtained from these normal vectors.

Recall that for two independent $X_i \sim N_1(\mu_i, \sigma_i^2), i = 1, 2$, we have $X_1 \pm X_2 \sim N_1(\mu_1 \pm \mu_2, \sigma_1^2 + \sigma_2^2)$. A similar relation holds for the multivariate normal. More precisely, we have the following.

Proposition 9.1.

(a) *Let* $\mathbf{X}_i \sim N_p(\boldsymbol{\mu}_i, \boldsymbol{\Sigma})$, $i = 1, \ldots, n$ *be independent normal vectors with same covariance matrix. Then*

$$\mathbf{W}_1 = \sum_{j=1}^{n} a_j \mathbf{X}_j \sim N_p \left(\sum_{j=1}^{n} a_j \boldsymbol{\mu}_j, \left(\sum_{j=1}^{n} a_j^2 \right) \boldsymbol{\Sigma} \right). \tag{9.5}$$

Similarly, let $\mathbf{W}_2 = \sum_{j=1}^{n} b_j \mathbf{X}_j \sim N_p(\sum_{j=1}^{n} b_j \boldsymbol{\mu}_j, (\sum_{j=1}^{n} b_j^2)\boldsymbol{\Sigma})$. *Then*

$$\mathbf{W}_3 = \begin{bmatrix} \mathbf{W}_1 \\ \mathbf{W}_2 \end{bmatrix} \sim N_{2p} \left(\begin{pmatrix} \boldsymbol{\mu}_a \\ \boldsymbol{\mu}_b \end{pmatrix}, \begin{bmatrix} \left(\sum_{j=1}^{n} a_j^2 \right) \boldsymbol{\Sigma} & \mathbf{b}^t \mathbf{a} \boldsymbol{\Sigma} \\ \mathbf{b}^t \mathbf{a} \boldsymbol{\Sigma} & \left(\sum_{j=1}^{n} b_j^2 \right) \boldsymbol{\Sigma} \end{bmatrix} \right). \tag{9.6}$$

     Hence, $\mathbf{W}_1$ *and* $\mathbf{W}_2$ *are independent if* $\sum_{i=1}^{n} a_i b_i = 0$.

(b) *More generally, let* $\mathbf{X}_i \sim N_p(\boldsymbol{\mu}_i, \boldsymbol{\Sigma}_i)$, $i = 1, \ldots, n$, *be independent, and* $\{a_i\}$, $i = 1, \ldots, n$, *be constants. Then*

$$\sum_{j=1}^{n} a_j \mathbf{X}_j \sim N_p \left(\sum_{j=1}^{n} a_j \boldsymbol{\mu}_j, \sum_{j=1}^{n} a_j^2 \boldsymbol{\Sigma}_j \right).$$

Figure 9.2 shows the density of the sum of two independent normal vectors,

$$\mathbf{X}_1 : \boldsymbol{\mu}_1 = \begin{pmatrix} 1.5 \\ 1 \end{pmatrix}; \quad \boldsymbol{\Sigma}_1 = \begin{pmatrix} 1.8 & 0.2147 \\ 0.2147 & 1 \end{pmatrix},$$

$$\mathbf{X}_2 : \boldsymbol{\mu}_2 = \begin{pmatrix} 1.5 \\ 3 \end{pmatrix}; \quad \boldsymbol{\Sigma}_2 = \begin{pmatrix} 2.8 & -1.2147 \\ -1.2147 & 1.6 \end{pmatrix},$$

$$\mathbf{X}_3 = \mathbf{X}_1 + \mathbf{X}_2 \sim N_2(\boldsymbol{\mu}_1 + \boldsymbol{\mu}_2; \boldsymbol{\Sigma}_1 + \boldsymbol{\Sigma}_2)$$

$$= N_2 \left(\begin{bmatrix} 3.0 \\ 4.0 \end{bmatrix}; \begin{bmatrix} 4.6 & -1.0 \\ -1.0 & 2,6 \end{bmatrix} \right)$$

(see this graph in the animated section of our web page).

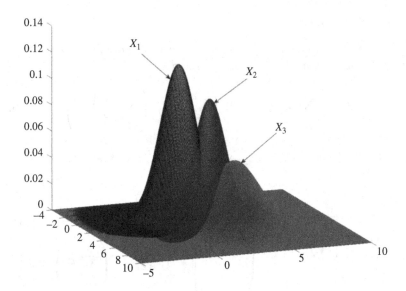

Fig. 9.2   Sum of two independent bivariate normal distributions.

9.7.2   *Matrix transformations*

Let $\mathbf{A}\,(q \times p)$ be a matrix of scalars, representing a linear transformation on the normal vector $\mathbf{X}$. Then we have the following:

(a)

$$\mathbf{A}\,(q \times p), \mathbf{X} \sim N_p\left(\boldsymbol{\mu}, \boldsymbol{\Sigma}\right) \Rightarrow \mathbf{AX} \sim N_q\left(\mathbf{A}\boldsymbol{\mu}, \mathbf{A}\boldsymbol{\Sigma}\mathbf{A}^t\right). \quad (9.7)$$

Equation (9.7) shows that any subset of q out of p of these rv's, with $q < p$, will be jointly Gaussian.

(b) **Block partition of a normal vector:** One of the most important properties of the multivariate normal $\mathbf{X} \sim N_p(\boldsymbol{\mu}, \boldsymbol{\Sigma})$ is, however, the sub-division into sub-groups of vectors that also have normal distributions, with derived parameters.

Let

$$\mathbf{X} = \begin{pmatrix} \mathbf{Y} \\ \mathbf{Z} \end{pmatrix}, \boldsymbol{\mu} = \begin{pmatrix} \boldsymbol{\mu}_{\mathbf{Y}} \\ \boldsymbol{\mu}_{\mathbf{Z}} \end{pmatrix}, \boldsymbol{\Sigma} = \begin{bmatrix} \boldsymbol{\Sigma}_{\mathbf{YY}} & \boldsymbol{\Sigma}_{\mathbf{YZ}} \\ \boldsymbol{\Sigma}_{\mathbf{ZY}} & \boldsymbol{\Sigma}_{\mathbf{ZZ}} \end{bmatrix}, \quad (9.8)$$

with appropriate dimensions $p = q + r$. Then for $\boldsymbol{\Sigma}_{\mathbf{YY}}$ and $\boldsymbol{\Sigma}_{\mathbf{ZZ}}$ positive definite, we have

$$\mathbf{Y} \sim N_q\left(\boldsymbol{\mu}_{\mathbf{Y}}, \boldsymbol{\Sigma}_{\mathbf{YY}}\right), \quad \mathbf{Z} \sim N_r\left(\boldsymbol{\mu}_{\mathbf{Z}}, \boldsymbol{\Sigma}_{\mathbf{ZZ}}\right).$$

If $\Sigma_{YZ} = \Sigma_{ZY} = 0$, Y and Z are independent and $|\Sigma_X| = |\Sigma_{YY}| \cdot |\Sigma_{ZZ}|$.

An immediate consequence is as follows.

Corollary 9.1. *Each component of* $X \sim N_p(\mu, \Sigma)$ *is univariate normal, i.e.* $X_i \sim N_1(\mu_i, \sigma_i^2)$, $1 \leq i \leq p$, *with* $\sigma_i^2 = \Sigma_{ii}$.

Other special properties of the multivariate normal are that the conditional distributions are also normal.

9.7.3 *Conditionally normal distributions*

In the above division into blocks, $X = \begin{bmatrix} Y \\ Z \end{bmatrix}$ the conditional distribution of Y, given Z (i.e. $Z = z_0$) is also normal. We have

$$[Y|Z = z_0] \sim N_p\left(\mu_Y + \Sigma_{YZ}\Sigma_{ZZ}^{-1}(z_0 - \mu_Z), \Sigma_{YY \cdot Z}\right), \qquad (9.9)$$

with

$$\Sigma_{YY \cdot Z} = \Sigma_{YY} - \Sigma_{YZ}\Sigma_{ZZ}^{-1}\Sigma_{ZY}. \qquad (9.10)$$

Hence, the conditional mean is a linear function of Z while the conditional variance matrix does not depend on Z. Here, we can find the earlier result in the bivariate case. The matrix $\Sigma_{YZ}\Sigma_{ZZ}^{-1}$ is the matrix of regression coefficients of Y on Z.

9.7.4 *Quadratic forms in normal variables*

Quadratic forms play a very important role in the study of the normal distribution because in the expression of the density of $X \sim N_p(\mu, \Sigma)$, $Q(x) = (x - \mu)^t \Sigma^{-1}(x - \mu)$ is a quadratic form in scalar x. Then, if we consider the random vector $X \sim N_p(\mu, \Sigma)$, $\Delta_X^2 = (X - \mu)^t \Sigma^{-1}(X - \mu)$ is a χ_p^2 central random variable, whereas $\Delta_{X*}^2 = (X)^t \Sigma^{-1}(X)$ is a non-central $\chi_p^2(\delta)$, with non-centrality parameter: $\delta = \mu^t \Sigma^{-1} \mu$.

9.8 Computation of the cdf of the Multinormal

We consider the following facts:

(a) In one dimension, we have the normal density, and the cdf, under their reduced forms:

$$\phi(x) = \frac{1}{\sqrt{2\pi}} \exp\left(-\frac{x^2}{2}\right), \quad -\infty < x < \infty,$$

$$\Phi(x) = \frac{1}{\sqrt{2\pi}} \int_{-\infty}^{x} \exp\left(-\frac{t^2}{2}\right) dt, \quad -\infty < x < \infty.$$

Both have to be computed numerically.

(b) In two dimensions, using the standard form again and the cdf as the double integral of that density, the formula given in the literature is, depending on the sign of ρ in $(-1, 1)$ (see Tong (1989, p. 14)),

$$F(x_1, x_2) = \int_{-\infty}^{\infty} \Phi\left(\frac{\sqrt{|\rho|}z + a_1}{\sqrt{1 - |\rho|}}\right) \Phi\left(\frac{\delta_\rho \sqrt{|\rho|}z + a_2}{\sqrt{1 - |\rho|}}\right) \phi(z)\, dz,$$

$$(9.11)$$

$\rho \in (-1, 1)$, $\delta_\rho = 1$, if $\rho \geq 0$, and $\delta_\rho = -1$ if $\rho < 0$. Here, $a_i = \frac{x - \mu_i}{\sigma_i}$, $i = 1, 2$, and Φ is the cdf of the univariate standard normal. But this single integral has to be computed by quadrature.

(c) In $\mathbb{R}^p$, let $A = \{x \in \mathbb{R}^p, b_i \leq x_i \leq a_i, i = 1, \dots, p\}$. As an extension of Eq. (9.11), we have

$$\mathbb{P}(\mathbf{X} \in A) = \int_{-\infty}^{\infty} \prod_{i=1}^{p} \left[\Phi\left(\frac{\sqrt{\rho}z + \frac{(a_i - \mu_i)}{\sigma_i}}{\sqrt{1 - \rho}}\right) \right.$$

$$\left. - \Phi\left(\frac{\sqrt{\rho}z + \frac{(b_i - \mu_i)}{\sigma_i}}{\sqrt{1 - \rho}}\right) \right] \phi(z)\, dz.$$

There is not yet a method, or software, that would perform satisfactorily in all cases.

(d) Methods presented in the statistical literature depend on the domain of integration, the approximation methods, and the errors acceptable. For $p \geq 4$, integration of the normal integral can be difficult, partly due to the "curse of dimensionality". Genz and Bretz (2009) have realized a survey of several methods to effectively compute the integral based on different hypotheses: orthant probability, limited domain, conditions on correlations, reduction formulas, Monte Carlo simulation, quasi-Monte Carlo simulation, approximations to the problem and to the integral, etc. Curnow and Dunnett (1962) discussed several approaches used earlier.

(e) Most works have been carried out using the standard multivariate normal, with density

$$f(\mathbf{x}) = \frac{|\mathbf{R}|^{-1/2}}{(2\pi)^{p/2}} \exp\left(-\frac{\mathbf{x}^t \mathbf{R}^{-1} \mathbf{x}}{2}\right),$$

where $\mathbf{R}$ is the correlation matrix. We define

$$\Phi(h_1, \ldots, h_p; \mathbf{R}) = \mathbb{P}\left(\bigcap_{j=1}^{p}(X_j \leq h_j)\right)$$

$$= \frac{|\mathbf{R}|^{-1/2}}{(2\pi)^{p/2}} \int_{-\infty}^{h_p} \cdots \int_{-\infty}^{h_1} \exp\left(-\frac{\mathbf{x}^t \mathbf{R}^{-1} \mathbf{x}}{2}\right)$$

$$\times \, dx_1 \ldots dx_p. \quad (9.12)$$

9.8.1 *Some specific approaches*

Let $f(\mathbf{x}, \boldsymbol{\mu}, \boldsymbol{\Sigma})$ be the density of $\mathbf{X}$ and Ω be a subset of $\mathbb{R}^p$. We wish to compute $I(f, \Omega) = \mathbb{P}(\mathbf{X} \in \Omega) = \int_{\Omega} f(\mathbf{x}; \boldsymbol{\mu}, \boldsymbol{\Sigma}) \, d\mathbf{x}$. Depending on the complexity of this integral, it can be evaluated accurately, or approximatively, by different methods. A short list follows.

(a) **Simulation:** For highly complex Ω, a simulation approach is best suited. The proportion of points inside Ω wrt the total number of points generated would give an approximate value of I. Care is to be taken so that all parts of Ω could receive simulation points.

(b) **Reduction in dimensions method:** Plackett (1954) pioneered this approach by proving that

$$\frac{\partial}{\partial \rho_{ij}} f(\mathbf{x}; \boldsymbol{\mu}, \boldsymbol{\Sigma}) = \frac{\partial^2}{\partial x_i \partial x_j} f(\mathbf{x}; \boldsymbol{\mu}, \boldsymbol{\Sigma}),$$

which is an extension of his equality (7.38) (in Section 7.18). Going to the cdf, we have

$$F(a_1, \ldots, a_p; \boldsymbol{\mu}, \boldsymbol{\Sigma}) = \int_{-\infty}^{a_1} \cdots \int_{-\infty}^{a_p} f(\mathbf{x}; \boldsymbol{\mu}, \boldsymbol{\Sigma}) \, dx_1 \ldots dx_p.$$

Interchanging differentiation and integration, we now have two integrals, one on the marginal f_1, and one on the conditional $f_{2|1}$, so that

$$\frac{\partial}{\partial \rho_{ij}} F(\mathbf{a}; \mathbf{0}, \boldsymbol{\Sigma}) = f_1(a_1, a_2, \mathbf{0}, \boldsymbol{\Sigma}_{11})$$

$$\times \int_{-\infty}^{a_3} \cdots \int_{-\infty}^{a_p} f_{2|1}(x_2; \boldsymbol{\mu}_{2|1}, \boldsymbol{\Sigma}_{22 \cdot 1}) \, dx_3 \ldots dx_p.$$

Hence, the probability integral $F(\mathbf{a}; \mathbf{0}, \boldsymbol{\Sigma})$ is expressed as a single integral on ρ_{ij} and another integral on the remaining variables.

(c) **Orthant probability:** In general, orthant probability means $\mathbb{P}\left(\bigcap_{i=1}^{p}(X_i > 0)\right)$ for $\mathbf{X} \sim N_p(\boldsymbol{\mu}, \boldsymbol{\Sigma})$. It is often, but not always, sufficient to consider the standard form of the multinormal, i.e. to suppose that all means zero and all variances one, and the distribution depends only on the matrix of correlations $\mathbf{R}$. Setting $\Omega = \mathbb{P}\left(\bigcap_{i=1}^{p}(X_i \le a_i)\right)$, for scalar $\{a_i\}$, $i = 1, \ldots, p$, we can write

$$\Omega = \mathbb{P}\left(\bigcap_{i=1}^{p} \{X_i \le a_i\}\right) = \mathbb{P}_R\left(\bigcap_{i=1}^{p} \{Z_i \le (a_i - \mu_i)/\sigma_i\}\right)$$

and bring the value of Ω to the standard normal. Similar considerations apply to $\Omega^* = \mathbb{P}\left(\bigcap_{i=1}^{p} \{X_i \ge b_i\}\right)$.

Note: Several approaches have been used by different authors to compute orthant probabilities. For $\{a_i = 0\}$, $i = 1, \ldots, p$, writing $I_p = \mathbb{P}_R\left(\bigcap_{i=1}^{p} \{X_i \le 0\}\right)$, we have closed-form formulas:

$$I_2 = \tfrac{1}{4} + \tfrac{\arcsin \rho}{2\pi} \text{ (see Sheppard (1898))},$$

$$I_3 = \tfrac{1}{8} + \tfrac{\arcsin \rho_{12} + \arcsin \rho_{13} + \arcsin \rho_{23}}{4\pi}$$

(see David (1953), using inclusion–exclusion relation).

For $p \geq 4$, some conditions on $\{\rho_{ij}\}$ are required to have a formula for I_p, but for $\rho_{ij} = 1/2$, $\forall i, j$, we have $I_p = \frac{1}{p+1}$ for any p.

(d) **Condition imposed on the correlation coefficients:** For $\mathbf{X} \sim N_p(\boldsymbol{\mu}, \boldsymbol{\Sigma})$, under the condition that the correlation coefficients are such that

$$\rho_{ij} = \lambda_i \lambda_j, \quad \lambda_i \in [-1, 1], \quad 1 \leq i \neq j \leq p,$$

for $\Omega = \{\mathbf{x} \in \mathbb{R}^p : b_i \leq x_i \leq a_i, 1 \leq i \leq p\}$, we have

$$\mathbb{P}(\mathbf{X} \in \Omega) = \int_{-\infty}^{\infty} \prod_{i=1}^{p} \left[\Phi \left(\frac{\frac{a_i - \mu_i}{\sigma_i} + \lambda_i z}{\sqrt{1 - \lambda_i^2}} \right) \right.$$

$$\left. - \Phi \left(\frac{\frac{b_i - \mu_i}{\sigma_i} + \lambda_i z}{\sqrt{1 - \lambda_i^2}} \right) \right] \phi(z) \, dz. \quad (9.13)$$

The last integral is a single integral, hence simpler to deal with, but it still needs final normal quadrature.

(e) **Domain of integration: Product of intervals**

- For the bivariate distribution, we have the corresponding expression applicable to any bivariate distribution:

$$\mathbb{P}(a < X \leq b, c < Y \leq d) = F_{X,Y}(b, d) - F_{X,Y}(a, d)$$

$$- F_{X,Y}(b, c) + F_{X,Y}(a, c).$$

Hence, we can determine the probability in any rectangle of the plane.

- For the trivariate case,

$$\mathbb{P}(x_{11} < X_1 \leq x_{12}; x_{21} < X_2 \leq x_{22}; x_{31} < X_3 \leq x_{32})$$

$$= F(x_{12}, x_{22}, x_{32}) - F(x_{11}, x_{21}, x_{32}) - F(x_{11}, x_{22}, x_{31})$$

$$- F(x_{12}, x_{21}, x_{31}) - F(x_{12}, x_{22}, x_{31}) + F(x_{12}, x_{21}, x_{32})$$

$$+ F(x_{11}, x_{22}, x_{32}) - F(x_{11}, x_{21}, x_{31}).$$

- For the p-variate distribution,

$$F\left(x_1, \ldots, x_p\right) = \int_{-\infty}^{x_p} \int_{-\infty}^{x_{p-1}} \cdots \int_{-\infty}^{x_1} f\left(y_1, \ldots, y_p\right) dy_1 \ldots dy_p$$

$$\text{(9.14)}$$

and the above relation becomes much more complicated: Following Thomasian (1969), we have the probability in any p-sided rectangular block:

$$\mathbb{P}\left(x_1^{(1)} < X_1 \le x_1^{(0)}, \ldots, x_p^{(1)} < X_p \le x_p^{(0)}\right) = \Delta F(\mathbf{x}^{(1)}, \mathbf{x}^{(0)}),$$

where

$$\Delta F\left(\mathbf{x}^{(1)}, \mathbf{x}^{(0)}\right) = \sum_{\varepsilon_1=0}^{1} \sum_{\varepsilon_2=0}^{1} \cdots \sum_{\varepsilon_p=0}^{1} (-1)^{\sum_{j=1}^{p} \varepsilon_j}$$

$$\times F\left(x_1^{(\varepsilon_1)}, x_2^{(\varepsilon_2)}, \ldots, x_p^{(\varepsilon_p)}\right). \quad \text{(9.15)}$$

Hence, $\Delta F\left(\mathbf{x}^{(1)}, \mathbf{x}^{(0)}\right)$ equals the sum of 2^p terms

$$\pm F\left(x_1^{(\varepsilon_1)}, x_2^{(\varepsilon_2)}, \ldots, x_p^{(\varepsilon_p)}\right),$$

with the sign being positive or negative, depending on whether an even or odd number of superscripts are (1).

9.9   Tables

(a) It is obvious that published tables are very voluminous and cumbersome. For practical purposes, we better use a software instead of tables. MATLAB, for example, has the routine **mvncdf** that computes the cdf of a multinormal distribution, which is quite easy to use.

(b) Tables for the binormal issued by the National Bureau of Standards can be used manually, but several relations need to be observed (see Section 7.16).

(c) For relatively simple tables, for limited use, one can consult those given by Tong (1989), which can be explained as follows.

The tables use quadrature formulas by Stroud and Secrest (1966) to compute the integrals

$$F_{\mathbf{X}}(\mathbf{a}) = \mathbb{P}\left(\bigcap_{i=1}^{p}\{X_i \leq a_i\}\right) = \int_{-\infty}^{\infty}\prod_{j=1}^{p}\Phi\left(\frac{\frac{a_i}{\sigma_i}+z\sqrt{\rho}}{\sqrt{1-\rho}}\right)\phi(z)\,dz$$

and

$$F_{|\mathbf{X}|}(\mathbf{a}) = \mathbb{P}\left(\bigcap_{i=1}^{p}\{|X_i| \leq a_i\}\right)$$

$$= \int_{-\infty}^{\infty}\prod_{j=1}^{p}\left(\Phi\left(\frac{\frac{a_i}{\sigma_i}+z\sqrt{\rho}}{\sqrt{1-\rho}}\right) - \Phi\left(\frac{\frac{-a_i}{\sigma_i}+z\sqrt{\rho}}{\sqrt{1-\rho}}\right)\right)\phi(z)\,dz.$$

The special case of $\sigma_1 = \cdots = \sigma_p = 1$ and $a_1 = \cdots = a_p = a$ is considered. There are four sections of the tables:

- **Section A: One-sided percentage points**
  A value c is provided, so that $\mathbb{P}\left(\bigcap_{i=1}^{p}\{X_i \leq c\}\right) = \gamma$ for given values of ρ, p and γ for $\mathbf{X} \sim N_p(\mathbf{0}, \mathbf{\Sigma})$.
- **Section B: Two-sided percentage points**
  A value c is provided, so that $\mathbb{P}\left(\bigcap_{i=1}^{p}\{|X_i| \leq c\}\right) = \gamma$ for given values of ρ, p and γ for $\mathbf{X} \sim N_p(\mathbf{0}, \mathbf{\Sigma})$.
- **Section C: One-sided probability integrals**
  The value of $\mathbb{P}\left(\bigcap_{i=1}^{p}\{X_i \leq b\}\right)$ is given for various values of b, ρ, p for $\mathbf{X} \sim N_p(\mathbf{0}, \mathbf{\Sigma})$.
- **Section D: Two-sided probability integrals**
  The value of $\mathbb{P}\left(\bigcap_{i=1}^{p}\{|X_i| \leq b\}\right)$ is given for various values of b, ρ, p for $\mathbf{X} \sim N_p(\mathbf{0}, \mathbf{\Sigma})$.

Although the tables seem to cover only a few cases of the whole vast topic of cdf, the author gives several important applications, such as confidence probabilities and sample size determination, test of hypotheses for means, and upper bounds for correlation coefficients. They form the Appendix of Tong (1989, pp. 229–259).

9.10   Generation of Observations from the Multinormal

The following result relates the multinormal to the χ_p^2, and the uniform on $[0, 2\pi]$, for simulation and cdf computation purposes (see Tong (1989)).

$\mathbf{X} \sim N_p(\boldsymbol{\mu}, \boldsymbol{\Sigma})$, with $\boldsymbol{\Sigma} = \mathbf{CC}^t$ iff $\mathbf{X}$ is the random vector corresponding to the following experiment:

(a) Let $V = \mathbb{R}^2$ with a χ_p^2 distribution and an observed value of it.
(b) Given $\mathbf{V} = v$, choose a point at random on the p-dim sphere with radius $r = \sqrt{v}$ and identify this point as $\mathbf{Z} = (Z_1, \ldots, Z_p)$.
(c) Obtain $\mathbf{X} = \boldsymbol{\mu} + \mathbf{CZ}$ as an observation from $N_p(\boldsymbol{\mu}, \boldsymbol{\Sigma})$.

See also Section 7.32.3.

9.11   Checking on Multinormality

There are a few methods suggested in the literature. However, none of these has been recognized as efficient and versatile. We mention *Mardia's method* (see Mardia (1970)) based on multivariate measures of skewness and kurtosis and *Cox and Small's approach* (see Cox and Small (1978)), which are coordinate-dependent.

The practice now is to test normality for marginal distributions, which is a necessary but not sufficient condition for multivariate normality. Bivariate and trivariate combinations of components can be tested too and any failure makes multivariate normality suspect. Gnanadesikan (1976) discussed several interesting approaches.

Transformations to reach normality: For multivariate data, the univariate procedure can be applied to each dimension, with a separate λ. We have the generalization of Box–Cox as follows:

- Let $\lambda = (\lambda_1, \ldots, \lambda_p)$ and the corresponding $x_i^{(\lambda)} = \left(x_{i1}^{(\lambda_1)}, \ldots, x_{ip}^{(\lambda_p)}\right)^t$. Let

$$\widehat{\boldsymbol{\Sigma}} = \sum_{i=1}^{n} \left(\mathbf{x}_i^{(\lambda)} - \overline{\mathbf{x}_i^{(\lambda)}}\right) \left(\mathbf{x}_i^{(\lambda)} - \overline{\mathbf{x}_i^{(\lambda)}}\right)^t \Big/ n. \qquad (9.16)$$

- We choose $\lambda = \widehat{\lambda}$ which maximizes

$$L = -\frac{n}{2} \ln\left(\left|\widehat{\boldsymbol{\Sigma}}\right|\right) + \sum_{j=1}^{p} (\lambda_j - 1) \sum_{i=1}^{n} \ln(x_{ij}).$$

For more details, we recommend Rencher and Christensen (2012).

In the following sections, we present the results associated with multivariate statistics, but concentrate on the confidence regions.

9.12 Multivariate Normal Sampling

In this section, we discuss several probability/statistics problems related to the mean values of normal random vectors. We will not go into the details concerning the statistical approaches to be adopted since they will take much more space.

We recall that in the univariate case we distinguish between two cases, σ^2 known, where we use Z and the normal distribution, and σ^2 unknown, where we use t and the Student t-distribution. We have similar considerations for the multivariate case.

We consider the vector $\mathbf{X} = \begin{bmatrix} X_1 \\ \cdots \\ X_p \end{bmatrix}$, with $\mathbf{X} \sim N_p(\boldsymbol{\mu}, \boldsymbol{\Sigma})$ and a random sample of size n, $\{\mathbf{X}_1, \ldots, \mathbf{X}_n\}$, with $n - 1 > p$. Then the maximum likelihood estimates of $\boldsymbol{\mu}$ and $\boldsymbol{\Sigma}$ are $\overline{\mathbf{X}} = \sum_{i=1}^n \mathbf{X}_i/n$ and $\widehat{\boldsymbol{\Sigma}} = \sum_{i=1}^n (\mathbf{X_i} - \overline{\mathbf{X}})^t (\mathbf{X_i} - \overline{\mathbf{X}})/n$, with the unbiased estimator being $\mathbf{S} = \frac{n}{n-1}\widehat{\boldsymbol{\Sigma}}$. Moreover, we have $\overline{\mathbf{X}}$ and $\mathbf{S}$ independent, with $\overline{\mathbf{X}} \sim N_p(\boldsymbol{\mu}, \boldsymbol{\Sigma}/n)$ and $\mathbf{S} \sim W(n-1, \boldsymbol{\Sigma}/(n-1))$, the Wishart distribution which is a generalization to matrices of the chi-square (see Chapter 12).

9.13 Test on the Population Mean Vector

For the multivariate case, to test $H_0 : \boldsymbol{\mu} = \boldsymbol{\mu}_0$, we distinguish two cases.

Case a: $\boldsymbol{\Sigma}$ is known $(= \boldsymbol{\Sigma}_0)$

We know that $\sqrt{n}(\overline{\mathbf{X}} - \boldsymbol{\mu}) \sim N_p(\mathbf{0}, \boldsymbol{\Sigma}_0)$, hence $n(\overline{\mathbf{X}} - \boldsymbol{\mu})^t \boldsymbol{\Sigma}_0^{-1}(\overline{\mathbf{X}} - \boldsymbol{\mu}) \sim \chi_p^2$.

To test $H_0 : \boldsymbol{\mu} = \boldsymbol{\mu}_0$, we use the critical region

$$n(\overline{\mathbf{X}} - \boldsymbol{\mu}_0)^t \boldsymbol{\Sigma}_0^{-1}(\overline{\mathbf{X}} - \boldsymbol{\mu}_0) \geq \chi_{p,1-\alpha}^2,$$

with $\chi_{p,1-\alpha}^2$ being the upper α-percentile (right-hand side test). The $(100)(1-\alpha)\%$ confidence region for $\boldsymbol{\mu}$ is

$$\left\{ \boldsymbol{\mu} : n(\boldsymbol{\mu} - \overline{\mathbf{X}})^t \boldsymbol{\Sigma}_0^{-1}(\boldsymbol{\mu} - \overline{\mathbf{X}}) \leq \chi_{p,1-\alpha}^2 \right\}.$$

This is the surface and interior of an ellipsoid centered at $\overline{\mathbf{X}}$.

Case b: Σ is unknown

In one dimension statistics, we test $H_0 : \mu = \mu_0$ vs $H_a : \mu \neq \mu_0$ using a small sample of size n from a normal population $N\left(\mu, \sigma^2\right)$. Supposing both parameters unknown, we use

$$t = \frac{\left(\overline{X} - \mu_0\right)}{s/\sqrt{n}},$$

where

$$\overline{X} = \frac{1}{n}\sum_{j=1}^{n} X_j, \quad S^2 = \frac{1}{n-1}\sum_{j=1}^{n}\left(X_j - \overline{X}\right)^2,$$

and t is the Student statistic t with $(n-1)$ dof.

Using observed values of $\{X_j\}$, we reject H_0 if the value of $|t|$ is too extreme, i.e. exceeds a specified percentile of the t-distribution, determined by α, the degree of signification adopted.

The same statistic can be used to derive a $(1-\alpha)\,100\%$ confidence interval for the value of μ:

$$\left\{\overline{X} - t_{(n-1),1-(\alpha/2)}\frac{S}{\sqrt{n}}; \overline{X} + t_{(n-1),1-(\alpha/2)}\frac{S}{\sqrt{n}}\right\}.$$

If we use the square of this statistic, i.e. $\left(t_{n-1}\right)^2$, we can carry out the same operations if we make the proper adjustments.

Hence, Student's t- statistic becomes Hotelling's T^2, for the multivariate normal case, defined by

$$T^2 = \left(\overline{\mathbf{X}} - \boldsymbol{\mu}_0\right)^t \left(\frac{1}{n}\mathbf{S}\right)^{-1}\left(\overline{\mathbf{X}} - \boldsymbol{\mu}_0\right) = n\left(\overline{\mathbf{X}} - \boldsymbol{\mu}_0\right)^t \mathbf{S}^{-1}\left(\overline{\mathbf{X}} - \boldsymbol{\mu}_0\right),$$

$$(9.17)$$

with $\boldsymbol{\mu}_0 = \begin{bmatrix} \mu_{01} \\ \cdots \\ \mu_{0p} \end{bmatrix}$.

9.14 Confidence Intervals for Mean Vectors and their Components

9.14.1 *One multinormal population*

On confidence intervals for the population mean $\boldsymbol{\mu} = (\mu_1, \ldots, \mu_p)^t$ and for its components, we consider these sub-topics:

(a) Hotelling statistic and the p-dimensional region of confidence and its projections;
(b) Individual independent confidence intervals for components of $\boldsymbol{\mu}$ and joint confidence level;
(c) Simultaneous confidence intervals;
(d) Separate confidence intervals;
(e) Large sample results;
(f) Bonferroni confidence intervals and their applications.

Frequently, the reader is confronted with the computation of confidence intervals and the associated confidence levels for each component of the mean and collectively.

Note: If $t_{n-1,\alpha/(2p)}$ is the upper $\frac{\alpha}{2p}$ percentile of t_{n-1}, then it corresponds to $t_{n-1,1-\alpha/(2p)}$ in this book's notation.

On the above-mentioned topics, we have the following:

(a) We can prove that the Hotelling statistic considered above behaves like a multiple of a Fisher–Snedecor F-variable,

$$T^2 = \frac{(n-1)\,p}{n-p} F_{p,n-p}.$$

Hence, once we have the values of a random sample $\{\mathbf{X}_1, \ldots, \mathbf{X}_n\}$, we can compute $\mathbf{S}_0$ and $\overline{\mathbf{X}}_0$, and for the multivariate case, to test $H_0 : \boldsymbol{\mu} = \boldsymbol{\mu}_0$, the critical region is

$$n\left(\overline{\mathbf{X}}_0 - \boldsymbol{\mu}_0\right)^t \mathbf{S}_0^{-1} \left(\overline{\mathbf{X}}_0 - \boldsymbol{\mu}_0\right) \geq T^2_{n,1-\alpha}.$$

The $100\,(1-\alpha)\,\%$ confidence region for the mean vector $\boldsymbol{\mu}$ is the ellipsoid determined by all values $\boldsymbol{\mu}$ such that

$$n\left(\overline{\mathbf{X}}_0 - \boldsymbol{\mu}\right)^t \mathbf{S}_0^{-1} \left(\overline{\mathbf{X}}_0 - \boldsymbol{\mu}\right) \leq \frac{(n-1)\,p}{(n-p)} F_{p,n-p,1-\alpha}, \qquad (9.18)$$

with $F_{p,n-p,1-\alpha}$ being the upper percentile. This confidence region is an ellipsoid in $\mathbb{R}^p$, centered at $\overline{\mathbf{X}}_0$, with axes on the eigenvectors of $\mathbf{S}_0$. Using $\overline{\mathbf{X}}_0$, $\mathbf{S}_0$ instead of μ, Σ, we have, approximately,

$$\text{Vol}\left\{\mathbf{X} : (\mathbf{X} - \overline{\mathbf{X}}_0)^t \mathbf{S}_0^{-1}(\mathbf{X} - \overline{\mathbf{X}}_0) \le c^2\right\}$$

$$= k_p c^p \sqrt{|\mathbf{S}_0|}, \quad k_p = \frac{2(\pi)^{p/2}}{p\Gamma(p/2)}.$$

Note: The determinant of $\mathbf{S}$, denoted $\det(\mathbf{S})$ or $|\mathbf{S}|$, is called generalized sample variance (see Section 12.7). Its volume is related to the volume Ω of the solid defined by p deviation vectors $\{\mathbf{d}_1, \ldots, \mathbf{d}_p\}$ defined by $\mathbf{d}_1 = \mathbf{y}_1 - \overline{x}_1 \mathbf{1}, \ldots, \mathbf{d}_p = \mathbf{y}_p - \overline{x}_p \mathbf{1}$. We have the relation: $|\mathbf{S}| = \Omega/(n-1)^p$ or $\sqrt{|\mathbf{S}|} = \sqrt{\Omega}/\sqrt{(n-1)^p}$ and the relation to the volume of the above ellipsoid is

$$\text{Vol}\left\{\mathbf{X} : (\mathbf{X} - \overline{\mathbf{X}})^t \mathbf{S}^{-1}(\mathbf{X} - \overline{\mathbf{X}}) \le c^2\right\}$$

$$= 2\left(\frac{\pi}{n-1}\right)^{p/2} \frac{c^p}{p\Gamma(p/2)}\sqrt{\Omega}. \tag{9.19}$$

(b) On the components of μ, the projection of this ellipsoid into the coordinate axis of a component gives the confidence interval for that component. We can then see that, upon obtaining a $(1-\alpha)\,100\%$ confidence region for the vector μ, all intervals obtained by projection in that region are valid simultaneously, with joint confidence level $(1-\alpha)$. These projections could be, however, difficult to obtain mathematically.

(c) We first prove that for a sample $\{\mathbf{X}_1, \ldots, \mathbf{X}_n\}$, and any scalar vector $\mathbf{a}$, the value $\mathbf{a}^t \mu$ is in the interval:

$$\mathbf{a}^t \overline{\mathbf{X}} \pm \sqrt{\frac{p(n-1)}{n(n-p)}F_{p,n-p,1-\alpha}}\sqrt{\mathbf{a}^t \mathbf{S}\mathbf{a}},$$

with probability $(1-\alpha)$.

Proof. The proof is based on the maximum of a quadratic form on a sphere.

Taking the value of **a** appropriately, we have

$$\mu_j \in \left\{ \bar{x}_j \pm \sqrt{\frac{p(n-1)}{n-p}} F_{p,n-p,1-\alpha} \sqrt{\frac{s_{jj}}{n}} \right\}. \qquad (9.20)$$

These are simultaneous confidence intervals for all components of $\boldsymbol{\mu}$, which hold together at the $(1-\alpha)$ confidence level.

(d) Separate confidence intervals for the components of $\boldsymbol{\mu}$: We consider each component separately and compute its $100(1-\alpha)\%$ confidence interval, using t-distribution already encountered. We have

$$\mu_i \in \left\{ \bar{x}_i \pm t_{n-1,1-\alpha/2} \sqrt{s_{ii}/n} \right\}, \quad 1 \leq i \leq p.$$

If the components are independent, we have the joint degree of confidence $\mathbb{P}\left(\bigcap \{\mu_i \in I_i\}\right) = (1-\alpha)^p$, which can be quite different, and smaller, than $(1-\alpha)$. There is a trade-off between interval lengths and their confidence levels. The CI obtained here are often shorter than the ones in (c).

(e) For large sample results, we have

$$n\left\{ (\overline{\mathbf{X}} - \boldsymbol{\mu})^t \mathbf{S}^{-1} (\overline{\mathbf{X}} - \boldsymbol{\mu}) \right\} \approx \chi_p^2$$

and

$$\mathbb{P}\left(n\left\{ (\overline{\mathbf{X}} - \boldsymbol{\mu})^t \mathbf{S}^{-1} (\overline{\mathbf{X}} - \boldsymbol{\mu}) \right\} \leq \chi_{p,1-\alpha}^2 \right) = 1 - \alpha.$$

When $n - p$ is large, we have

$$\mathbf{a}^t \boldsymbol{\mu} \in \mathbf{a}^t \overline{\mathbf{X}} \pm \sqrt{\chi_{p,1-\alpha}^2} \sqrt{\frac{\mathbf{a}^t \mathbf{S} \mathbf{a}}{n}},$$

for each scalar vector **a** simultaneously, leading to

$$\mu_j \in \bar{x}_j \pm \sqrt{\chi_{p,1-\alpha}^2} \sqrt{\frac{s_{jj}}{n}}, \quad 1 \leq j \leq p, \quad \text{simultaneously.} \quad (9.21)$$

For n large, (9.21) is very close to (9.20).

(f) In the Bonferroni approach, we control the overall sum of errors $\sum_{j=1}^{p} \alpha_j$, in the relation on confidence intervals (CI):

$$\mathbb{P}\,(\text{all CI true}) = 1 - \mathbb{P}(\text{at least there is one CI false}).$$

We have Bonferroni's inequality:

$$\mathbb{P}\left(\bigcap_{i=1}^{k} A_i\right) \geq 1 - \sum_{i=1}^{k} \mathbb{P}\left(A_i{}^C\right),$$

which leads to

$$\mathbb{P}\,(\text{all CI true}) = 1 - (\alpha_1 + \cdots + \alpha_p).$$

Hence, take

$$\left\{\bar{x}_i \pm t_{n-1,1-\alpha/(2p)} \sqrt{\frac{s_{ii}}{n}}\right\} \tag{9.22}$$

as confidence interval for μ_i. With $\alpha_i = \alpha/(2p)$, we have the required condition. This last method is often simpler and gives shorter intervals at the same time. □

9.14.2   *Two multinormal populations and more*

9.14.2.1   *Case of two multivariate normal distributions*

In univariate statistics, it is well known that when comparing two means, we use Student's t-statistic, which, however, depends on whether the two population variances are equal or not. For more than two means, we perform an analysis of variance first and then make several tests and evaluations to have confidence intervals for linear combinations of means, using the contrast theory.

The same approach is adopted when we consider observations coming from two multivariate normal populations, and two mean vectors can be compared using the T^2 confidence intervals for differences of components of these mean vectors.

(1) **Multivariate observations are paired:** Let $\{\mathbf{X}_{ijk}\}, i = 1, 2,$ $j = 1, ..., n, k = 1, ..., p$, be two random samples of same size n, each containing vectors of p dimensions. Let $\mathbf{D}_j = \begin{pmatrix} D_{j1} \\ ... \\ D_{jp} \end{pmatrix}$ be the

*j*th *sample of differences,* $j = 1, \ldots, n$. Suppose that

$$\mathbb{E}(\mathbf{D}_j) = \boldsymbol{\delta} = \begin{pmatrix} \delta_1 \\ \cdots \\ \delta_p \end{pmatrix}, \quad \mathrm{Cov}(\mathbf{D}_j) = \boldsymbol{\Sigma}.$$

Setting

$$\overline{\mathbf{D}} = \frac{1}{n} \sum_{i=1}^{n} \mathbf{D}_i, \mathbf{S} = \frac{1}{n-1} \sum_{i=1}^{n} (\mathbf{D}_i - \overline{\mathbf{D}})(\mathbf{D}_i - \overline{\mathbf{D}})^t,$$

we have

$$n(\overline{\mathbf{D}} - \boldsymbol{\delta})^t \mathbf{S}^{-1} (\overline{\mathbf{D}} - \boldsymbol{\delta}) = T^2 = \frac{(n-1)p}{n-p} F_{p,n-p},$$

which can serve for hypothesis testing and confidence intervals computation, as seen previously.

(2) **Unpaired multivariate observations:** We distinguish between two cases (see Johnson and Wichern (1992)):

(a) **Equal population covariance matrices,** $\boldsymbol{\Sigma}_1 = \boldsymbol{\Sigma}_2$: We then pool the two sample covariance matrices to form a new one:

$$\mathbf{S}_{\mathrm{pool}} = \frac{1}{n_1 + n_2 - 2} [(n_1 - 1)\mathbf{S}_1 + (n_2 - 1)\mathbf{S}_2]. \quad (9.23)$$

We can now construct a confidence region for $\boldsymbol{\mu}_1 - \boldsymbol{\mu}_2$, which is an ellipse, or ellipsoid, when the difference $(\overline{\mathbf{X}}_1 - \overline{\mathbf{X}}_2)$ is taken as variable and this region will, in turn, help to compute simultaneous confidence regions for components differences by using either the above methods or the Bonferroni approach. The region of confidence for the difference of mean vectors is

$$[(\overline{\mathbf{X}}_1 - \overline{\mathbf{X}}_2) - (\boldsymbol{\mu}_1 - \boldsymbol{\mu}_2)]^t \left[\left(\frac{1}{n_1} + \frac{1}{n_2} \right) \mathbf{S}_{\mathrm{pool}} \right]^{-1}$$

$$\times [(\overline{\mathbf{X}}_1 - \overline{\mathbf{X}}_2) - (\boldsymbol{\mu}_1 - \boldsymbol{\mu}_2)] \leq \frac{p(n_1 + n_2 - 2)}{n_1 + n_2 - p - 1}$$

$$\times F_{p,n_1+n_2-p-1,1-\alpha},$$

which is an ellipsoid with probability $1 - \alpha$.

Note: T^2 can be shown to be equivalent to the Mahalanobis distance between the two mean vectors. Its statistic is defined by

$$D^2 = \left(\overline{\mathbf{X}}_1 - \overline{\mathbf{X}}_2\right)^t \frac{(n_1 - 1)\mathbf{S}_1 + (n_2 - 1)\mathbf{S}_2}{n_1 + n_2 - 2} \left(\overline{\mathbf{X}}_1 - \overline{\mathbf{X}}_2\right).$$

(b) **Unequal population covariance matrices, $\boldsymbol{\Sigma}_1 \neq \boldsymbol{\Sigma}_2$:** The two sample covariance matrices are joined to form a new one by using all observations. We now use an approximate approach since this multivariate *Behrens–Fisher* problem does not have a satisfactory solution yet:

$$\left[\left(\overline{\mathbf{X}}_1 - \overline{\mathbf{X}}_2\right) - (\boldsymbol{\mu}_1 - \boldsymbol{\mu}_2)\right]^t \left[\left(\frac{1}{n_1}\mathbf{S}_1 + \frac{1}{n_2}\mathbf{S}_2\right)^{-1}\right]$$
$$\times \left[\left(\overline{\mathbf{X}}_1 - \overline{\mathbf{X}}_2\right) - (\boldsymbol{\mu}_1 - \boldsymbol{\mu}_2)\right] \sim \chi_p^2, \tag{9.24}$$

and we can derive the $100\,(1 - \alpha)\,\%$ simultaneous confidence intervals for all linear combinations $\mathbf{a}^t\,(\boldsymbol{\mu}_1 - \boldsymbol{\mu}_2)$.

Note: The hypothesis $H_0 : \boldsymbol{\Sigma}_1 = \boldsymbol{\Sigma}_2$ can be tested using $\mathbf{S}_1$, $\mathbf{S}_2$ before we decide to use one of the two above approaches.

9.14.2.2   *More than two multinormal populations*

Multivariate analysis of variance (MANOVA): Using the same premises as the univariate analysis of variance, where the goal is to compare several scalar mean values, MANOVA has a similar model to construct the MANOVA table. It is exactly the same table as the univariate ANOVA, only now, scalar data is replaced by vector data. Appropriate changes in computations, and in using the related statistics, have to be made.

(a) The model is, in MANOVA,

$$\mathbf{X}_{ij} = \boldsymbol{\mu} + \boldsymbol{\tau}_i + \mathbf{e}_{ij}, \;\; i = 1, \dots, g, \;\; j = 1, \dots, n_i, \mathbf{e}_{ij} \sim N\,(\mathbf{0}, \boldsymbol{\Sigma}).$$

Here, $\boldsymbol{\mu}$ is the overall mean vector and $\boldsymbol{\tau}_i$ is the ith treatment effect vector. We suppose that $\sum_{i=1}^{g} n_i \boldsymbol{\tau}_i = 0$.

Table 9.1   The MANOVA table.

Variations	Matrix sums of squares	Degree of freedom
Treatment	$\mathbf{B} = \sum\limits_{i=1}^{g} n_i \left(\mathbf{x}_i - \overline{\mathbf{x}}\right)\left(\mathbf{x}_i - \overline{\mathbf{x}}\right)^t$	$g - 1$
Residuals	$\mathbf{SSB} = \mathbf{W} = \sum\limits_{i=1}^{g}\sum\limits_{j=1}^{n_i} \left(\mathbf{x}_{ij} - \overline{\mathbf{x}}_i\right)\left(\mathbf{x}_{ij} - \overline{\mathbf{x}}_i\right)^t$	$\sum\limits_{i=1}^{g} n_i - g$
Totals	$\mathbf{SSW} = \mathbf{S} = \mathbf{B} + \mathbf{W} = \sum\limits_{i=1}^{g}\sum\limits_{j=1}^{n_i} \left(\mathbf{x}_{ij} - \overline{\mathbf{x}}\right)\left(\mathbf{x}_{ij} - \overline{\mathbf{x}}\right)^t$	$\sum\limits_{i=1}^{g} n_i - 1$

The vector of observations may be decomposed, as in the univariate case, as

$$\mathbf{X}_{ij} = \overline{\mathbf{X}} + \left(\overline{\mathbf{X}}_i - \overline{\mathbf{X}}\right) + \left(\mathbf{X}_{ij} - \overline{\mathbf{X}}_i\right),$$

where the MANOVA table, with matrix outputs, $\mathbf{B}, \mathbf{W}, \mathbf{S}$ is given in Table 9.1.

But, for testing the equality of means, while univariate ANOVA uses ratios of chi-squares, and Fisher–Snedecor statistic F, in classical MANOVA, we usually compute the ratio of determinants, called Wilks's statistic, $\Lambda = \frac{|\mathbf{W}|}{|\mathbf{W}+\mathbf{B}|}$, which is related to the likelihood ratio criterion but corresponds, however, to the beta variable in univariate ANOVA and not the Fisher–Snedecor statistic. This point has been raised in Pham-Gia (2008) and in subsequent publications of the AMSARG, where we call for the use the matrix variate F-distribution, which is here

$$F = \frac{|\mathbf{B}|}{|\mathbf{W}|}. \tag{9.25}$$

This statistic has been shown to perform very well and could be used in MANOVA, just like Wilks's statistic Λ.

(b) The exact expression of Wilks's statistic was given in Pham-Gia (2008), in terms of a Meijer G-function, since it is the determinant of the central beta matrix distribution. Hence, distribution-wise, it is also a product of independent univariate betas of the first kind. It is to be noted that for the two population cases, Wilks's Λ becomes a multiple of Fisher–Snedecor statistic F, since

$$
\left(\frac{\sum\limits_{i=1}^{g} n_i - p - 1}{p} \right) \left(\frac{1 - \Lambda}{\Lambda} \right) \sim F_{p, \sum\limits_{i=1}^{g} n_i - p - 1}, \quad p \geq 1.
$$

Other approximate relations concerning other cases exist.

(c) For questions related to the power of tests performed in MANOVA, we have to deal with non-central matrix betas and non-central matrix F-distributions, which have expressions in terms of hypergeometric distributions of one or two matrix arguments. The non-central beta matrices are present under four forms, as given in Gupta and Nagar (2000), see also Chapter 13. For the non-central F, we have considered the linear and planar cases only for the non-centrality parameter matrix and expressed the density in closed forms using only common mathematical functions (here the Legendre polynomials) instead of zonal polynomials. As a consequence, the matrix variate F-statistics, central and non-central, can be used for MANOVA. Numerical testing shows that they perform equally well as the central and non-central Wilks statistics.

9.15   Correlation Coefficients

There are several notions of correlation when we are in the presence of more than two rv's. In fact, in everyday's language, the term *correlation* means simply a relation between two entities.

Zero covariance implies that the corresponding components are independent for normal variables, but only if they are binormal. There are several examples of dependent univariate normal variables with zero correlation.

There are several other correlation coefficients that the reader can come into contact with in his/her reading or research. In this section, we deal briefly with univariate observations only. But the same ideas apply to multivariate observations, with more complexity. Also, the population coefficients of various types are reviewed here, but the corresponding sample coefficients, and their distributions, are often not for lack of space. Correlation coefficients, in general, measure *the degrees of association* between sets of random variables.

9.15.1 *Correlation between two univariate rv's*

If X, Y are two univariate random variables, then

$$Cov\,(X,Y) = \mathbb{E}\left[(X - \mu_X)\,(Y - \mu_Y)\right]$$

and the **correlation coefficient** $\rho_{XY} = Cov(X,Y)/\sigma_X\sigma_Y$.

It is proven that $|\rho_{XY}| \leq 1$ and

$$(X \text{ and } Y \text{ independent}) \Rightarrow (\rho_{XY} = 0),$$

but the converse is untrue, except for the binormal case (see Chapter 7).

The sample correlation coefficient is given by

$$r = \frac{\displaystyle\sum_{j=1}^{n} (x_j - \overline{x})\,(y_j - \overline{y})}{\sqrt{\displaystyle\sum_{j=1}^{n} (x_j - \overline{x})^2}\,\sqrt{\displaystyle\sum_{j=1}^{n} (y_j - \overline{y})^2}}, \tag{9.26}$$

associated with the scattergram of the sample, where $0 \leq |r| \leq 1$, with its absolute value getting closer to 0, or 1, if the scattergram shows very little linear tendency or has a pronounced linear shape, respectively.

Under the hypothesis that the couple has a binormal distribution, the distribution of r can be studied, but depends on whether the population coefficient ρ is zero or different from zero, and inferences can be made on ρ depending on the value of r.

9.15.2 *The multiple correlation coefficient in practice*

Definition 9.1. The *coefficient of multiple correlation*, denoted R, is defined as the Pearson correlation coefficient between the predicted and actual values of the dependent variable in a linear regression model that includes an intercept.

Hence, it measures how well a given variable can be predicted using a linear function of a set of other variables. It is the correlation between the variable's values and the best predictions that can be computed linearly from the predictive variables. The following remarks apply:

(a) The coefficient of multiple correlation takes values between 0 and 1; a higher value indicates a better predictability of the dependent variable from the independent variables, with a value of 1 indicating that the predictions are exactly correct and a value of 0 indicating that no linear combination of the independent variables is a better predictor than is the fixed mean of the dependent variable.

(b) It is computed as the positive square root of D, the coefficient of determination, $R = +\sqrt{D}$, but under the particular assumptions that an intercept is included and that the best possible linear predictors are used.

(c) **Computation in a sample:** In a random sample, the squared coefficient of multiple correlation can be computed as the fraction of variation of the dependent variable that is explained by the independent variables, which in turn is 1 minus the unexplained fraction. The unexplained fraction can be computed as the sum of squared residuals, i.e. the sum of the squares of the prediction errors, divided by the sum of the squared deviations of the values of the dependent variable from its expected value.

Considering the residuals, $\eta_{1\cdot23\ldots p} = \xi_1 - \beta_{12}\xi_2 - \cdots - \beta_{1p}\xi_p = \xi_1 - \sigma_1$, where $\sigma_1 = \beta_{12}\xi_2 - \cdots - \beta_{1p}\xi_p$ is the best linear estimate of ξ_1, we have

$$\rho_{1(23\ldots p)} = \frac{\mathbb{E}\left(\xi_1 \sigma_1\right)}{\sqrt{\mathbb{E}\left(\xi_1^2\right)\mathbb{E}\left(\sigma_1^2\right)}}. \tag{9.27}$$

Hence, it measures the correlation between ξ_1 on the one side and all other variables on the other side, with σ_1 having the maximum ordinary correlation coefficient with ξ_1, as given by Eq. (9.26). In a sample, R^2 gives the proportion of the total variations in the y_j's attributable to the predictor variables $\{x_1, \ldots, x_n\}$:

$$R^2 = 1 - \frac{\sum\limits_{i=1}^{n} \hat{\varepsilon}_i^2}{\sum\limits_{i=1}^{n} (y_i - \bar{y})^2} = \frac{\sum\limits_{i=1}^{n} (\hat{y}_i - \bar{y})^2}{\sum\limits_{i=1}^{n} (y_i - \bar{y})^2}.$$

9.15.3   *Coefficient of partial correlation*

It measures the degree of association between two random variables Y_1, Y_2, with the effect of a set of controlling random variables removed.

We have

$$\rho_{Y_1 Y_2 \cdot \mathbf{Z}} = \frac{\sigma_{Y_1 Y_2 \cdot \mathbf{Z}}}{\sqrt{\rho_{Y_1 Y_1 \cdot \mathbf{Z}}}\sqrt{\rho_{Y_2 Y_2 \cdot \mathbf{Z}}}}.$$

Formally, the partial correlation between X and Y, given a set of n controlling variables $\mathbf{Z} = \{Z_1, Z_2, \ldots, Z_n\}$, written $\rho_{XY \cdot Z}$, is the correlation between the residuals R_X and R_Y resulting from the linear regression of X with $\mathbf{Z}$ and of Y with $\mathbf{Z}$, respectively:

$$r_{Y_1 Y_2 \cdot \mathbf{Z}} = \frac{s_{Y_1 Y_2 \cdot \mathbf{Z}}}{\sqrt{s_{Y_1 Y_1 \cdot \mathbf{Z}}}\sqrt{s_{Y_2 Y_2 \cdot \mathbf{Z}}}}.$$

Computation using linear regression: A simple way to compute the sample partial correlation for some data is to solve the two associated linear regression problems, get the residuals, and calculate the correlation coefficient between the residuals. Then

$$\widehat{\rho}_{XY \cdot Z} = \frac{n \sum_{i=1}^{n} e_{X,i} e_{Y,i} - \sum_{i=1}^{n} e_{X,i} \sum_{i=1}^{n} e_{Y,i}}{\sqrt{n \sum_{i=1}^{n} e_{X,i}^2 - \left(\sum_{i=1}^{n} e_{X,i}\right)^2} \sqrt{n \sum_{i=1}^{n} e_{Y,i}^2 - \left(\sum_{i=1}^{n} e_{Y,i}\right)^2}},$$

$$(9.28)$$

where $\{e_{X,i}, e_{Y,i}\}_{i=1}^{n}$ are the residuals. When Z is a single variable, we have

$$\rho_{XY \cdot Z} = \frac{\rho_{XY} - \rho_{XZ} \rho_{ZY}}{\sqrt{1 - \rho_{XZ}^2}\sqrt{1 - \rho_{ZY}^2}}. \qquad (9.29)$$

9.15.4   *Semi-partial correlation or part correlation*

In the computation of the semi-partial correlation between X and Y, we hold the third variable Z constant for either X or Y. The absolute value of this coefficient is always less than the partial correlation.

9.15.5   *Some theory on correlations*

Correlation coefficients, in the population, are related to conditional distributions. Recall that if

$$\mathbf{X} = \begin{bmatrix} \mathbf{X}_1 \\ \mathbf{X}_2 \end{bmatrix} \begin{array}{l} (q \times 1), \\ ((p-q) \times 1), \end{array}$$

then the conditional distribution

$$\mathbf{X}_1 \,|\, \mathbf{x}_2 \sim N_q \left(\boldsymbol{\mu}_1 + \boldsymbol{\Sigma}_{12}\boldsymbol{\Sigma}_{22}^{-1} (\mathbf{x}_2 - \boldsymbol{\mu}_2); \boldsymbol{\Sigma}_{11} - \boldsymbol{\Sigma}_{12}\boldsymbol{\Sigma}_{22}^{-1}\boldsymbol{\Sigma}_{21} \right). \tag{9.30}$$

The matrix $\boldsymbol{\Sigma}_{12}\boldsymbol{\Sigma}_{22}^{-1}$ is the matrix of *partial regression* coefficients of $\mathbf{X}_1$ on $\mathbf{X}_2$. If $q = 1$, as it is often the case, the regression equation is $\mathbb{E}\left(X_1 \,|\, \mathbf{X}_2 = \mathbf{x}_2\right) = \alpha + \sum_{i=2}^{p} \beta_i x_i$, often referred to as *the general linear model*. Then the scalar measure of correlation between $\mathbf{X}_1$ and $\mathbf{X}_2$ is the *multiple correlation coefficient*, given by $\theta = \frac{\sigma_{12}^t \cdot \boldsymbol{\Sigma}_{22}^{-1} \cdot \sigma_{12}}{\sqrt{\sigma_{11}}}$. Also, in Eq. (9.29), the coefficient of correlation between any two elements of $\mathbf{X}_1$ is a *partial correlation coefficient*. On the other hand, the non-zero characteristic roots of $\boldsymbol{\Sigma}_{11}^{-1}\boldsymbol{\Sigma}_{12}\boldsymbol{\Sigma}_{22}^{-1}\boldsymbol{\Sigma}_{21}$ are called *canonical correlations* between $\mathbf{X}_1$ and $\mathbf{X}_2$.

9.15.6   *Correlation matrix*

For a random vector $\mathbf{X} = (X_1, \ldots, X_p)^t$, the correlation between the component X_i and X_j is

$$\rho_{ij} = \frac{\mathbb{E}\left[(X_i - \mu_i)(X_i - \mu_i)\right]}{\sqrt{\mathbb{E}\left[(X_i - \mu_i)^2\right]}\sqrt{\mathbb{E}\left[(X_j - \mu_j)^2\right]}}.$$

The associated matrix is the correlation matrix $\Xi = [\rho_{ij}]$, with 1 on the diagonal, wrt which it is symmetrical. We have

$$\Xi = \left[\mathbf{V}^{1/2}\right]^{-1}\boldsymbol{\Sigma}\left[\mathbf{V}^{1/2}\right]^{-1},$$

with

$$\mathbf{V}^{1/2} = \begin{bmatrix} \sqrt{\sigma_{11}} & \cdots & 0 \\ & \cdots & 0 \\ sym. & & \sqrt{\sigma_{pp}} \end{bmatrix}.$$

The density of Ξ has been studied by several authors (see Kshirsagar (1972)).

For $\mathbf{X}$ multinormal, some results and applications are given by Pham-Gia and Choulakian (2014). For a random sample, we have the corresponding sample variables $\mathbf{R}, \mathbf{D}, \mathbf{S}$, with

$$\mathbf{R} = \left[\mathbf{D}^{1/2}\right]^{-1} \mathbf{S} \left[\mathbf{D}^{1/2}\right]^{-1}$$

and

$$|\mathbf{S}| = (s_{11}...s_{pp}) \, |\mathbf{R}| \, .$$

Remark: In general, random variables may be uncorrelated but statistically dependent. But if a random vector is multivariate normal, then any two (or more) of its components that are uncorrelated are independent.

9.16 Simulation of the Multivariate Normal

It is important to know how to generate a sample of observations from a normal distribution of any dimension.

(a) The random numbers we usually generate are pseudo-random numbers, and the simplest approach is using the modulo method, starting with a seed X_0 and choosing positive integers a, c, m and then letting

$$X_{n+1} = aX_n + c \, (\text{mod} \, m)$$

for $n \geq 0$. Each X_n is either $(0, 1, \ldots, m-1)$ and X_n/m is taken as a random value in $(0, 1)$.

(b) For simulation of continuous random variables, the basic approach is the inverse transformation: For any continuous cdf $F(.)$, if we define the rv X by $X = F^{-1}(U)$, where U is a uniform rv on $(0, 1)$, then X has cdf $F(.)$. Other methods include the *rejection method*, the *hazard rate method*.

(c) We start with the standard normal $N(0, 1)$ on $\mathbb{R}$. We can use the central limit theorem for an approximate simulation.

Algorithm: To have a sample $\{x_i\}_{i=1}^{N}$ from $N_1\,(0,1)$, we draw a sample $\{u_i\}_{i=1}^{N}$ from the uniform distribution on $[0,1]$ and compute $x_i = \left(\sum_{i=1}^{N} u_i - \frac{N}{2}\right)\Big/\sqrt{\frac{N}{12}}$, with $N \geq 12$. For $N = 12$, we have $x_i = \sum_{i=1}^{12} u_i - 6$.

(d) **Box–Mueller method:** This approach is much more efficient than the previous one. We obtain two independent samples of $N\,(0,1)$.

Algorithm: Generate two independent observations from the uniform on $[0,1]$. Then the two values

$$X = \sqrt{-2\log U}\,\sin(2\pi V), \quad Y = \sqrt{-2\log U}\,\cos(2\pi V)$$

are independent and each has a $N\,(0,1)$ distribution. At the next step, to obtain samples from the univariate $W \sim N\left(\mu, \sigma^2\right)$, it suffices to compute $W = \mu + X\sigma$.

This approach presented now is based on the polar coordinates

$$R^2 = X^2 + Y^2, \quad \Theta = \tan^{-1}\frac{Y}{X},$$

which can be shown to be independent, with an exponential and uniform distribution, respectively. If we do not want to use cos and sin, we can use

$$X = \sqrt{-2\log R^2}\,\frac{V_1}{R} = \sqrt{-\frac{2\log S}{S}}V_1,$$

$$Y = \sqrt{-2\log R^2}\,\frac{V_2}{R} = \sqrt{-\frac{2\log S}{S}}V_2$$

as two independent unit normals, where $S = R^2 = V_1^2 + V_2^2$. So, our algorithm is as follows:

(i) Generate U_1 and U_2.
(ii) Compute $V_1 = 2U_1 - 1$, $V_2 = 2U_2 - 1$, $S = V_1^2 + V_2^2$.
(iii) If $S > 1$, return step (i).

(iv) Return the independent unit normal: $\sqrt{-\frac{2\log S}{S}}V_1$, $\sqrt{-\frac{2\log S}{S}}V_2$.

(e) Similarly, to generate a sample from a multivariate normal $N_p\,(\boldsymbol{\mu}, \boldsymbol{\Sigma})$, where $\boldsymbol{\mu} \in \mathbb{R}^p$ and $\boldsymbol{\Sigma}$ is positive, symmetric $(p \times p)$, we draw a sample $\{y_i\}_{i=1}^N$ from $N(0,1)$ and make the transformation: $\mathbf{x} = \boldsymbol{\mu} + \boldsymbol{\Sigma}^* \mathbf{y}$, where $\boldsymbol{\Sigma}^*$ is obtained from $\boldsymbol{\Sigma}$ (see Chapter 7).

9.17   An Example from MATLAB

- **Syntax**

$$R = \text{normrnd}(\text{mu}, \text{sigma})$$

- **Description**

R = normrnd(mu, sigma) generates random numbers from the normal distribution with mean parameter mu and standard deviation parameter sigma. mu and sigma can be vectors, matrices, or multidimensional arrays that have the same size, which is also the size of R. A scalar input for mu or sigma is expanded to a constant array with the same dimensions as the other input.

Example

$$n1 = \text{normrnd}(1:6, 1./(1:6))$$

output

$$n1 = 2.1650 \ 2.3134 \ 3.0250 \ 4.0879 \ 4.8607 \ 6.2827$$

9.18   Conclusion

This chapter presents the main results, theoretical and applied, related to the normal distribution in p dimensions. Many more theoretical results exist and provide grounds for the development of other interesting properties of the multivariate normal. They are not treated here, and we refer the reader to other specialized works listed in the bibliography section.

Bibliography

Cox, D.R. and Small, N.J.H. (1978). Testing multivariate normality, *Biometrika*, **65**, 263–272.

Cramer, H. (1961). *Mathematical Methods of Statistics* (Princeton University Press, New Jersey).

Csaki, F. (1970). A concise proof of Sylvester's theorem, *Periodica Polytechnica Electrical Engineering (Archives)*, **14**(2), 105–112. Available at: https://pp.bme.hu/ee/article/view/5099.

Curnow, R.N. and Dunnett, C.W. (1962). The numerical evaluation of certain multivariate normal integrals, *Annals of Mathematical Statistics*, **33**, 571–579.

David, F.N. (1953). A note on the evaluation of the multivariate normal integral, *Biometrika*, **40**, 458–459.

Genz, A. and Bretz, F. (2009). *Computation of Multivariate Normal and t Probabilities*, *Lecture Notes in Statistics* 195 (Springer-Verlag, Berlin, Heidenberg).

Gnanadesikan, R. (1976). *Methods for Statistical Data Analysis of Multivariate Observations* (John Wiley and Sons, New York).

Gupta, A.K. and Nagar, D.K. (2000). *Matrix Variate Distributions* (Chapman and Hall/CRC, New York).

Johnson, R.A. and Wichern, D.W. (1992). *Applied Multivariate Statistical Analysis*, 3rd ed. (Prentice-Hall, Englewoods Cliffs, New York).

Kshirsagar, A.M. (1972). *Multivariate Analysis* (Marcel Dekker, New York).

Mardia, K.V. (1970). Measures of multivariate skewness and kurtosis with applications, *Biometrika*, **57**, 510–530.

Pham-Gia, T. (2008). Exact expression of Wilks's statistic and applications, *Journal of Multivariate Analysis*, **99**(8), 1698–1716.

Pham-Gia, T. and Choulakian, V. (2014). Distribution of the sample correlation matrix and applications, *Open Journal of Statistics*, **4**(5), 330–344.

Pierce, D.A. and Dykstra, R.L. (1969). Independence and the normal distribution, *The American Statistician*, **23**(4), 39.

Plackett, R.L. (1954). A reduction formula for normal multivariate integrals, *Biometrika*, **41**, 351–360.

Rencher, A.C. and Christensen, W.F. (2012). *Methods of Multivariate Analysis*, 3rd ed. (John Wiley and sons, New Jersey).

Sheppard, W.F. (1898). On the geometric treatment of the normal curve of statistics with special reference to correlation and to the theory of errors, *Proceedings of the Royal Society of London*, **62**, 170–173.

Stroud, A.H. and Secrest, D. (1966). *Gaussian Quadrature Formulas* (Prentice-Hall, Englewood Cliffs, New Jersey).

Thomasian, A.J. (1969). *The Structure of Probability Theory with Applications* (McGraw Hill, New York).

Tong, Y.L. (1989). *The Multivariate Normal Distribution* (Springer, New York).

Chapter 10

Some Applications of the Multivariate Normal Distribution in Science and Engineering

10.1 Introduction

In this chapter, we present some applications particular to the two-dimensional normal vector, with graphs drawn in $\mathbb{R}^3$, something we will not be able to do in higher dimensions. In particular, we will present the following:

(a) An application in engineering: Bayesian stress–strength, using a multinormal model.
(b) An application in imagery: Eye recognition approach in subject identification.
(c) Three applications in pattern recognition and statistics: Discrimination, classification and clustering. Variables considered here go from the bivariate to the quadrivariate. The maximum function approach is introduced. Clustering of probability densities is also treated.
(d) An application of the Mahalanobis distance in biology/wildlife management.

For each application, we will present the abstract, the purpose, and the main results. Technical details are left out, and so are

311

many supporting results and arguments, but the interested reader can always find them in the publications that we clearly indicate. All applications use, or are based on, the multinormal distribution directly or indirectly.

10.2   Bayesian Stress–Strength Multinormal Model

Abstract: Multivariate models offer much more realistic applications to real-life reliability studies than univariate models (see Chapter 5), since components of strength $\mathbf{X}$ are now considered individually, and also collectively, through their dependence on each other, as measured by a covariance matrix. A linear combination of these components, $\sum_{j=1}^{k_1} a_j X_j$, with $a_j > 0$, $1 \leq j \leq k_1$, now represents their combined effects, with the weights a_j, $1 \leq j \leq k_1$, reflecting their relative importances. Similar considerations apply for stress $\mathbf{Y}$, of dimension k_2, and we have now, for the reliability of the system, $\Re = \mathbb{P}\left(\sum_{j=1}^{k_1} a_j X_j > \sum_{j=1}^{k_2} b_j Y_j\right)$. We can see that, when $\mathbf{X}$ and $\mathbf{Y}$ are random vectors, with specified distributions, $\Re$ is a constant. But when these distributions have parameters that, themselves, have prior probability distributions, as in the Bayesian approach, $\Re$ becomes a random variable, with its distribution to be determined.

A Bayesian approach is adopted for $\Re = \mathbb{P}\left(\mathbf{B}^t \mathbf{W} \geq 0\right)$, where $\mathbf{W}$ is multinormal, i.e. $\mathbf{W} \sim N_k\left(\boldsymbol{\mu}, \mathbf{T}\right)$, with the unknown mean vector $\boldsymbol{\mu}$ and the unknown precision matrix $\mathbf{T}$ having a joint normal–Wishart prior distribution (see Chapter 6). Some computations are carried out by simulation, an approach which plays an important role in this problem.

The concept of highest posterior density (hpd) region has been introduced in Chapter 6. Here, we have the two hpd regions for stress and for strength. Several conclusions can be drawn from these regions and their overlap (see Fig. 10.1).

Note: Due to the complexity of the arguments used in this research, we invite the reader to consult Pham-Gia and Turkkan (2006) directly.

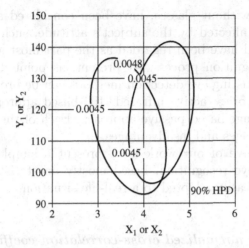

Fig. 10.1 95% highest posterior density intersection regions of strength and stress (multinormal case).

10.3 Application to Eye Recognition of a Subject

Abstract: In image processing, the area around the eyes can be used efficiently to recognize an individual within a large database. In this chapter, we use the statistical two sample t-test (or the normal test) and the statistical cross-correlation coefficient to that purpose.

Matching technique, based on the "cross-correlation coefficient", is widely used in image processing to locate an image within a larger one.

On the other hand, recognition of a subject based on a part of his/her face has been a very active research topic in the recent years. Several approaches have been suggested in the literature, going from the very sophisticated ones, using textures, colors, special points on the face, linear and nonlinear filters, to highly mathematical ones, and to the relatively simple ones, based on standard concepts. Using fingerprints to identify a subject seems to be a popular, more advanced approach, although not without its shortcomings. For pictures, they can vary widely because of either the lighting conditions under which they were taken or the positions and attitudes of the subjects being photographed. All these factors can influence widely the recognition capability of an approach.

The eyes, without glasses, have been considered as the part of the face least affected by the subject's attitude, such as smiling or grimacing, and have been regarded as the part most apt to be used in a face recognition process. At present, as pointed out by some researchers, existing eye detection methods can be broadly classified into two categories: active infrared (IR)-based approaches and the traditional image-based passive approach. Each of these approaches has its advantages and disadvantages.

The objective of our work is to present a simple, but robust, approach to eye recognition, based mostly on statistical methods, which can be easily adapted to a real-life situation.

10.3.1   *The normalized cross-correlation coefficient*

10.3.1.1   *Principle of template matching*

To identify an image within another one, we look at the distance between pixels of the two. Let f be an image to be processed, w a pattern to search for, also called template, which could be a subset of f. A formula which represents a good matching criterion describing a match between f and w located at a position (u, v) is

$$C_1(u, v) = \frac{1}{\max_{(i,j)\in V}\left\{\left|f\left(i+u, j+v\right) - w\left(i, j\right)\right| + 1\right\}},$$

where V is the set of all image pixels in the processed image. However, it could be difficult to use this criterion in practice since pixels are considered individually.

10.3.1.2   *Statistical cross-correlation coefficient*

Several versions of the statistical cross-correlation coefficient exist, most of them based on the Cauchy–Schwartz inequality in the mathematical function theory. The normalized coefficient presented here, also called the *statistical cross-correlation coefficient*, seems to be subject to a limited number of weaknesses only.

Let $f(x, y)$ be an image of size $M \times N$ and let $w(x, y)$ be a smaller image of size $J \times K$. We define the *normalized cross-correlation* between the two images at the point as

$$\gamma(x,y) = \frac{\sum_s \sum_t \left[f(s,t) - \overline{f}(s,t)\right]\left[w(x+s,y+t) - \overline{w}\right]}{\left\{\sum_s \sum_t \left[f(s,t) - \overline{f}(s,t)\right]^2 \sum_s \sum_t \left[w(x+s,y+t) - \overline{w}\right]^2\right\}^{1/2}},$$

where $\overline{w}$ is the average value of the pixels in w, while $\overline{f}(s,t)$ is the average value of f in the region coincident with the current location of w. The summations are taken here over the coordinates common to both f and w, and hence some appropriate *padding* is required for the computation of $\gamma(x,y)$, which has its value between -1 and $+1$, and is independent of the linear scale changes in the amplitude of f and w. We know that, as an image itself, $\gamma(x,y)$ has the highest (brightest) value where the best match between f and w is found. Also, we consider both cases: common variance (or covariance matrix) and different variances (or covariance matrices), so that the reader can see that one case provides better results. In a real statistical study, a test could be performed to decide which case applies.

10.3.1.3   *The Database*

The use of a template for object recognition, although very convenient, must be subject to the most stringent conditions. When a template is taken from a file image, the distribution of the pixels in the template is identical to that of one of the image. However, when a part of another digital photo of the same subject is used as a template, the difference in sizes, lightings, head position, facial expressions, and resolutions can render the template and file image quite different and matching will not be successful. For these reasons, every effort should be made to put a template and image on a comparable basis.

Due to the different nature of the problem we are addressing here, a new face database needs to be set up. They consist of a limited number of volunteers (65 in total) coming to be photographed at our laboratory at the Université de Moncton. Frontal face pictures with two different face expressions (no expression and smiling) were taken

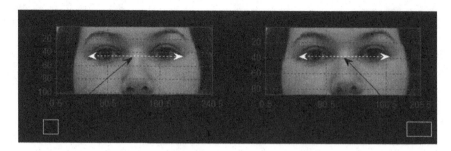

Fig. 10.2   Normalized and non-normalized distances between the eyes of a subject.

of these individuals, who come from different race backgrounds and also belong to varying age groups. Our basic database hence contains 130 pictures (Fig. 10.2).

To be precise in our measurements, an edge finder is used on all face sections to evidence the eye areas of all the subjects. This edge finder is based on the normal test, or the Student t-test, for two samples in statistics, which will enable us to detect the points where the edge is detected, that can be summarized as follows: Let $P(x,y)$ be any pixel. We consider the left and right neighborhoods of $P(x,y)$, N_L and N_R consisting of 15 pixels each, with P being on the right boundary of N_L (and in its middle) and similarly, on the left boundary of N_R. An edge height factor δ is now added to N_L and deducted from N_R. Its value is determined rather empirically, but its optimal value can be found by experimentation.

Let $\overline{X}_L$ and s_L^2 be the mean and variance of the pixel values in N_L. Similarly, we have $\overline{X}_R$ and s_R^2 for the right neighborhood. The ratio

$$t_{LR} = \frac{\overline{X}_L - \overline{X}_R}{S_p\sqrt{\frac{2}{15}}},$$

with $S_p^2 = 14\left(S_L{}^2 + S_R{}^2\right)/24$, will allow us to decide whether the mean to the left is different from the mean to the right by comparing it to 2.00. Similarly, we have $\overline{X}_T$ and s_T^2 for the top neighborhood of $P(x,y)$, and $\overline{X}_B$ and s_B^2 for its bottom neighborhood. We declare a pixel as belonging to an edge when t_{LR} or t_{TB} is larger than 2, taken as the approximate critical value of the t-distribution or of *the normal distribution* (see details in Hervet *et al.* (2009)).

It is nice to report that the software produced has allowed us to recognize most (about 96%) of the 65 subjects when they came to our lab for identification.

10.4 Application to Pattern Recognition: Discrimination, Classification, and Clustering

Abstract: The following three chapters treat these topics, using the classical approach, but the metric considered is the maximum function and its integral, as originally suggested by Glick (1972, 1973). We present the main results and invite the readers to consult these articles for all other details.

The *maximum function* is used for three purposes:

(1) **Discrimination:** To determine the intersection curves in $\mathbb{R}^3$ of the normal surfaces $g_i(x_1, x_2) = q_i f_i(x_1, x_2)$, the projections of which on the horizontal plane give the boundaries, linear or quadratic, between the groups, as established by the classical theory.

So, we define $g_{\max}(x_1, x_2) = \max_i \{g_i(x_1, x_2)\}$ and determine its definition domain. Also, its integral is related to a distance of L^1 type and an error of Bayes type. Extensions to $\mathbb{R}^p, p \geq 3$, uses the quasi-Monte Carlo method for integration and gives similar results (Pham-Gia *et al.*, 2008).

(2) **Classification:** Determine the function $g_{\max}(x_1, x_2)$ by identifying its regions of definition in $\mathbb{R}^2$ and use it to classify a new point. Extension to $g_{\max}(x_1, x_2, \ldots, x_p)$ is possible (see Pham-Gia *et al.* (2015)).

(3) **Clustering:** Using the width of a group of densities $\|f_1, f_2, \ldots, f_k\|_1$, which is related to $f_{\max}(x_1, \ldots, x_p)$, to cluster a set of densities or a set of points after computing the approximating densities (see Vovan and Pham-Gia (2010)).

10.4.1 *Background*

The function defined pointwise as the maximum of several given functions, denoted by $f_{\max}$, is occasionally encountered in statistics. It usually provides a theoretical upper bound for these functions, from which other results can be derived. When the functions considered are probability densities,

$$f_{\max}(x) = \max\{f_1(x), \ldots, f_k(x)\}, \quad \forall x \in \mathbb{R}^n$$

is directly associated with the maximum of the two, or several, likelihoods, either normal or non-normal. Since it is at the root of the discriminant principle,

$$g_{\max}(x) = \max\{q_1 f_1(x), \ldots, q_k f_k(x)\}, \sum_{j=1}^{k} q_j = 1, \quad \forall x \in \mathbb{R}^n,$$

is a nonlinear transform of $f_{\max}$ and can have a particularly interesting role, as an efficient classifier, once its definition regions are clearly determined. The boundaries of these regions have linear or quadratic expressions in the normal case, and quite different algebraic expressions in other cases, but $g_{\max}$ can still fulfill its role.

Furthermore, via the value of its integral, $g_{\max}$ provides the value of the Bayes error in classification, which is the theoretical minimum misclassification probability. Also, for the case of two populations, clear relations exist between its integral and the L^1-distance between the two densities, and also with the overlapping coefficient, which measures the common area between them.

We will first establish the basic properties of $g_{\max}$ and evidence its role as a classifier. An integrated computer software, developed by our group AMSARG, extends and generalizes the MATLAB function BayesGauss (see Gonzalez *et al.* (2004, p. 493)), which is based on the same decision principles.

10.4.2   *Discrimination between two populations*

Let $f_1(x)$ and $f_2(x)$ be two functions on $\mathbb{R}^n$ and let $f_{\max}(x) = \max\{f_1(x), f_2(x)\}$ and $f_{\min}(x) = \min\{f_1(x), f_2(x)\}$ be two functions defined from them.

The L^1-distance between two functions is well defined by

$$\|f_1 - f_2\|_1 = \int_{\mathbb{R}^n} |f_1(x) - f_2(x)| dx = \int_{\mathbb{R}^n} f_{\max}(x) dx - \int_{\mathbb{R}^n} f_{\min}(x) dx.$$

$$(10.1)$$

Setting $\int_{\mathbb{R}^n} f_{\min}(x)dx = \lambda_{12}$, the measure of the overlap region, we have the following.

Properties: For densities $f_1(x)$ and $f_2(x)$, we have

$$\int_{\mathbb{R}^n} f_{\max}(x)dx + \int_{\mathbb{R}^n} f_{\min}(x)dx = \int_{\mathbb{R}^n} f_1(x)dx + \int_{\mathbb{R}^n} f_2(x)dx = 2,$$

$$\|f_1 - f_2\|_1 = 2\left(1 - \lambda_{1,2}\right). \tag{10.2}$$

10.4.3 *Relations between $f_{\max}$ and some measures in statistical discrimination analysis*

(1) When no prior probability is given to f_1 and f_2, classification of an observation x_0 is made by considering the likelihood ratio f_1/f_2 at x_0 and by assigning x_0 to f_1 if this ratio is higher than 1. Since $\lambda_{1,2}$ represents the minimum value of the classification error, also called the Bayes error, and denoted by Pe, we have

$$\int_{\mathbb{R}^n} f_{\max}(x)dx = 2 - Pe = \|f_1 - f_2\|_1 + Pe$$

or

$$\|f_1 - f_2\|_1 = \int_{R^n} f_{\max}(x)dx - Pe.$$

Classification by the Pe rule, as the most effective approach, is discussed by Ben Bassat (1982).

(2) With prior probabilities q and $1-q$ given to f_1 and f_2, the optimal classification rule assigns an observation x_0 to the class with maximum posterior probability at x_0. The intersection area between the two functions qf_1 and $(1-q)f_2$, denoted by $\lambda_{1,2}^{[q]}$, now represents the minimum value of the classification error $Pe^{[q]}$, with

$$Pe^{[q]} = \lambda_{1,2}^{[q]} = \int_{\mathbb{R}^n} \min\left\{qf_1(x), (1-q)f_2(x)\right\}dx. \tag{10.3}$$

The function of interest is now $\max\{f_1^*(x), f_2^*(x)\}$, where $f_i^*(x)$ are posterior densities. But since

$$f_1^*(x) = \frac{qf_1(x)}{qf_1(x) + (1-q)f_2(x)},$$

and similarly for $f_2^*(x)$, we can consider the function $g_{\max}^{[q]}(x) = \max\{qf_1(x), (1-q)f_2(x)\}$, which can serve for discrimination. We also have

$$Pe^{[q]} = \int\limits_{R^n} \min\{f_1^*(x), f_2^*(x)\}d\mu(x), \tag{10.4}$$

where $d\mu(x) = [qf_1(x) + (1-q)f_2(x)]\,dx$.

10.4.4   *Discrimination between k populations, k > 2*

Several of the above relations become invalid when there are more than two populations. Let us consider k densities $f_i(x)$, $i = 1, \ldots, k$, with $k \geq 3$ and let

$$f_{\max}(x) = \max\{f_1(x), f_2(x), \ldots, f_k(x)\}, \quad \forall x \in \mathbb{R}^n.$$

An L^1-distance between all k densities taken at the same time cannot really be defined, and the closest to it is a weighted sum of pairwise L^1-distances. However, using $f_{\max}$, we can devise a measure which could be considered as generalized L^1-distance between these k functions, since it is consistent with other considerations related to distances in general. This measure is

$$\|f_1, f_2, \ldots, f_k\|_1 \equiv \int\limits_{\mathbb{R}^n} f_{\max}(x)dx - 1, \tag{10.5}$$

and is slightly different than the case $k = 2$ given by Eq. (10.1). We have the double inequality:

$$\max_{i<j}\|f_i - f_j\|_1 \leq 2\|f_1, f_2, \ldots, f_k\|_1 \leq \sum_i \sum_j \|f_i - f_j\|_1. \tag{10.6}$$

But, as pointed out by Glick (1972), this distance $\|f_1, f_2, \ldots, f_k\|_1$ cannot be expressed as a function of pairwise distances, although it is bounded below by the max of these distances and above by their sum, as seen from the above relation.

Let us consider the case where the prior probability for f_i is q_i, $1 \leq i \leq k$, with $\sum_{i=1}^{k} q_i = 1$, and let $g_i(x) = q_i f_i(x)$ and $(q) = (q_1, \ldots, q_k)$. As an extension of Eq. (10.3), we have the following.

Definition 10.1. We define

$$g_{\max}(x) = \max \{g_1(x), g_2(x), \ldots, g_k(x)\}$$

and, similarly, for $g_{\min}(x)$.

10.4.5 *Relations between distances and Bayes error*

The relations between Bayes error $Pe_{1,2,\ldots,k}^{[(q)]}$, the generalized distance $\|f_1, f_2, \ldots, f_k\|_1$, $g_{\max}(x)$ and overlapping coefficients of different orders $\lambda_{i,j,\ldots,l}$, for any subset $\{i, j, \ldots, l\} \subset \{1, 2, \ldots, k\}$, become more complex, but the following results parallel those related to two distributions.

Using the inclusion–exclusion principle, we have

$$f_{\max} = \sum_{k=1}^{i} f_k - \sum_{k<l} \min(f_l, f_k) + \sum_{k<l<m} \min(f_l, f_k, f_m) + \cdots.$$

$$(10.7)$$

Integrating the above relation, we obtain

$$\int_R f_{\max}(x)dx = k - \sum_{k<l} \lambda_{kl} + \sum_{k<l<m} \lambda_{klm} + \cdots + \qquad (10.8)$$

and

$$\|f_1, f_2, \ldots, f_k\|_1 = \int_{R^n} f_{\max}(x)dx - 1$$

$$= (k-1) - \sum_{k<l} \lambda_{kl} + \sum_{k<l<m} \lambda_{klm}$$

$$+ \cdots + (-1)^{i-1} \lambda_{1,2\ldots k}. \qquad (10.9)$$

Since we have $\lambda_{kl} = 1 + (\|f_k - f_l\|_1/2)$, relation (10.9) follows.

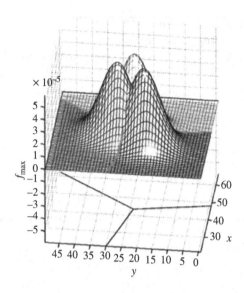

Fig. 10.3   Same covariance matrix, different means.

10.4.6 *Intersections of normal surfaces*

We give some graphical representations of intersections of three and four normal surfaces, and their projections on the horizontal plane in the case of equal covariance matrix (Figs. 10.3–10.5). We can see why this case is simple to treat. The case of unequal covariance matrices is much more complex (see Fig. 10.11). Let's consider five normal surfaces given by

$$\boldsymbol{\mu}_1^t = (40, 20), \ \boldsymbol{\mu}_2^t = (48, 24), \ \boldsymbol{\mu}_3^t = (43, 32),$$

$$\boldsymbol{\mu}_4^t = (38, 28), \ \boldsymbol{\mu}_5^t = (50, 30). \tag{10.10}$$

$$\boldsymbol{\Sigma}_1 = \boldsymbol{\Sigma}_2 = \boldsymbol{\Sigma}_3 = \boldsymbol{\Sigma}_4 = \boldsymbol{\Sigma}_5 = \begin{pmatrix} 20 & 5 \\ 5 & 30 \end{pmatrix}. \tag{10.11}$$

10.4.7 *Equivalence between classical approach boundaries and boundaries provided by $g_{\max}$*

We recall here first the basic theory related to normal distributions (see Johnson and Wichern (1998)) and establish, in two cases, the

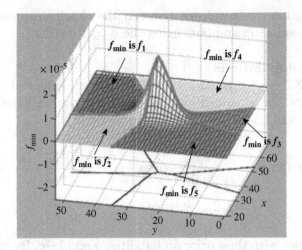

Fig. 10.4   Minimum function (same covariance matrix).

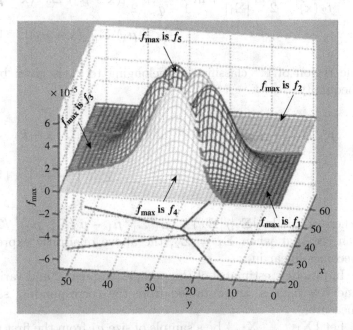

Fig. 10.5   Maximum function (same covariance matrix).

above-mentioned equivalence between the classical approach using the likelihood ratio and ours using the g_{max} function.

(A) Two populations

Let $\mathbf{X}_i \sim N_p(\boldsymbol{\mu}_i, \boldsymbol{\Sigma}_i), i = 1, 2$ be two independent normal vectors. In classification, we consider the ratio of the two densities:

$$f_i(\mathbf{x}) = \frac{1}{(2\pi)^{p/2}|\boldsymbol{\Sigma}_i|^{1/2}} \exp\left[-\frac{(\mathbf{x} - \boldsymbol{\mu}_i)^t \boldsymbol{\Sigma}_i^{-1}(\mathbf{x} - \boldsymbol{\mu}_i)}{2}\right],$$

$$i = 1, 2, \tag{10.12}$$

together with their prior probabilities q and $1-q$. By considering the log-likelihood ratio,

$$\ln \frac{f_1(\mathbf{x})}{f_2(\mathbf{x})} = \frac{1}{2}\ln\frac{|\boldsymbol{\Sigma}_2|}{|\boldsymbol{\Sigma}_1|} + \ln\frac{q}{1-q} + \frac{1}{2}\left[(\mathbf{x} - \boldsymbol{\mu}_2)^t\boldsymbol{\Sigma}_2^{-1}(\mathbf{x} - \boldsymbol{\mu}_2)\right.$$

$$\left. -(\mathbf{x} - \boldsymbol{\mu}_1)^t\boldsymbol{\Sigma}_1^{-1}(\mathbf{x} - \boldsymbol{\mu}_1)\right], \tag{10.13}$$

the region of $\mathbb{R}^n$ classified as belonging to π_1 is given by the inequality

$$\left[\boldsymbol{\mu}_1{}^t\boldsymbol{\Sigma}_1{}^{-1} - \boldsymbol{\mu}_2{}^t\boldsymbol{\Sigma}_2{}^{-1}\right]\mathbf{x} - \frac{1}{2}\mathbf{x}^t\left[\boldsymbol{\Sigma}_1{}^{-1} - \boldsymbol{\Sigma}_2{}^{-1}\right]\mathbf{x} - k$$

$$\geq \ln\left(\frac{q}{1-q}\right), \tag{10.14}$$

where $k = \left[\ln(|\boldsymbol{\Sigma}_1|/|\boldsymbol{\Sigma}_2|) + (\boldsymbol{\mu}_1{}^t\boldsymbol{\Sigma}_1^{-1}\boldsymbol{\mu}_1 - \boldsymbol{\mu}_2{}^t\boldsymbol{\Sigma}_1^{-1}\boldsymbol{\mu}_2)\right]/2$. We can see that when $\boldsymbol{\Sigma}_1 = \boldsymbol{\Sigma}_2$, the above quadratic expression becomes linear in $\mathbf{x}$.

In practice, estimates of the population means, and variances and covariances, are obtained from the corresponding sample measures.

Let $\{\mathbf{X}_{11}, \ldots, \mathbf{X}_{1n_1}\}$ be a sample of size n_1 from the first population, and similarly, we have $\{\mathbf{X}_{21}, \ldots, \mathbf{X}_{2n_2}\}$ from the second. We can then compute the sample mean vector $\bar{\mathbf{x}}_1, \bar{\mathbf{x}}_2$ and sample covariance matrices S_1 and S_2. We distinguish now between the two cases:

(a) **Equal covariance matrices: $\Sigma_1 = \Sigma_2 = \Sigma$**

The decision rule is then: For a new vector x_0, allocate it to π_1 if

$$(\bar{x}_1 - \bar{x}_2)^t (S^*)^{-1} x_0 - \frac{1}{2}(\bar{x}_1 - \bar{x}_2)^t (S^*)^{-1} (\bar{x}_1 + \bar{x}_2)$$

$$\geq \ln\left(\frac{q}{1-q}\right), \quad (10.15)$$

and to π_2, otherwise. Here, S^* is the estimate of the common variance matrix Σ and can be obtained by pooling S_1 and S_2.

We can see that the discriminant function $ld(x)$ is linear in x, since

$$ld(x) = (\bar{x}_1 - \bar{x}_2)^t (S^*)^{-1} x - A, \quad (10.16)$$

where $A = \frac{1}{2}(\bar{x}_1 - \bar{x}_2)^t (S^*)^{-1}(\bar{x}_1 + \bar{x}_2) + \ln(1 - q/q)$ and $x_0 \in \pi_1$ if $ld(x_0) \geq 0$.

But equation $ld(x_0) = 0$ precisely determines the linear boundary of the two regions, where $g_{max} = qf_1$ and $g_{max} = (1 - q)f_2$, and hence, the two approaches are equivalent in this case.

(b) **Different covariance matrices: $\Sigma_1 \neq \Sigma_2$**

For a new vector x_0, we consider the quadratic discrimination function, $qd(x)$, and allocate it to π_1 if

$$qd(x_0) = \left[\bar{x}_1{}^t(S_1)^{-1} - \bar{x}_2{}^t(S_2)^{-1}\right] x_0 +$$

$$- \frac{1}{2}x_0^t \left[(S_1)^{-1} - (S_2)^{-1}\right] x_0 - k \geq \ln\left(\frac{1-q}{q}\right), \quad (10.17)$$

and to π_2, otherwise, where

$$k = \frac{\ln(|S_1|/|S_2|) + \left(\bar{x}_1{}^t S_1^{-1}\bar{x}_1 - \bar{x}_2{}^t S_2^{-1}\bar{x}_2\right)}{2}.$$

Again, the two approaches are equivalent, since Eq. (10.17) also determines the quadratic boundary of the region, where $g_{max} = qf_1$.

(B) More than two populations

By considering each normal density, we obtain the associated quadratic discriminant score:

$$d_i^Q(\mathbf{x}) = -\ln|\mathbf{S}_i| - \left[(\mathbf{x} - \bar{\mathbf{x}}_i)^t(\mathbf{S}_i)^{-1}(\mathbf{x} - \bar{\mathbf{x}}_i)\right] + 2\ln(p_i),$$

$$i = 1, 2, \ldots, k, \tag{10.18}$$

and assign $\mathbf{x}_0$ to the population having the highest value $d_i^Q(\mathbf{x}_0)$. Since $g_{max} = \max\left(\exp\left[d_i^Q(\mathbf{x})\right]\right)$, $\forall \mathbf{x} \in \mathbb{R}^n$, we have equivalent results by considering g_{max}.

When all $\boldsymbol{\Sigma}_i$ are equal, $d_i^Q(\mathbf{x})$ becomes a linear discriminant score

$$d_i^L(\mathbf{x}) = 2(\bar{\mathbf{x}}_i)^t\mathbf{S}^{-1}\mathbf{x} - (\bar{\mathbf{x}}_i)^t\mathbf{S}^{-1}\bar{\mathbf{x}}_i + \ln p_i. \tag{10.19}$$

Since Eqs. (10.18) and (10.19) also determine the different quadratic or linear boundaries for g_{max}, there is equivalence of the two approaches.

10.4.8   *Numerical applications*

We will use Fisher's well-known data on iris to carry out discriminant analysis with g_{max}, by considering progressively one, two, and all four variables under the normal model, which has been verified to apply in each case.

Data: The three varieties of Iris, namely *Setosa* (Se), *Versicolor* (Ve), and *Virginica* (Vi), have data in four attributes: $x_1 =$ sepal length, $x_2 =$ sepal width, $x_3 =$ petal length, $x_4 =$ petal width. In the notation $x_{ij}, 1 \leq i \leq 3, 1 \leq j \leq 50$, the first index refers to the Iris variety and the second one to the observation. We can see that there are several combinations of iris varieties $\left(\binom{3}{2} + \binom{3}{3}\right)$ with iris attributes $\left(\binom{4}{1} + \binom{4}{2} + \binom{4}{3} + \binom{4}{4}\right)$, with two cases of covariance matrices in data analysis. Let us consider the following cases:

(A) Univariate case

(i) **Common variance:**
   Numerical example 10.1: We consider (x_1) in (Se), (Ve), and (Vi). We have $\bar{x}_{11} = 5.006$ and $\bar{x}_{21} = 5.936$ as the sample means

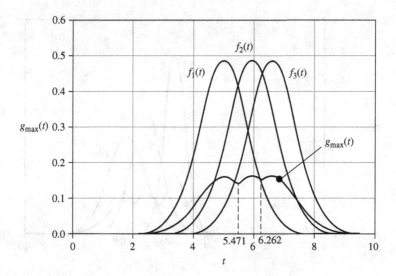

Fig. 10.6   Univariate case: $g_{\max}$ for three groups, same dispersion.

for (Se) and (Ve), respectively, and $\overline{x}_{31} = 6.588$. The total sample size for (x_1) is 150, which gives a common variance 0.825^2. Taking as $f_i, i = 1, \ldots, 3$, the density of $X_i \sim N\left(\overline{x}_{i1}, 0.825^2\right)$, we have the following: For $q_1 = q_2 = q_3 = 1/3$, we have $g_{\max} = \max\{f_1, f_2, f_3\}/3$. This is a curve with three equal maxima at the three mean values and two minima at their mid-points, i.e. at 5.471 and 6.262 (Fig. 10.6). Hence, a new value x_0 will be classified as belonging to f_1, f_2 or f_3, depending on whether $x_0 < 5.471$, or $5.471 \le x_0 < 6.262$, or $x_0 \ge 6.262$. We have

$$Pe_{1,2}^{[(1/3)]} = 1 - \int_{\mathbb{R}} g_{\max}(x)dx = 0.422,$$

which clearly shows that this univariate model is not adequate.

(ii) **Different variances:** This is a more complex case, which can lead to different results. Suppose that $\mu_1 < \mu_2 < \mu_3$ and also $\sigma_1 < \sigma_2 < \sigma_3$. Then, each couple $f_i, f_j, 1 \le i, j \le 3$ has two intersection points. In general, only one of the two intersection points is within the three standard deviation range, and the corresponding function $g_{\max}$ usually has three visible modes only.

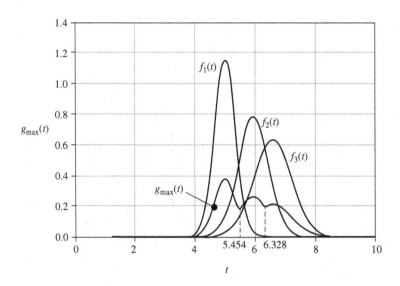

Fig. 10.7   Univariate case: g_{max} for three groups, different dispersions.

Numerical example 10.2: (Se), (Ve), and (Vi) now have different variances, estimated separately $s_1^2 = 0.349^2$, $s_2^2 = 0.511^2$, and $s_3^2 = 0.629^2$.

For $q_1 = q_2 = q_3 = 1/3$ (unless we have better information on $\{q_i\}$, we have g_{max} as a curve with three local max at the three mean values, and two minima at 5.454 and 6.328 (Fig. 10.7), and $Pe_{1,2,3}^{[(1/3)]} = 1 - \int_{\mathbb{R}} g_{max}(x)dx = 0.278$, which is better than the previous model.

We can see that, if (x_1) is chosen as the sole classification criterion, then for either two groups, or three groups, the different variance case is better, which is reflected by a smaller value of Pe.

The multivariate cases are more complicated although they follow the same steps.

(B) Bivariate case
Two approaches will be used, both based on the g_{max} function.

(1) **The parametric approach:** We consider the cases of two groups, supposed normally distributed, and in each case, bivariate (x_1, x_3) and quadrivariate distributions (all x_i, $i = 1, \ldots, 4$), with common or different dispersions, are considered. Results on these variables only are reported here due to space restrictions,

but all cases have been studied. Also, we only consider the case $q_i = 1/k, i = 1, \ldots, k$, where k is the number of groups, i.e. prior probabilities are taken as equal. But the general case, for $(q) = (q_1, q_2, \ldots, q_k)$, with $\sum_{j=1}^{k} q_j = 1$, and $g_j = q_j f_j$ can be dealt with in a similar way.

(2) **The non-parametric approach:** This approach uses density approximation, as presented in the article.

Numerical example 10.3: We have three estimates for (Se), (Ve), and (Vi): $S_1 = \begin{pmatrix} 0.122 & 0.016 \\ 0.016 & 0.03 \end{pmatrix}$, $S_2 = \begin{pmatrix} 0.261 & 0.179 \\ 0.179 & 0.216 \end{pmatrix}$, $S_3 = \begin{pmatrix} 0.396 & 0.297 \\ 0.297 & 0.298 \end{pmatrix}$.

For $q_1 = q_2 = q_3 = 1/3$, we have g_{max} defined in complementary regions T_1 to T_5 of $\mathbb{R}^2$, which are determined by the quadratic curves Λ_1 and Λ_1', Λ_2, and Λ_3 as follows: In T_1, $g_{max} = f_1/3$, in T_2, $g_{max} = f_2/3$, in T_3, $g_{max} = f_2/3$, in T_4, $g_{max} = f_3/3$, and in T_5, $g_{max} = f_2/3$. The three density surfaces, with different shapes, intersect along three space curves, Ω_1, Ω_2 and Ω_3 in $\mathbb{R}^3$ (Fig. 10.8), whose projections into the (x_1, x_3)-plane provide the boundaries, Λ_1 and Λ_1', Λ_2 and Λ_3, for discrimination between the three groups (Figs. 10.9

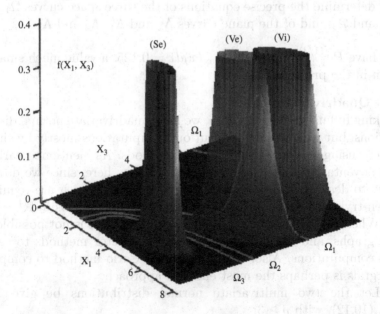

Fig. 10.8 Bivariate case: $f(x_1, x_3)$ for three groups, different dispersions in $\mathbb{R}^2$.

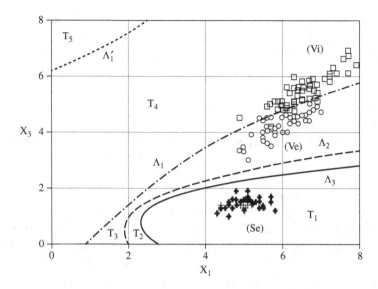

Fig. 10.9   Bivariate case: Parametric quadratic discriminant curves for three groups in $\mathbb{R}^2$.

and 10.10). Here, as in the previous case, we can analytically determine the precise equations of the three space curves Ω_1, Ω_2 and Ω_3, and of the plane curves Λ_1 and Λ_1', Λ_2 and Λ_3.

We have $Pe_{1,2,3}^{[(1/3)]} = 1 - \int_{\mathbb{R}^2} g_{\max}(x)dx = 0.125$, a value much smaller than in the previous case.

(C) Quadrivariate case

Making full use of Fisher's data, we have quadrivariate normal distributions, but we have to carry out our computations mostly by simulation, using the quasi-Monte Carlo methods. As mentioned earlier, the advantages of using $g_{\max}$ become obvious here, since we do not have to deal with complex analytic expressions, which are required when trying to determine different classification regions in $\mathbb{R}^4$.

When the number of variables is 3 or higher, it is not possible to use graphs, and we also have to resort to vector methods to carry out computations. Also, the quasi-Monte Carlo method to compute integrals is perhaps the most effective approach.

Let the two multivariate normal distributions be given by Eq. (10.12), with $p \geq 3$:

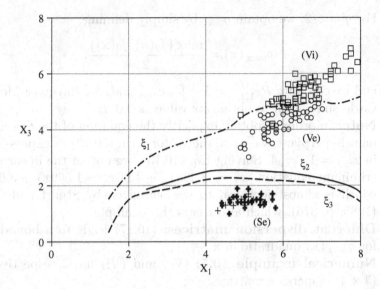

Fig. 10.10 Bivariate case: Non-parametric discriminant curves by kernel estimated densities in $\mathbb{R}^2$.

(a) **Same dispersion matrix:** For mean vectors, we now have $\overline{x}_1 = (\overline{x}_{11}, \overline{x}_{12}, \overline{x}_{13}, \overline{x}_{14})$ and similarly for and $\overline{x}_3$. Let us consider the last two groups, and suppose that they have a common (4×4) dispersion matrix $\mathbf{T}$.

By (10.16), we have a linear expression of $\mathbf{x}$ as the boundary for $g_{\max}(\mathbf{x})$ related to these two groups.

Numerical example 10.4: For this example, and the following one, let us consider (Ve) and (Vi), instead of (Ve) and (Se), so that the same numerical example already worked out in Stuart and Ord (1983, p. 374) can be verified. We hence use the complete set of data on Iris, in their usual order, with all four attributes (x_1, x_2, x_3, x_4). We have $\overline{x}_2 = (5.936, 2.77, 4.26, 1.326)^t$, $\overline{x}_3 = (6.588, 2.974, 5.552, 2.026)^t$, and the (4×4) *common dispersion matrix* is

$$\mathbf{T} = \begin{pmatrix} 0.435 & 0.121 & 0.449 & 0.165 \\ & 0.11 & 0.141 & 0.079 \\ & & 0.675 & 0.286 \\ & & & 0.179 \end{pmatrix}.$$

For $q = 1/2$, we obtain g_{max} by simply defining

$$g_{max}(\mathbf{x}) = \frac{\max\{f_2(\mathbf{x}), f_3(\mathbf{x})\}}{2},$$

and we compute $Pe_{1,2}^{(1/2)} = 1 - \int_{\mathbb{R}^4} g_{max}(x)dx$ by the quasi-Monte Carlo simulation to obtain its value as 0.027.

Note: In order to obtain explicitly the equation of the discriminant hyperplane, we write the vector form (10.12) as expressions in $x_i, i = 1, \ldots, 4$. Solving Eq. (10.13), we obtain the linear discriminant equation, $x_1 = 3.483x_4 + 1.96x_3 - 1.569x_2 + 3.685$, which is almost identical to the one given by Stuart and Ord (1983, p. 376), which also treats this example.

(b) **Different dispersion matrices:** (10.17) leads to a boundary for $g_{max}(\mathbf{x})$ quadratic in $\mathbf{x}$.

Numerical example 10.5: (Ve) and (Vi) have, respectively, (4×4) dispersion matrices:

$$S_{Ve} = \begin{pmatrix} 0.261 & 0.083 & 0.179 & 0.055 \\ & 0.097 & 0.081 & 0.040 \\ & & 0.216 & 0.072 \\ & & & 0.038 \end{pmatrix},$$

$$S_{Vi} = \begin{pmatrix} 0.396 & 0.092 & 0.297 & 0.048 \\ & 0.102 & 0.070 & 0.047 \\ & & 0.298 & 0.048 \\ & & & 0.074 \end{pmatrix},$$

while (Se) has as dispersion

$$S_{Se} = \begin{pmatrix} 0.122 & 0.097 & 0.016 & 0.010 \\ & 0.141 & 0.011 & 0.009 \\ & & 0.030 & 0.006 \\ & & & 0.011 \end{pmatrix}.$$

For $q = 1/2$, we have $g_{max}(\mathbf{x})$ defined as before. Similar to the preceding example, we can also obtain explicit quadratic expressions relating x_1 to the two other variables, which would determine the quadratic hypersurfaces separating different groups.

10.5   Application to Statistical Classification

Abstract: The maximum of k numerical functions defined on $\mathbb{R}^p$, $p \geq 1$, by $f_{\max}(\mathbf{x}) = \max\{f_1(\mathbf{x}), \ldots, f_k(\mathbf{x})\}$, $\forall \mathbf{x} \in \mathbb{R}^p$ is used here in statistical classification. We present first some theoretical results, then its application in classification using a computer program we have developed. This approach leads to clear decisions, even in cases where the extension to several classes of Fisher's linear discriminant function fails to be effective (Fig. 10.11).

10.5.1   *Our approach*

For normal surfaces of different means and covariance matrices, in the Bayesian approach, we classify a new value $\mathbf{x}_0$ into the class j_0, such that $\phi_{j_0}(\mathbf{x}_0) = \max_j\{\phi_j(\mathbf{x}_0)\}$. It corresponds to Eq. (10.18).

In the common Bayesian approach, to classify a new observation, we have the choice between the following:

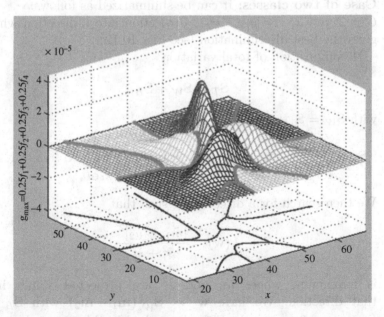

Fig. 10.11   Maximum of five binormal distributions with different covariance matrices.

(1) One against all, using the $(C-1)$ discriminant functions, with the decision each time: $\mathbf{x}_0$ is in group j or not in group j. As pointed out by several authors, this method can lead to regions not clearly assignable to any group.

(2) Two at a time, using $C(C-1)/2$ comparisons with regions delimited by straight lines or quadratic curves, each expression classifies new data as in C_i or C_j.

In our approach, we use the second method and compile all results so that $\mathbb{R}^2$ is now divided into disjoint sub-regions, each having a surface atop of it. Then, for a new observation $\mathbf{x}_0$, to classify it, we just use the max-classification function $\varphi\{g_i\}$ formally defined in the article.

10.5.2 *Fisher's approach*

It is the original method suggested first in discrimination and then in classification.

(a) **Case of two classes:** It can be summarized as follows:
$C = 2$, $p = 2$, $r = p - 1 = 1$: Projection into a direction which gives the best discrimination (see Fig. 10.12).
   Decomposition of total variation

$$\mathbf{S}_T = \mathbf{S}_W + \mathbf{S}_B,$$

with $\mathbf{S}_W = \mathbf{S}_1 + \mathbf{S}_2$ and

$$\mathbf{S}_i = \sum_{\mathbf{x} \in D_i} (\mathbf{x} - \mathbf{m}_i)(\mathbf{x} - \mathbf{m}_i)^t, \quad \mathbf{m}_i = \frac{1}{n_i} \sum_{\mathbf{x}_k \in D_i} \mathbf{x}_k, \quad i = 1, 2.$$

We then search for a direction $\mathbf{w}$ such that

$$J(\mathbf{w}) = \frac{|\tilde{m}_1 - \tilde{m}_2|^2}{\tilde{s}_1^2 + \tilde{s}_2^2}$$

is maximum, where $\tilde{m}_i$ and $\tilde{s}_i^2$ are projected values into that direction. We have $\mathbf{w} = \mathbf{S}_W^{-1}(\mathbf{m}_1 - \mathbf{m}_2)$ with $\mathbf{S}_B = (\mathbf{m}_1 - \mathbf{m}_2)(\mathbf{m}_1 - \mathbf{m}_2)^t$. Fisher's method in this case reduces to the discriminant method if we suppose the population normal.

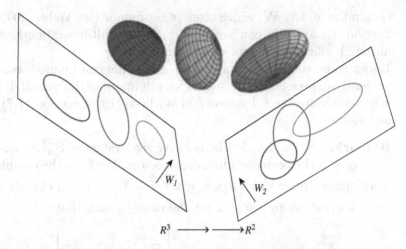

$$R^3 \longrightarrow \longrightarrow R^2$$

Fig. 10.12   Projections can be different according to the direction. The first direction separates the projections better.

(b) **Fisher's multilinear method (extension of the above approach due to Rao):** C classes of dimension p and $r = C - 1 < p$.
Projection into space of dimension r:

Decomposition of total variation in original space:

$$\mathbf{S}_T = \sum (\mathbf{x} - \mathbf{m})(\mathbf{x} - \mathbf{m})^t, \quad \mathbf{m} = \frac{1}{n} \sum_{i=1}^{C} n_i \mathbf{m}_i,$$

$$\mathbf{S}_T = \mathbf{S}_W + \mathbf{S}_B,$$

$$\mathbf{S}_W = \sum_{i=1}^{C} \mathbf{S}_i = \sum_{i=1}^{C} \sum_{\mathbf{x} \in D_i} (\mathbf{x} - \mathbf{m}_i)(\mathbf{x} - \mathbf{m}_i)^t,$$

$$\mathbf{S}_B = \sum_{i=1}^{C} n_i (\mathbf{m}_i - \mathbf{m})(\mathbf{m}_i - \mathbf{m})^t.$$

The projection from a p-dim space to a $(C - 1)$-dim space is done with a matrix $\mathbf{W}$ and we have $\mathbf{y} = \mathbf{W}^t \mathbf{x}$. Using $\mathbf{y}$, let the projected quantities be $\tilde{\mathbf{S}}_W = \mathbf{W}^t \mathbf{S}_W \mathbf{W}$, $\tilde{\mathbf{S}}_B = \mathbf{W}^t \mathbf{S}_B \mathbf{W}$.

We want to find the matrix $\mathbf{W}$ so that the ratio $J(\mathbf{W}) = |\tilde{\mathbf{S}}_B| / |\tilde{\mathbf{S}}_W|$ is maximum. Solving $|\mathbf{S}_B - \lambda_i \mathbf{S}_W| = 0$ to obtain λ_i and then solving $(\mathbf{S}_B - \lambda_i \mathbf{S}_W) w_i = 0$ to have eigenvectors w_i, we obtain the matrix $\mathbf{W}$, which often is not unique (see Webb (2002, Section 4.3.3)). We can see that Fisher's multilinear method, although efficient, can get quite complicated.

Under some conditions, the classification rule can be more conveniently expressed as the search for minimum of individual discriminant values (see Johnson and Wichern (1998, Section 11.7)) as follows:

Remark: Let $\{\lambda_1, \ldots, \lambda_s\}$ be positive eigenvalues of $\mathbf{S}_W^{-1}\mathbf{S}_B$ and $\{\mathbf{e}_1, \ldots, \mathbf{e}_s\}$ the corresponding eigenvectors. We form the sample discriminant functions $\left\{\hat{\mathbf{a}}_j^t\mathbf{x} = \hat{\mathbf{e}}_j^t\mathbf{x}\right\}, j = 1, \ldots, s$. To classify a new observation $\mathbf{x}_0$, we look for the class k_0 such that

$$\sum_{j=1}^{s} \mathbf{a}_{k_0}^t(\mathbf{x}_0 - \overline{\mathbf{x}}_{k_0})^2 = \min_i \left\{ \sum_{j=1}^{s} \mathbf{a}_j^t(\mathbf{x}_0 - \overline{\mathbf{x}}_i)^2 \right\}$$

and classify $\mathbf{x}_0$ in class k_0.

Several other approaches to classification have been proposed in the literature. Since the Bayes error is used as a yardstick, against which other errors are measured, the maximum function, as developed above, comes in very handy to deal with applied and computational aspects.

10.6   Application to Clustering

The same theory can be applied to clustering of probability densities.

Abstract: Using theoretical results on the maximum of several functions already established, and the definition of the joint distance of k probability densities, we derive new algorithms for cluster analysis based on cluster widths.

10.6.1   *Clustering with a bound on the width of a cluster*

We establish an algorithm so that each cluster is optimal and the joint distance of elements within a cluster does not exceed a chosen value.

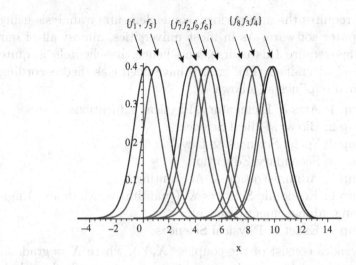

Fig. 10.13   Graph of the nine probability density functions and their $f_{\max}(x)$.

Numerical example 10.6

(a) Clustering univariate densities

We have nine populations with univariate normal probability densities (Fig. 10.13)

$$f_i(x) = \frac{1}{\sigma_i \sqrt{2\pi}} \exp\left[-\frac{(x-\mu_i)^2}{2\sigma_i^2}\right], \quad i = 1, \ldots, 9,$$

with specific parameters:

$$\sigma_1 = \sigma_2 = \sigma_3 = \sigma_4 = \sigma_5 = \sigma_6 = \sigma_7 = \sigma_8 = \sigma_9 = 1,$$

$$\mu_1 = 0.3, \quad \mu_2 = 4, \quad \mu_3 = 8.6, \quad \mu_4 = 9.7, \quad \mu_5 = 1,$$

$$\mu_6 = 5.3, \quad \mu_7 = 3.6, \quad \mu_8 = 8, \quad \mu_9 = 4.8.$$

With our approach presented, we obtain three clusters as given here.

(b) Clustering bivariate data

We will use real data in this application, and consider the clustering of groups taking the introductory statistics course at our university. According to our records, seven groups took our STAT 2853 in the year 2002, a first course in applied statistics, designed to introduce them to basic notions of probability and statistics.

It also requires them to perform some elementary analysis using a computer software. As in most universities, almost all of our programs require this course, and hence, its clientele is quite large and diversified. The groups have been classified according to their disciplines as follows:

- Group 1: Arts & Elementary/Physical Education,
- Group 2: Bio & Medical Sciences,
- Group 3: Social Sciences & Law,
- Group 4: Secondary Education,
- Group 5: Administration & Accounting,
- Group 6: Economic Sciences & Quantitative Methods (Engineering, Management Science),
- Group 7: Exact & Physical Sciences.

Final grades consist of the couples (X, Y), where X = grade of final exam and Y = grade of mid-term exam (the final grade is a weighted mean of these two grades, with weights determined by each individual instructor).

Table 10.1 and Fig. 10.14 give the seven samples of size 20 each, taken from the department records.

Our objective is to find the best way to reduce the number of groups, using the student's performance (X, Y) as criterion. Here, X and Y have the same weight.

Our approach is particularly suited to this kind of problem where entities to be clustered are groups of elements and not single elements. Also, these groups of discrete data must be estimated by probability densities, which is not the case of categorical data, for example. The kernel estimate of each density has the following form:

$$f(x, y) = \frac{1}{nh_x h_y} \sum_{j=1}^{m} \sum_{i=1}^{n} K\left(\frac{x - x_i}{h_x}, \frac{y - y_j}{h_y}\right),$$

where $n = 10$ is the number of elements in each group, and $K(z_1, z_2)$ is the normal kernel,

$$K(z_1, z_2) = \frac{1}{2\pi} \exp\left(-\frac{z_1^2 + z_2^2}{2}\right).$$

Here, we have seven sets of 20 points each. Using different bivariate distributions (bivariate beta, bivariate uniform, bivariate exponential and bivariate normal), we obtain different

Table 10.1   Grades (X, Y) for seven samples of STAT 2853.

No.	Group 1 (Arts & Elementary/Physical Education)		Group 2 (Bio & Medical Sciences)		Group 3 (Social Sciences & Law)		Group 4 (Secondary Education)		Group 5 (Administration & Accounting)		Group 6 (Economic Sciences & Quantitative Methods)		Group 7 (Exact/Physical Sciences)	
	X	Y	X	Y	X	Y	X	Y	X	Y	X	Y	X	Y
1	55.9	56.4	79.9	80.2	64.3	51.7	84.2	63.9	62.9	72.7	72.9	76.0	85.7	86.1
2	56.5	60.3	81.2	81.5	67.6	61.0	86.4	63.9	66.9	76.3	74.4	70.5	87.6	79.2
3	52.4	59.3	78.1	74.9	67.2	53.6	89.4	61.5	64.9	76.8	72.2	67.1	86.5	87.8
4	56.7	56.3	83.9	84.8	67.4	57.0	83.2	60.6	63.3	82.2	78.9	76.4	83.6	93.5
5	56.1	58.5	76.1	83.4	64.4	54.9	84.7	66.1	60.8	81.0	69.0	72.8	83.0	77.5
6	58.4	56.9	81.3	79.2	67.2	57.6	86.9	59.9	65.3	73.9	75.7	66.9	87.2	82.7
7	51.6	58.9	81.0	74.1	65.8	53.6	86.9	62.7	65.3	72.2	77.3	69.9	84.0	89.2
8	52.8	58.0	79.4	81.6	67.3	59.7	89.0	62.8	59.9	82.3	72.2	75.3	83.8	87.8
9	60.9	58.3	83.7	86.3	67.3	47.9	87.5	67.5	59.1	79.5	72.4	70.9	89.0	91.0
10	54.8	58.3	77.3	83.1	68.7	59.4	86.7	63.5	64.0	76.8	72.1	68.3	85.0	78.5
11	59.7	60.3	78.1	73.6	65.6	55.2	89.0	62.9	68.9	77.0	76.7	66.9	85.4	82.5
12	56.0	60.9	81.0	86.5	73.7	53.7	87.9	61.6	65.6	74.3	73.4	75.0	82.8	74.6
13	59.0	62.8	78.4	78.2	68.1	59.8	88.8	67.0	65.5	78.6	73.5	76.1	86.6	85.4
14	54.6	56.1	81.7	79.5	64.0	57.8	86.5	64.0	67.3	79.4	74.4	67.8	84.6	91.7
15	55.6	67.8	84.2	80.6	65.0	52.1	86.4	66.3	64.6	81.2	75.6	71.8	82.0	81.8
16	54.2	60.5	87.0	84.5	67.8	52.4	85.8	60.5	66.2	77.3	75.7	74.0	85.3	84.0
17	53.8	70.4	81.9	77.1	65.8	58.8	85.7	63.8	66.5	74.6	70.3	68.9	87.7	80.1
18	54.4	58.4	79.5	75.9	72.7	57.3	88.3	68.1	67.4	71.6	75.8	75.9	84.4	84.6
19	54.5	61.8	76.7	78.8	66.2	55.0	87.1	64.7	66.8	69.3	73.5	68.5	82.3	89.8
20	55.1	62.1	80.8	82.4	67.7	53.3	85.6	58.4	64.8	78.8	74.5	68.1	85.5	92.0

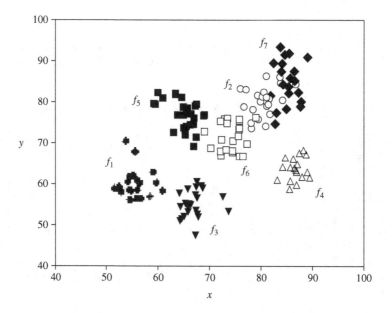

Fig. 10.14   Seven samples as points in $\mathbb{R}^2$ for STAT 2853.

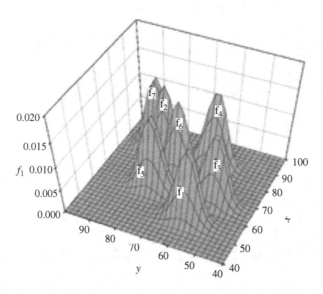

Fig. 10.15   Non-parametric kernel estimated densities.

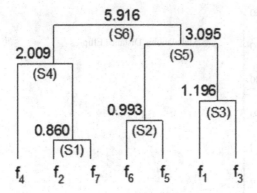

Fig. 10.16   Dendrogram of densities estimated from discrete data in $\mathbb{R}^2$.

density forms. The above normal kernel gives estimates of the seven densities, which are drawn in Fig. 10.15.

The clustering of these densities, according to our approach, is given by Fig. 10.16. Numbers on the graph gives distances, or widths, of progressive clusters (see Vovan and Pham-Gia (2010)).

10.7   Application of the Mahalanobis Square Distance in Landscape Analysis

We have seen the Mahalanobis square distance in Chapter 9, associated with the exponent of the density of a multivariate normal distribution.

Jenness Enterprises of Arizona, USA, has used this square distance and computed the spectrum of its values, as given in the joined graph (Fig. 10.17). Furthermore, when the two variables are identified in their domain of variation, we can obtain a grid of Mahalanobis values, as explained in the following section on landscape analysis.

10.7.1   *Applications to Landscape Analysis*[1]

For example, suppose that we have a raster of elevation values and a raster of slope values, and we are interested in identifying those

---

[1]This section has been reproduced with permission from Mr. J. Jenness of Jenness Enterprises, AZ, USA.

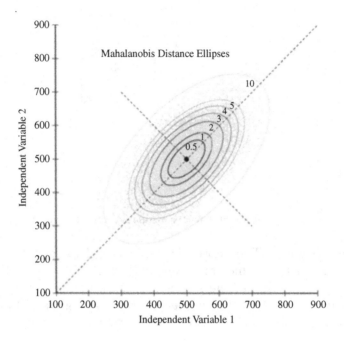

Fig. 10.17   Mahalanobis square distance ellipses.

regions on the landscape that have similar slopes and elevations to a mean slope and elevation preferred by some species of interest. Furthermore, we want to analyze the slope and elevations in combination so that if our species likes steep slopes at low elevations but shallow slopes at high elevations, then we won't inadvertently select steep slopes at high elevations or shallow slopes at low elevations.

Assume that the niche of our species of interest can be described in terms of elevation and slope with the following parameters:

$$\text{Vector of Mean Values} = \begin{pmatrix} \text{Elevation} = 2{,}121 \\ \text{Slope} = 18 \end{pmatrix}$$

$$\text{Covariance Matrix} = \begin{pmatrix} 1{,}931 & -54 \\ -54 & 87 \end{pmatrix}.$$

We can then enter the elevation and slope rasters directly into the Mahalanobis equation to produce a raster of Mahalanobis distance

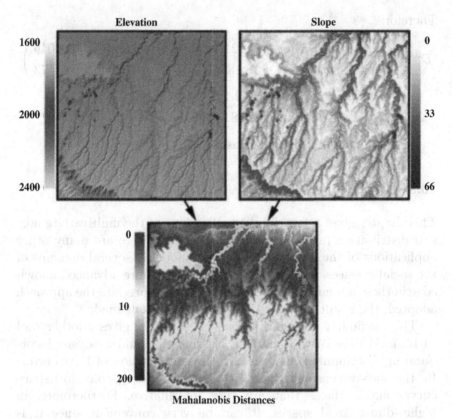

Fig. 10.18 Combined rasters of Mahalanobis distances related to elevation and slope.

values (Fig. 10.18): Given that

$$D^2 = (\mathbf{x} - \mathbf{m})^t \mathbf{C}^{-1} (\mathbf{x} - \mathbf{m}),$$

where $\mathbf{x}$ is the vector of variable values, $\mathbf{m}$ is the vector of mean values, and $\mathbf{C}$ is the covariance matrix.

Then

$$(\mathbf{x} - \mathbf{m}) = \begin{pmatrix} [\text{Elevation Raster}] - 2{,}121.41667 \\ [\text{Slope Raster}] - 18.18997 \end{pmatrix}$$

$$\mathbf{C}^{-1} = \begin{pmatrix} 1{,}931 & -54 \\ -54 & 87 \end{pmatrix}^{-1} = \begin{pmatrix} 0.00074 & 0.00046 \\ 0.00046 & 0.01173 \end{pmatrix}.$$

Therefore,

$$D^2 = \begin{pmatrix} \text{[Elevation Raster]} - 2{,}121.41667 \\ \text{[Slope Raster]} - 18.18997 \end{pmatrix}^t \times \begin{pmatrix} 0.00074 & 0.00046 \\ 0.00046 & 0.01173 \end{pmatrix}$$

$$\times \begin{pmatrix} \text{[Elevation Raster]} - 2{,}121.41667 \\ \text{[Slope Raster]} - 18.18997 \end{pmatrix}$$

$$= \text{[Mahalanobis Distance Raster]}.$$

10.8   Conclusion

This chapter gives a few simple applications of the multivariate normal distribution in science and engineering. There are many other applications of the multivariate normal model in several domains of the social sciences. Most of them, however, require advanced knowledge in these domains themselves in order to appreciate the approach adopted, the results obtained, and the conclusions made.

The maximum function, as presented above, gives another tool to be used in statistical pattern classification and analysis, incorporating discriminant analysis and the computation of Bayes error. In the two-dimensional case, it also provides graphs for space curves and surfaces that are very informative. Furthermore, in higher-dimensional spaces, it can be very convenient since it is machine-oriented and can free the analyst from complex analytic computations related to the determination of different domains of definition of the discriminant function.

Bibliography

Ben Bassat, M. (1982). *Use of Distance Measures, Information Measures and Error Bounds in Feature Evaluation*, in: Krishnaiah, P.R., Kanal, L.N., eds., *Handbook of Statistics*, Vol. 2 (Amsterdam, North-Holland), pp. 773–791.

Glick, N. (1972). Sample-based classification procedures derived from density estimators, *Journal of the American Statistical Association*, **67**, 116–122.

Glick, N. (1973). Separation and probability of correct classification among two or more distributions, *Annals of the Institute of Statistical Mathematics*, **25**, 373–382.

Gonzalez, R.C., Woods, R.E. and Eddins, S.L. (2004). *Digital Image Processing with MATLAB* (Prentice-Hall, New York).

Hervet, E., Kardouchi, M., Nyiumgere, F. and Pham-Gia, T. (2009). Eye recognition using a statistical approach, *Advances in Statistics and Applications*, **12**(2), 209–223.

Johnson, R. and Wichern, D. (1998). *Applied Multivariate Statistical Analysis*, 4th ed., (Prentice-Hall, New York).

Pham-Gia, T. and Turkkan, N. (2006). Stress–strength reliability: The multivariate case, *IEEE Transactions on Reliability*, **56**(1), 115–124.

Pham-Gia, T., Nhat, N.D. and Phong, N.V. (2015). Statistical classification using the maximum function, *Open Journal of Statistics*, **5**(7), 665–679.

Pham-Gia, T., Turkkan, N. and Vovan, T. (2008). Statistical discrimination analysis using the maximum function, *Communications in Statistics — Simulation and Computation*, **37**(2), 320–336.

Stuart, A. and Ord, K. (1983). *Kendall's Advanced Theory of Statistics*, Vol. 3, 4th ed., (Macmillan, London).

Vovan, T. and Pham-Gia, T. (2010). Clustering probability densities, *Journal of Applied Statistics*, **37**(11), 1891–1910.

Webb, A.T. (2002). *Statistical Pattern Recognition*, 2nd ed. (John Wiley and Sons, New York).

Chapter 11

Functions of Matrices and Jacobians of Transformations

11.1 Introduction

In multivariate analysis, we frequently have to consider matrices, instead of vectors, as variables. If they are random and have a probability distribution, special rules apply to operations on these distributions. Furthermore, one richness of matrix theory is the various decompositions of a matrix into simpler, or special, matrices to make operations easier to handle and use already established results. But decomposition combined with variable changes can also make operations quite complicated. One basic question we may wish to ask is whether we can divide in this algebra, or equivalently, whether we can find multiplicative inverses for matrices. For example, in matrix algebra, any matrix with non-zero determinant has an inverse, so there are some matrices that we really can divide by. However, there are also matrices which are not invertible. So, in general, you cannot divide two matrices. A difficulty related to deriving Jacobians for matrix transformations is that these Jacobians can have forms not at all expected or familiar.

After reviewing basic matrix properties and discussing about Jacobians, we concentrate on two basic matrix operations: differentiation and integration. For differentiation, we recall, without proofs, several important results. For integration, various suppositions will be made on the matrix in order to be able to derive results.

347

Skew-symmetric matrix component, spectral and singular value decompositions will be considered.

11.2 Review on Matrices

11.2.1 *Linear transformations and matrices*

Consider the matrix $\mathbf{A}$ which can have special forms.

$\mathbf{A} = \begin{bmatrix} c & 0 \\ 0 & c \end{bmatrix} = c\mathbf{I}_2$, this matrix stretches every vector by the vector c, with the result that the whole space expands or contracts, or goes to the opposite side when c is negative.

$\mathbf{A} = \begin{bmatrix} 0 & -1 \\ 1 & 0 \end{bmatrix}$ is a rotation which turns all vectors through $90°$, mapping $(x, y) \to (-y, x)$. If the angle is different from $90°$, we have cases in Chapter 7.

$A = \begin{bmatrix} 0 & 1 \\ 1 & 0 \end{bmatrix}$ is a reflection wrt $45°$ line, sending $(x, y) \to (y, x)$.

Finally, the matrix $\mathbf{A} = \begin{bmatrix} 1 & 0 \\ 0 & 0 \end{bmatrix}$ is a projection and takes the whole space into a lower-dimensional subspace.

For all real numbers α, β and vectors $\mathbf{x}, \mathbf{y}$, the rule of linearity of transformations T is

$$T(\alpha\mathbf{x} + \beta\mathbf{y}) = \alpha T(\mathbf{x}) + \beta T(\mathbf{y}),$$

and any transformation that satisfies this requirement is a linear transformation. Matrix multiplication is a linear transformation in $\mathbb{R}^p$.

11.2.2 *Basic properties*

(a) We suppose that the basic knowledge on matrices is known. Let $\mathbf{A}$ and $\mathbf{B}$ be two matrices of the same dimensions $(m \times n)$.

Addition and difference of $\mathbf{A}$ and $\mathbf{B}$: Operations are made on each entry, resulting in $\mathbf{A} + \mathbf{B}$ and $\mathbf{A} - \mathbf{B}$ of the same dimensions.

Multiplication of $\mathbf{A}$ and $\mathbf{B}$: $\mathbf{A} \times \mathbf{B}$ can only be made if $\mathbf{A}(n \times p)$ and $\mathbf{B}(p \times r)$ and $\mathbf{AB}$ is $(n \times r)$, while $\mathbf{BA}$ is defined only if $r = n$. The product is hence not always commutative.

(b) $\mathbf{A}^{-1}$, the inverse of $\mathbf{A}$, is the matrix, s.t. $\mathbf{A}.\mathbf{A}^{-1} = \mathbf{A}^{-1}.\mathbf{A} = \mathbf{I}_p$, where $\mathbf{I}_p$ is the square identity matrix of size p, i.e. it has 1 on the diagonal and 0 everywhere else. $\mathbf{A}^{-1}$ can exist or not. Often, only the left or right inverse exists. We define $\mathbf{A}/\mathbf{B} = \mathbf{A}\mathbf{B}^{-1}$, when the second operation can be done. Generalized inverses can also be defined.

(c) An orthogonal matrix has its inverse equal its transpose, i.e. $\mathbf{Q}^{-1} = \mathbf{Q}^t$. Then, for the ith row q_i^t, the product $q_i{}^t q_i = 1$ (if orthonormal), while $q_i{}^t q_j = 0, i \neq j$, and similarly for columns.

(d) **Eigenvalues and eigenvectors:** For a square matrix $\mathbf{A}$, λ is an eigenvalue, with corresponding eigenvector $\mathbf{x} \neq 0$ if $\mathbf{A}\mathbf{x} = \lambda\mathbf{x}$. In other words, $\mathbf{x}$ is a fixed point direction for the linear transformation performed by $\mathbf{A}$, denoted $\mathbf{e}_j$. We set $\mathbf{x}^t\mathbf{x} = 1$ to normalize $\mathbf{x}$. For a $(k \times k)$ matrix, we then have up to k pairs $\{\lambda_i, \mathbf{e}_i\}_{i=1}^k$, with $\mathbf{e}_j{}^t\mathbf{e}_j = 1$. They form an orthonormal basis. Also, the eigenvalues are unique if there are no multiple values. If we write $\mathbf{A} = \sum_{j=1}^k \lambda_j \mathbf{e}_j \mathbf{e}_j{}^t$, we have made a *spectral decomposition* of the matrix $\mathbf{A}$ into k column eigenvectors (see (f)).

For $\mathbf{A}(n \times n)$, the eigenvalues of $\mathbf{A}^k$, $k = 1, \ldots, m$ are λ_i^k, $i = 1, \ldots, n$, with same eigenvectors. For $\mathbf{A}^{-1}$, it is λ^{-1}. Also, eigenvalues of $\mathbf{A}$ equal eigenvalues of $\mathbf{A}^t$.

We have trace $= \lambda_1 + \cdots + \lambda_n = a_{11} + \cdots + a_{nn}$ and $\det\mathbf{A} = \lambda_1 \times \cdots \times \lambda_n$.

(e) **Positive definite matrix and quadratic form:** Let $\mathbf{A}$ be a $(k \times k)$ symmetric matrix. Then $Q_k(x) = x^t\mathbf{A}x$ is a quadratic form. If for any k-vector different from 0, we have $0 \leq x^t\mathbf{A}x$, $\mathbf{A}$ is called non-negative definite. If $0 < x^t\mathbf{A}x$, it is positive definite. This condition for a matrix is equivalent to a real number being strictly positive. It can be shown that a $(k \times k)$ matrix is positive definite iff all its eigenvalues are positive or that all sub-determinants taken along the diagonal are positive.

(f) **Square root of a matrix:** Let $\mathbf{A}$ be a $(k \times k)$ positive definite matrix. We can write

$$\mathbf{A} = \sum_{i=1}^k \lambda_i e_i e_i{}^t = \mathbf{P}\mathbf{\Lambda}\mathbf{P}^t$$

$$= [e_1 \ e_2 \ \ldots \ e_k] \, diag(\lambda_1, \ldots, \lambda_k)[e_1 \ e_2 \ \ldots \ e_k]^t.$$

We have $\mathbf{A}^{-1} = \mathbf{P}\mathbf{\Lambda}^{-1}\mathbf{P}^t$. We define $\mathbf{\Lambda}^{1/2} = diag(\sqrt{\lambda_1}, \ldots, \sqrt{\lambda_k})$, and the square root matrix $\mathbf{A}^{1/2} = \mathbf{P}\mathbf{\Lambda}^{1/2}\mathbf{P}^t$, which has many properties similar to those of the scalar square root.

11.2.3 *Important decompositions of a matrix*

(a) In a factorization $\mathbf{A} = \mathbf{LU}$, we seek the decomposition of a matrix $\mathbf{A}$ into the product of two triangular matrices, the lower one $\mathbf{L}$ and the upper one $\mathbf{U}$.

In general, systems of equations involving triangular coefficient matrices are easier to deal with. Indeed, the whole point of Gaussian elimination is to replace the coefficient matrix with one that is triangular.

Equation $\mathbf{AX} = \mathbf{B}$ becomes $\mathbf{LUX} = \mathbf{B}$ or $\mathbf{UX} = \mathbf{L}^{-1}\mathbf{B}$, where $\mathbf{U}$ is upper triangular and easier to solve. By writing down the equations relating the entries a_{ij}, l_{ij}, u_{ij}, we can easily find these entries values.

For example,

$$A = \begin{bmatrix} 3 & 1 & 6 \\ -6 & 0 & -16 \\ 0 & 8 & -17 \end{bmatrix} = \begin{bmatrix} 1 & 0 & 0 \\ -2 & 1 & 0 \\ 0 & 4 & 1 \end{bmatrix} \times \begin{bmatrix} 3 & 1 & 6 \\ 0 & 2 & -4 \\ 0 & 0 & -1 \end{bmatrix}.$$

A positive definite matrix always has a $\mathbf{LU}$ decomposition, but not every matrix has. A reordering of the rows of an invertible matrix can lead to this decomposition.

(b) Factorization $\mathbf{A} = \mathbf{QR}$. We have to first obtain an orthonormal basis for $\mathbf{A}$, whose column vectors will give $\mathbf{Q}$. This can be done in several ways, one of which is is the Gram–Schmidt process.

Gram–Schmidt process: From a set of independent vectors, this process leads progressively to an orthonormal basis by following these operations:

- Suppose that V is a k dim space with basis $B = \{\mathbf{b}_1, \ldots, \mathbf{b}_k\}$.
- Create progressively

$$\mathbf{u}_1 = \mathbf{b}_1,$$

$$\mathbf{u}_2 = \mathbf{b}_2 - \mathrm{proj}(\mathbf{b}_{2\,\mathrm{onto}}\,\mathbf{u}_1),$$

$$\mathbf{u}_3 = \mathbf{b}_3 - \text{proj}(\mathbf{b}_{3\,\text{onto}}\,\mathbf{u}_1) - \text{proj}(\mathbf{b}_{3\,\text{onto}}\,\mathbf{u}_2),$$
$$\cdots$$

$$\mathbf{u}_k = \mathbf{b}_k - \text{proj}(\mathbf{b}_{k\,\text{onto}}\,\mathbf{u}_1) - \cdots - \text{proj}(\mathbf{b}_{k\,\text{onto}}\,\mathbf{u}_{k-1}).$$

- $\mathbf{v}_i = \mathbf{u}_i/\|\mathbf{u}_i\|, i = 1,\ldots,k$. Let set $\mathbf{Q} = \begin{bmatrix} \mathbf{v}_1\ \mathbf{v}_2\ \cdots\ \mathbf{v}_k \end{bmatrix}$. $\mathbf{Q}$ is orthogonal, with $\mathbf{Q}\mathbf{Q}^t = \mathbf{Q}^t\mathbf{Q} = \mathbf{I}$.

Every $(m \times n)$ matrix with independent columns can be decomposed into $\mathbf{A} = \mathbf{Q}\mathbf{R}$, where the columns of $\mathbf{Q}$ are orthonormal and $\mathbf{R}$ is upper triangular and invertible. When $m = n$, $\mathbf{Q}$ becomes an orthogonal matrix.

(c) A matrix $\mathbf{A}$ is skew-symmetric if $\mathbf{A}^t = -\mathbf{A}$, with all zeros on the diagonal. Writing $\mathbf{A} = \frac{1}{2}(\mathbf{A} + \mathbf{A}^t) + \frac{1}{2}(\mathbf{A} - \mathbf{A}^t)$, we see that any matrix is the sum of a symmetric matrix and a skew-symmetric matrix.

(d) Spectral decomposition of a matrix is seen in (d) of the previous section, and an application in whitening is seen in Chapter 9.

(e) **Cholesky decomposition**: If $\mathbf{A}$ is a positive definite square matrix, there exists a unique lower (or upper) triangular matrix $\mathbf{T}$, $t_{ij} = 0$, $i < j$, (or $t_{ij} = 0, i > j$) with positive diagonal elements such that $\mathbf{A} = \mathbf{T}\mathbf{T}^t$.

(f) **Singular value decomposition of a rectangular matrix:** Let $\mathbf{A}$ be a $(m \times k)$ matrix. Then there exist matrices $\mathbf{U}(m \times m)$ and $\mathbf{V}(k \times k)$, both orthogonal s.t. $\mathbf{A} = \mathbf{U}\Lambda\mathbf{V}^t$, where Λ is partially diagonal with entry, called *singular values*, with λ_{jj} positive, $1 \leq j \leq \min(m,k)$. This decomposition shows that the effects of a rectangular matrix on a vector can be decomposed into a rotation, a stretching or compressing, and another rotation. We will use some properties of this decomposition in later chapters. This decomposition can be considered as an extension of the spectral decomposition seen in Section 11.2.2(d).

This is a very important decomposition of a rectangular matrix and it has numerous applications in various domains. An example follows: Let $\mathbf{A}(2 \times 3) = \begin{bmatrix} -1 & 1 & 0 \\ 0 & -1 & 1 \end{bmatrix}$. Its **SVD** is

$$\mathbf{A} = \frac{1}{\sqrt{2}}\begin{bmatrix} -1 & 1 \\ 1 & 1 \end{bmatrix}\begin{bmatrix} \sqrt{3} & 0 & 0 \\ 0 & 1 & 0 \end{bmatrix}\begin{bmatrix} 1/\sqrt{6} & -2/\sqrt{6} & 1/\sqrt{6} \\ -1/\sqrt{2} & 0 & 1/\sqrt{2} \\ 1/\sqrt{3} & 1/\sqrt{3} & 1/\sqrt{3} \end{bmatrix}.$$

(g) **The generalized inverse:** For an $(n \times m)$ matrix $\mathbf{X}$, its Moore–Penrose generalized inverse is an $(m \times n)$ matrix $\mathbf{X}^+$ such that

$$\mathbf{X}\mathbf{X}^+\mathbf{X} = \mathbf{X},$$
$$\mathbf{X}^+\mathbf{X}\mathbf{X}^+ = \mathbf{X}^+,$$
$$(\mathbf{X}\mathbf{X}^+)^t = \mathbf{X}\mathbf{X}^+,$$
$$(\mathbf{X}^+\mathbf{X})^t = \mathbf{X}^+\mathbf{X}.$$

We know that $\mathbf{X}^+$ exists and is unique. Furthermore, if

$$\mathbf{X} = \mathbf{U}\begin{pmatrix} \mathbf{D} & \mathbf{0} \\ \mathbf{0} & \mathbf{0} \end{pmatrix}\mathbf{V}$$

is the singular value decomposition (SVD) of $\mathbf{X}$, then

$$\mathbf{X}^+ = \mathbf{V}^t\begin{pmatrix} \mathbf{D}^{-1} & \mathbf{0} \\ \mathbf{0} & \mathbf{0} \end{pmatrix}\mathbf{U}.$$

(h) In one variable, if the rv X has density $f(x)$, then its reciprocal, or inverse, $Y = 1/X$, has density $g(y) = \frac{1}{y^2}f\left(\frac{1}{y}\right)$ on an appropriate domain. We know that for $X \sim N_1(0,1)$, Y is bimodal.

   If the inverse of a matrix does not exist, we can consider its generalized Moore–Penrose pseudo-inverse and consider its distribution, as done by Zhang (2007), who presents the expressions of the pseudo-inverse multivariate normal distributions and the pseudo-inverse matrix variate normal distributions, among others.

Applications of the SVD: Let $\mathbf{A} = \mathbf{U}\boldsymbol{\Sigma}\mathbf{V}^t$ be the singular value decomposition of the matrix $\mathbf{A}$. It could be seen as a generalization of the eigenvalue–eigenvector decomposition, where $\mathbf{U}$ and $\mathbf{V}$ are now any two orthogonal matrices, not necessarily transposes of each other. The diagonal matrix $\boldsymbol{\Sigma}$ has eigenvalues from $\mathbf{A}^t\mathbf{A}$, denoted $\sigma_1, \ldots, \sigma_r$. The columns of $\mathbf{U}(m \times m)$ are eigenvectors of $\mathbf{A}^t\mathbf{A}$ while the columns of $\mathbf{V}(n \times n)$ are eigenvectors of $\mathbf{A}^t\mathbf{A}$. The r singular values on the diagonal of $\boldsymbol{\Sigma}(m \times n)$ are the square roots of the non-zero eigenvalues of both $\mathbf{A}^t\mathbf{A}$ and $\mathbf{A}\mathbf{A}^t$.

   The ratio $\sigma_{\max}/\sigma_{\min}$ is the condition number and can serve in many situations. An example is given in Strang (2006).

In sending information from space, a satellite can send the whole picture $\mathbf{A}$ of $(1{,}000 \times 1{,}000)$ or one million pixels. But if we know the svd of $\mathbf{A}$ and find that among $\sigma_1, \ldots, \sigma_r$, only 20 of them are significant, we can keep them and ignore the others. We send only the corresponding 20 columns of $\mathbf{U}$ and $\mathbf{UA} = \mathbf{U\Sigma V}^t$ gives $\mathbf{A} = \sum_{j=1}^{20} u_j \sigma_j v_j^t$, which is 20 times 2,000 numbers instead of one million.

(i) **Vinograd's theorem**: Let $\mathbf{A}(k \times m)$ and $\mathbf{B}(k \times n)$ be two real matrices with $m \le n$. Then $\mathbf{AA}^t = \mathbf{BB}^t$ iff there exists an orthogonal matrix $\mathbf{H}$, $\mathbf{HH}^t = \mathbf{I}_m$ such that $\mathbf{AH} = \mathbf{B}$.

One consequence is as follows: Let $\mathbf{A}$ be a $(n \times m)$ real matrix of rank m, $m \le n$. Then $\mathbf{A}$ can be written as $\mathbf{A} = \mathbf{H}\begin{bmatrix} \mathbf{I}_m \\ 0 \end{bmatrix}\mathbf{B}$, where $\mathbf{H}(n \times n)$ is orthogonal and $\mathbf{B}(m \times m)$ is positive definite.

11.3 Two Important Operations on Matrices

We consider first the *Vec operation* and the *Kronecker product* as follows:

(1) **The Vec operation**

Let $\mathbf{A}$ be a $(p \times q)$ matrix. Vec$(\mathbf{A})$ is the *column vector* obtained by appending the columns of $\mathbf{A}$ one atop of the other to form a $(1 \times pq)$ column vector.

Conversely, given a $[(n \times m) \times 1]$ column vector, it can be divided into n column vectors $(m, 1)$ and build a $(m \times n)$ matrix by putting these $(m, 1)$ column vectors side by side, and similarly for n line vectors $(1, n)$.

(2) **The Kronecker product**

Let $\mathbf{A}(p \times q)$ and $\mathbf{B}(m \times n)$ be two matrices. Their *Kronecker product* $\mathbf{A} \otimes \mathbf{B}$ is the matrix $\mathbf{C}(pm \times qn)$ obtained by multiplying each element of $\mathbf{A}$ by $\mathbf{B}$:

$$\mathbf{C} = \begin{bmatrix} a_{11}\mathbf{B} & a_{12}\mathbf{B} & \cdots & a_{1q}\mathbf{B} \\ a_{21}\mathbf{B} & a_{22}\mathbf{B} & \cdots & a_{2q}\mathbf{B} \\ \vdots & \vdots & \ddots & \vdots \\ a_{p1}\mathbf{B} & a_{p2}\mathbf{B} & \cdots & a_{pq}\mathbf{B} \end{bmatrix}.$$

This product is not commutative, i.e. $\mathbf{A} \otimes \mathbf{B} \ne \mathbf{B} \otimes \mathbf{A}$, but enjoys several properties:

(a) $|\mathbf{I} \otimes \mathbf{A}| = |\mathbf{A}|^q$ since

$$\mathbf{I} \otimes \mathbf{A} = \begin{bmatrix} \mathbf{A} & 0 & \cdots & 0 \\ 0 & \mathbf{A} & \cdots & 0 \\ \vdots & \vdots & \ddots & \vdots \\ 0 & 0 & \cdots & \mathbf{A} \end{bmatrix}.$$

(b) For any non-zero scalars α, β, $(\alpha \mathbf{A}) \otimes (\beta \mathbf{B}) = \alpha \beta (\mathbf{A} \otimes \mathbf{B})$.
(c) For $\mathbf{A}$ and $\mathbf{B}$ of same dimensions, $(\mathbf{A} + \mathbf{B}) \otimes \mathbf{C} = (\mathbf{A} \otimes \mathbf{C}) + (\mathbf{B} \otimes \mathbf{C})$.
(d) $(\mathbf{A} \otimes \mathbf{B}) \otimes \mathbf{C} = \mathbf{A} \otimes (\mathbf{B} \otimes \mathbf{C})$.
(e) $(\mathbf{A} \otimes \mathbf{B})^t = (\mathbf{A}^t \otimes \mathbf{B}^t)$.
(f) $\mathrm{tr}(\mathbf{A} \otimes \mathbf{B}) = \mathrm{tr}(\mathbf{A})\mathrm{tr}(\mathbf{B})$ if $\mathbf{A}$ and $\mathbf{B}$ are square and have same dimensions.
(g) For non-singular matrices, $(\mathbf{A} \otimes \mathbf{B})^{-1} = (\mathbf{A}^{-1} \otimes \mathbf{B}^{-1})$.
(h) For $\mathbf{A}(m \times m)$ and $\mathbf{B}(n \times n)$ square,

$$\det(\mathbf{A} \otimes \mathbf{B}) = (\det \mathbf{A})^n (\det \mathbf{B})^m.$$

(i) Relations between eigenvalues of $\mathbf{A}, \mathbf{B}$ and $\mathbf{A} \otimes \mathbf{B}$ exist. For example, if $\{\lambda_1, \ldots, \lambda_m\}$ and $\{\mu_1, \ldots, \mu_n\}$ are the two sets of eigenvalues for $\mathbf{A}$ and $\mathbf{B}$, respectively, then the eigenvalues of $\mathbf{A} \otimes \mathbf{B}$ are $\{\lambda_i \mu_j, \ i = 1, m, j = 1, n\}$.

Theorem 11.1. *Let* $\mathbf{A}(p \times q)$, $\mathbf{X}(q \times r)$, *and* $\mathbf{B}(r \times s)$ *be three matrices. Then we have the equality:*

$$Vec(\mathbf{AXB}) = (\mathbf{B}^t \otimes \mathbf{A})Vec(\mathbf{X})$$

for the $(ps \times 1)$ *vector* $Vec(\mathbf{AXB})$.

Proof. The proof is quite involved.   □

11.4   Jacobians of Transformations Between Special Matrices

Distributions concerning a unidimensional random variable are encountered in basic statistics. But we soon find the need to deal with vector variables, which have multivariate distributions. Here, besides the marginal distributions of the components, we also have to consider the covariances between two components. The multivariate normal, seen in the previous chapters, is a well-known example. Now, how about matrices as variates? We can see that the Jacobians of transformations between special matrices can be challenging to derive.

11.4.1 Matrices of functions and functions of matrices

Both topics make sense. On the one hand, we can have a matrix with functions of one or several variables as entries, and, on the other hand, we can define, for example, the exponential of a matrix $\mathbf{A}$ as $\exp(\mathbf{A}) = \sum_{j=0}^{\infty} \frac{\mathbf{A}^j}{j!}$, with $\rho(\mathbf{A}) < 1$, where this condition on the spectral radius $\rho(\mathbf{A})$ ensures convergence.

11.4.1.1 Matrices of functions

Let $\mathbf{X}$ be a $(m \times n)$ matrix with entries of scalar-valued functions. Then continuity, derivative, and integral are defined component by component. For example,

$$\int_a^b \mathbf{X} = \left(\int_a^b x_{ij}(t)dt \right)$$

is a matrix of scalar values.

11.4.1.2 Functions of matrices

When $f(t)$ is a polynomial or rational function, with scalar coefficients and scalar arguments t, we can define $f(\mathbf{A})$ for a matrix $\mathbf{A}$ by substituting $\mathbf{A}$ for t, matrix inversion for division, and the identity matrix $\mathbf{I}_p$ for 1. For example,

$$f(t) = \frac{1+t^3}{1-t} \Rightarrow f(\mathbf{A}) = (\mathbf{I}_p + \mathbf{A}^3)(\mathbf{I}_p - \mathbf{A})^{-1}$$

if $1 \notin \Lambda(\mathbf{A})$, where $\Lambda(\mathbf{A})$ denotes the spectrum (set of eigenvalues) of $\mathbf{A}$.

We have

$$(\mathbf{I} - \mathbf{X})^{-1} = \sum_{k=0}^{\infty} \mathbf{X}^k$$

and

$$\cos \mathbf{X} = \sum_{k=0}^{\infty} (-1)^k \frac{\mathbf{X}^{2k}}{(2k)!}.$$

Matrix exponential

$$e^{\mathbf{A}t} = \mathbf{I} + \mathbf{A}t + \frac{(\mathbf{A}t)^2}{2!} + \frac{(\mathbf{A}t)^3}{3!} + \cdots.$$

This series always converges and has the usual properties

$$e^{\mathbf{A}t} e^{\mathbf{A}s} = e^{\mathbf{A}(t+s)}, \quad e^{\mathbf{A}t} e^{-\mathbf{A}t} = \mathbf{I}, \quad \text{and} \quad \frac{d}{dt}(e^{\mathbf{A}t}) = \mathbf{A} e^{\mathbf{A}t}.$$

For a square matrix $\mathbf{X}$,

$$\det(e^{\mathbf{X}}) = \exp(\operatorname{tr} \mathbf{X}).$$

When matrices are considered as input, we have either scalar functions of one or several matrix arguments or matrix-valued functions of the same variables. This has led us to distinguish between entries distribution, determinant distribution and eigenvalue distribution when talking about a matrix variate distribution (see Chapter 12). For scalar-valued matrix functions, they are usually based on symmetric functions of the entries or of the eigenvalues of the input square matrices. We recall here some basic notions of calculus on matrices, which are not so evident.

For scalar functions, we first have the notions of derivatives wrt to either a scalar variable, or a vector, or a matrix. These notions differ, naturally, and results can sometimes look completely different from the classical calculus results.

Hypergeometric functions: Hypergeometric functions of scalar arguments are well used in advanced probability and statistics. In multivariate analysis, we can have hypergeometric functions in one or several scalar arguments or in matrix arguments. A survey on these functions is made by Pham-Gia and Dinh (2016), where some difficulties to define the hypergeometric function when going from a scalar to a matrix are presented. In particular, the numerical computation of a hypergeometric function in one or several matrix arguments is still difficult.

11.4.2 *Transformations in* $\mathbb{R}^p$

Let $f(\mathbf{x}) = f(x_1, \ldots, x_p)$ be the density of the vector $\mathbf{X} = (X_1, \ldots, X_p)^t$ and $\mathbf{H}$ be the transformation $\mathbf{X} \to \mathbf{Y}$, determined by

$$\begin{cases} y_1 = h_1(x_1, \ldots, x_p) = h_1(\mathbf{x}), \\ \quad \cdots \\ y_p = h_p(x_1, \ldots, x_p) = h_p(\mathbf{x}). \end{cases}$$

Assume these transformations are one to one and continuously differentiable, so that the inverse transformations exist:

$$\begin{cases} x_1 = k_1(y_1, \ldots, y_p) = k_1(\mathbf{y}), \\ \quad \cdots \\ x_p = k_p(y_1, \ldots, y_p) = k_p(\mathbf{y}). \end{cases}$$

The Jacobian of this transformation is then

$$J\left(\frac{\mathbf{x}}{\mathbf{y}}\right) = \begin{vmatrix} \dfrac{\partial x_1}{\partial y_1} & \cdots & \dfrac{\partial x_1}{\partial y_p} \\ \vdots & \ddots & \vdots \\ \dfrac{\partial x_p}{\partial y_1} & \cdots & \dfrac{\partial x_p}{\partial y_p} \end{vmatrix} \neq 0,$$

with the rhs being an expression in $(y_1, \ldots, y_p)$. We have the density of $\mathbf{Y}$, $f(\mathbf{y}) = f(\mathbf{x}^*)\left|J(\frac{\mathbf{x}}{\mathbf{y}})\right|$, where $|\mathbf{A}|$ is the absolute value of $\mathbf{A}$. Here, $f(\mathbf{x}^*)$ is the density of $\mathbf{X}$, but expressed in y. So, when we perform the change of variable $\mathbf{X} \to \mathbf{Y}$, the final objective is to have the density of $\mathbf{Y}$ and its precise definition domain. To this purpose, we have to determine the Jacobian $J(\frac{\mathbf{x}}{\mathbf{y}}) = \left[J(\frac{\mathbf{y}}{\mathbf{x}})\right]^{-1}$. Recall that for one variable we have $f(y) = f(x(y)) \cdot \left|\frac{dx}{dy}\right|$. This notion will be used frequently in this chapter. The determination of the Jacobian can be quite complex, especially when the domains of $\mathbf{X}$, and of $\mathbf{Y}$, have to be sub-divided and matched so that the correspondence is 1–1 between them. Furthermore, multivariate Jacobians can often have unexpected forms, and Farrell (1985) remarked right at the beginning of the preface of his book: "I found the multivariate Jacobian calculations horrible and unbelievable ... I continually saw alternatives to the Jacobian and variable change method of computing probability density functions...". His book makes heavy use of differential forms and exterior algebras. Furthermore, there are transformations appropriate to some specific fields (differential equations, algebraic geometry, etc.), which give rise to Jacobians used in those fields. Here, we only consider cases encountered in probability and statistics.

11.4.3 *Jacobians in changing variables*

Although some change of variables can be straightforward when these variables are scalar, some can be particularly complicated when we deal with matrices and their decompositions. In this chapter, we deal with some classical cases in increasing degrees of complexity.

Our main goal is to find $d\mathbf{Y}$, or $(d\mathbf{Y})$, subsequent to a change of variable $\mathbf{X} \to \mathbf{Y}$. For matrices, we also use their decompositions, which can take various forms, and compute the Jacobians for each part, resulting in a final differential having a form, such as

$$d\mathbf{Y} = A\varphi(t_{ij})g(\mathbf{X})d\mathbf{T}d\mathbf{X}d\mathbf{W},$$

where $\mathbf{W}$ is an abstract group, and integration is now made separately on $\mathbf{T}$, $\mathbf{W}$, and $\mathbf{X}$ when they are independent to obtain the expression of the density of $\mathbf{Y}$.

There are decomposition strategies of matrices into well-known and simpler forms, with the related integrals already available. A good reference is Mathai (1997) or Seber (2007). Naturally, methods based exclusively on characteristic functions can also be used, but they are not necessarily simpler.

If exterior forms and exterior algebra are used (see Rudin (1976) for an introduction), some elegant and convenient results can be derived when considering differential forms, but the notions on Stiefel, and sometimes, Grassman, manifolds, on zonal polynomials, among others, are additional knowledge required from the reader. Exterior forms and wedge product are introduced because they are very useful when dealing with differentials. But they can be avoided, and we just recall their basic properties in the following section.

11.4.4   *Jacobian and exterior product*

In carrying out the required changes of variables mentioned above, we have to use Jacobians, and using wedge products $dx_1 \wedge dx_2 = -dx_2 \wedge dx_1$ and $dx_1 \wedge dx_1 = 0$ and exterior forms would be helpful. We have, for example, for

$$I = \int_A f(x_1, \ldots, x_m) dx_1 \wedge \cdots \wedge dx_m$$

and transforms $x_i = x_i(y_1, \ldots, y_m)$, $i = 1, \ldots, m$,

$$I = \int_A g(y_1, \ldots, y_m) dy_1 \wedge \cdots \wedge dy_m,$$

with

$$\overset{m}{\underset{i=1}{\wedge}} dx_i = \left| \det \left[\left(\frac{\partial x_i}{\partial y_j} \right) \right] \right| \overset{m}{\underset{i=1}{\wedge}} dy_i,$$

where the Jacobian of the transformation is the absolute value of the determinant, $\det \left[\left(\frac{\partial x_i}{\partial y_j} \right) \right]$. Hence, the formal procedure of multiplying differential forms is equivalent to calculating the Jacobian. An illustration is the transformation from rectangular coordinates $\{x_1, \ldots, x_m\}$ to polar coordinates $\{r, \theta_1, \ldots, \theta_{m-1}\}$. We then have (see Muirhead (1983))

$$\overset{m}{\underset{i=1}{\wedge}} dx_i = \left[r^{m-1} \sin^{m-2} \theta_1 \sin^{m-3} \theta_2 \ldots \sin \theta_{m-2} \right] .dr \overset{m-1}{\underset{i=1}{\wedge}} d\theta_i.$$

We can see immediately that for $m = 2$, we have

$$dx \wedge dy = r.dr.d\theta, \quad r \geq 0, 0 \leq \theta \leq 2\pi.$$

11.4.5   *The multigamma function*

Let X be a positive $(p \times p)$ matrix, where T is upper triangular with positive diagonal entries. Then the correspondence between the two matrices of differentials is

$$(d\mathbf{X}) = 2^p \left(\prod_{i=1}^{p} t_{ii}^{p+1-i} \right) (d\mathbf{T}) \quad \text{or} \quad \bigwedge_{i=1}^{p} dx_{ij}$$

$$= 2^p t_{11}^p \dots t_{22}^{p-1} \dots t_{pp} \bigwedge_{i \leq j} dt_{ij}. \tag{11.1}$$

If $f(\mathbf{X}) = \text{etr}(-\mathbf{X})|\mathbf{X}|^{a - \frac{p+1}{2}}$, where $\text{etr}(\mathbf{X})$ is the exponential of the trace of $\mathbf{X}$, with the domain of positive definite matrices, $\Omega = \{\mathbf{X} : \mathbf{X} > 0\}$, we have the multivariate gamma function

$$\Gamma_p(a) = \int_{\Omega} \text{etr}(-\mathbf{X})|\mathbf{X}|^{a - (p+1)/2} d\mathbf{X}.$$

Carrying out the integral as explained above, we obtain a product of ordinary gamma functions, i.e.

$$\Gamma_p(a) = \pi^{p(p-1)/4} \prod_{i=1}^{p} \Gamma\left(a - \frac{i-1}{2} \right).$$

Details of the proof are given in Theorem 11.2, but very often, the multivariate gamma function is just defined that way.

11.5   Derivatives of Vectors and Matrices

Some results are reviewed here without proof.

11.5.1   *Derivatives of a vector*

Let $\mathbf{X}(p \times 1)$ be a vector of real independent variables $\{x_1, \dots, x_p\}^t$, $\mathbf{a}^t = \{a_1, \dots, a_p\}$ be a scalar vector, and $f(\mathbf{X})$ be a scalar function of $\mathbf{X}$.

The derivative of $f(\mathbf{X})$ wrt $\mathbf{X}$ is a column vector of partial derivative operators:

$$\left(\frac{\partial f(.)}{\partial \mathbf{X}} \right)^t = \left(\frac{\partial f(.)}{\partial x_1}, \dots, \frac{\partial f(.)}{\partial x_p} \right).$$

We have, for some particular values of $f(\mathbf{X})$, e.g. $\mathbf{a}^t\mathbf{X}$ and $\mathbf{X}^t\mathbf{X}$:

$$\frac{\partial(\mathbf{a}^t\mathbf{X})}{\partial\mathbf{X}} = \frac{\partial(\mathbf{X}^t\mathbf{a})}{\partial\mathbf{X}} = \mathbf{a},$$

$$\frac{\partial(\mathbf{X}^t\mathbf{X})}{\partial\mathbf{X}} = 2\mathbf{X}.$$

11.5.2 *Derivatives of a matrix and of a function of a matrix*

(a) Let $\mathbf{X}$ be a matrix of real variables. We can take derivatives of $\mathbf{X}$ wrt a scalar entry, or a vector component of the matrix, or the matrix itself, and results are different. The derivative of $\mathbf{X}$ is as follows:

  (i) wrt a scalar variable: $\frac{\partial\mathbf{X}}{\partial u} = (\frac{\partial x_{ij}}{\partial u})$, with the rhs being the matrix of derivatives wrt u, of all entries x_{ij}, which are functions of u.

  (ii) wrt an entry: Let the entry have the (l, k) position in the matrix $\mathbf{X}$. Then $\frac{\partial\mathbf{X}}{\partial x_{lk}} = \Delta_{lk}$, with Δ_{lk} being the matrix with zero everywhere, except at (l, k) position, where the value is 1.

  (iii) wrt $\mathbf{X}$ itself:

$$\frac{\partial f(\mathbf{X})}{\partial\mathbf{X}} = \left(\frac{\partial f(\mathbf{X})}{\partial x_{ij}}\right),$$

  the rhs being the matrix of all partial derivatives wrt each entry (this is a definition).

(b) Examples of simple functions of a matrix are as follows:

  (b.1) $f(\mathbf{X}) = \text{tr}(\mathbf{X}) \Rightarrow \frac{\partial f(\mathbf{X})}{\partial\mathbf{X}} = \mathbf{I}_p$ (trace is a scalar, but we have here the identity matrix because of the above definition).

  (b.2) For $\mathbf{A}, \mathbf{X}(p \times q), f(\mathbf{X}) = \text{tr}(\mathbf{A}^t\mathbf{X}) \Rightarrow \frac{\partial f(\mathbf{X})}{\partial\mathbf{X}} = \mathbf{A}$, and for $\mathbf{A}, \mathbf{X}$ both $(p \times p)$ and $\mathbf{X}$ not symmetric, we have

$$f(\mathbf{X}) = \text{tr}(\mathbf{X}\mathbf{A}) \Rightarrow \frac{\partial f(\mathbf{X})}{\partial\mathbf{X}} = \mathbf{A}^t,$$

  and for $\mathbf{X}$ symmetric,

$$\frac{\partial f(\mathbf{X})}{\partial\mathbf{X}} = \mathbf{A} + \mathbf{A}^t - diag(\mathbf{A}).$$

(b.3) For a $(p \times p)$ matrix $\mathbf{A}$, $\mathbf{X}^t \mathbf{A} \mathbf{X} = \sum_{j=1}^{p} \sum_{i=1}^{p} a_{ij} x_i x_j$,
$\frac{\partial (\mathbf{X}^t \mathbf{A} \mathbf{X})}{\partial \mathbf{X}} = (\mathbf{A} + \mathbf{A}^t) \mathbf{X}$ ($= 2\mathbf{A}\mathbf{X}$ if $\mathbf{A}$ symmetric).

(b.4) The derivative wrt $\mathbf{X}$ and its transpose is

$$\frac{\partial^2 f(\mathbf{X})}{\partial \mathbf{X} \partial \mathbf{X}^t} = \begin{bmatrix} \dfrac{\partial^2 f(\mathbf{X})}{\partial x_1^2} & \cdots & \dfrac{\partial^2 f(\mathbf{X})}{\partial x_1 \partial x_p} \\ \cdots & \ddots & \cdots \\ \dfrac{\partial^2 f(\mathbf{X})}{\partial x_p \partial x_1} & \cdots & \dfrac{\partial^2 f(\mathbf{X})}{\partial x_p^2} \end{bmatrix}.$$

So, it is the matrix of partial derivatives of order 2 of $f(\mathbf{X})$, also known as the Hessian.

(c) **Derivative of a determinant wrt the matrix**
We have

$$\frac{\partial |\mathbf{X}|}{\partial \mathbf{X}} = |\mathbf{X}| (\mathbf{X}^{-1})^t,$$

and when $\mathbf{X}$ is symmetric i.e. $\mathbf{X} = \mathbf{X}^t$,

$$\frac{\partial |\mathbf{X}|}{\partial \mathbf{X}} = |\mathbf{X}| (2\mathbf{X}^{-1} - diag(\mathbf{X}^{-1})).$$

(d) **Derivatives of the trace of a square matrix**
In function of matrices, the trace of a square matrix has often the same role as x in scalar functions. For example, we have $\mathrm{etr}(\mathbf{A}) = \exp(\mathrm{tr}(\mathbf{A}))$, replacing $\exp(x)$.

For $\mathbf{X}(p \times p)$, a matrix of independent real entries, $\frac{\partial (\mathrm{tr}\mathbf{X}^2)}{\partial \mathbf{X}} = 2\mathbf{X}^t$.

If $\mathbf{X}(p \times q)$ and $\mathbf{A}(q \times q)$ matrix of constants,

$$\frac{\partial (\mathrm{tr}(\mathbf{X} \mathbf{A} \mathbf{X}^t)}{\partial \mathbf{X}} = \mathbf{X}(\mathbf{A} + \mathbf{A}^t).$$

We have

$$\frac{\partial (\mathrm{etr}(\mathbf{X}^2))}{\partial \mathbf{X}} = 2(\mathrm{etr}(\mathbf{X}^2)\mathbf{X}^t).$$

(e) **Derivatives of a power of the determinant of a square matrix X**

$$\frac{\partial |\mathbf{X}|^k}{\partial \mathbf{X}} = k|\mathbf{X}|^k (\mathbf{X}^{-1})^t$$

$$= k|\mathbf{X}|^k (2\mathbf{X}^{-1} - \text{diag}(\mathbf{X}^{-1})) \text{ if } \mathbf{X} = \mathbf{X}^t.$$

(f) **Derivatives of the logarithm of the determinant of a square matrix X**

$$\frac{\partial (\ln |\mathbf{X}|)}{\partial \mathbf{X}} = (\mathbf{X}^{-1})^t$$

$$= (2\mathbf{X}^{-1} - \text{diag}(\mathbf{X}^{-1})) \text{ if } \mathbf{X} = \mathbf{X}^t.$$

11.5.3 *Jacobian for the inverse matrix*

In one variable, we have a simple form for the derivative of a function. For a matrix, the formula can be different, depending on the nature of that matrix.

We set $d\mathbf{X} = \prod_{i=1}^{p} \prod_{j=1}^{q} dx_{ij}$, while for the matrix of differentials, we write $(d\mathbf{X}) = (dx_{ij})$.

We know that for certain nonlinear transformations, the two Jacobians are the same, i.e. $J(\mathbf{Y} : \mathbf{X}) = J(d(\mathbf{Y}) : (d\mathbf{X}))$ and that we can take differentials using the relations between the variables. For example, for $\mathbf{Y} = \mathbf{X}^{-1}$, we can write

$$\mathbf{XY} = \mathbf{I}_p \Rightarrow (d\mathbf{X})\mathbf{Y} + \mathbf{X}(d\mathbf{Y}) = 0 \Rightarrow (d\mathbf{Y}) = -\mathbf{X}^{-1}(d\mathbf{X})\mathbf{X}^{-1}.$$

Note: Let $\mathbf{X}$ be a $(p \times p)$ non-singular matrix of independent real entries and $\mathbf{Y} = \mathbf{X}^{-1}$. Then we have

$$d\mathbf{Y} = |\mathbf{X}|^{-2p} d\mathbf{X} \text{ for general } \mathbf{X}$$

$$= |\mathbf{X}|^{-(p+1)} d\mathbf{X} \text{ for } \mathbf{X} \text{ symmetric, } \mathbf{X} = \mathbf{X}^t$$

$$= |\mathbf{X}|^{-(p-1)} d\mathbf{X} \text{ for } \mathbf{X} \text{ skew-symmetric, } \mathbf{X} = -\mathbf{X}^t$$

$$= |\mathbf{X}|^{-(p+1)} d\mathbf{X} \text{ for } \mathbf{X} \text{ lower or upper triangular.}$$

11.5.4　*Powers of a square matrix*

Let $\mathbf{X}(p \times p)$ be positive definite, with eigenvalues $\{\lambda_1, \ldots, \lambda_p\}$. Then for a general $\mathbf{X}$, we have the following:

(a) $\mathbf{Y} = \mathbf{X}^2 \to d\mathbf{Y} = \left\{ \prod_{i=1}^{p} \prod_{j=1}^{p} (\lambda_i + \lambda_j) \right\} d\mathbf{X}$, which becomes

$d\mathbf{Y} = 2^p |\mathbf{X}| \left\{ \prod_{i<j=1}^{p} (\lambda_i + \lambda_j) \right\} d\mathbf{X}$ for $\mathbf{X} = \mathbf{X}^t$.

(b) $\mathbf{Y} = \mathbf{X}^m \to d\mathbf{Y} = m^p |\mathbf{X}|^{m-1} \left\{ \prod_{i<j} \left(\frac{\lambda_i^m - \lambda_j^m}{\lambda_i - \lambda_j} \right)^2 \right\} d\mathbf{X}$,

$|\mathbf{X}| = |\mathbf{Y}|^{1/m}$ for a general $\mathbf{X}$, which becomes $d\mathbf{Y} = m^p |\mathbf{X}|^{m-1} \left\{ \prod_{i<j} \left| \frac{\lambda_i^m - \lambda_j^m}{\lambda_i - \lambda_j} \right| \right\} d\mathbf{X}$ for $\mathbf{X} = \mathbf{X}^t$, with $|\mathbf{X}| = |\mathbf{Y}|^{1/m} = $ mth positive root of $|\mathbf{Y}|$.

Remark: Pham-Gia *et al.* (2019) have studied the distributions of powers of the central matrix variate beta distributions.

11.6　Differentials in Matrix Decompositions

In the remainder of this chapter, we examine some special matrix decompositions (such as Cholesky, spectral and singular values) and apply them to compute some matrix integrals.

In the case of a decomposition of Cholesky type, we have the following.

Theorem 11.2. *Let $\mathbf{X}$ be a $(p \times p)$ symmetric positive definite matrix of independent real variables. Then for $\mathbf{T} = (t_{ij})$ real lower triangular matrix, we have*

$$\mathbf{X} = \mathbf{TT}^t \Rightarrow d\mathbf{X} = 2^p \left\{ \prod_{j=1}^{p} t_{jj}^{p+1-j} \right\} d\mathbf{T}, \qquad (11.2)$$

where $d\mathbf{T} = $ differential on all t_{ij}.

Remark: For the upper triangular case,

$$\mathbf{X} = \mathbf{T}^t\mathbf{T} \Rightarrow d\mathbf{X} = 2^p \left\{ \prod_{j=1}^{p} t_{jj}^{j} \right\} d\mathbf{T}.$$

Proof. Using the matrix of differentials.　　□

Example 11.1. Theorem 11.2 can be used to calculate the multivariate Gamma function:

$$\Gamma_p(\alpha) = \int\limits_{\mathbf{X}>0} |\mathbf{X}|^{\alpha - \frac{p+1}{2}} \exp\left[-\mathrm{tr}(\mathbf{X})\right] d\mathbf{X}$$

$$= \pi^{\frac{p(p-1)}{4}} \Gamma(\alpha)\Gamma\left(\alpha - \frac{1}{2}\right) \cdots \Gamma\left(\alpha - \frac{p-1}{2}\right), \qquad (11.3)$$

also called the integral of Siegel.

Note: This integral is associated with the space of positive definite $(p \times p)$ matrices. Since a symmetric $(p \times p)$ matrix has $p(p+1)/2$ elements, the set of all such matrices is an Euclidean space of dimension $p(p+1)/2$. The set of positive definite matrices is a subset of this Euclidean space defined by the consecutive inequalities

$$a_{11} > 0, \begin{vmatrix} a_{11} & a_{12} \\ a_{21} & a_{22} \end{vmatrix} > 0, \ldots, |\mathbf{A}| > 0.$$

Also, the measure for the integral is the Lebesgue measure

$$d\mathbf{A} = da_{11} \wedge da_{12} \wedge \cdots \wedge da_{pq} = da_{11}da_{12}\ldots da_{pq}.$$

Proof. We have $\int_{-\infty}^{\infty} e^{-t^2} dt = \sqrt{\pi}$ and $\int_0^{\infty} e^{-x}x^{\alpha-1}dx = \Gamma(\alpha)$, which also gives

$$2\int\limits_0^{\infty} (t^2)^{\alpha - \frac{j}{2}} e^{-t^2} d(t^2) = \Gamma\left(\alpha - \frac{j-1}{2}\right), 1 \le j \le p.$$

Let $\mathbf{X} = \mathbf{TT}^t \Rightarrow d\mathbf{X} = 2^p \left\{\prod_{j=1}^p t_{jj}^{p+1-j}\right\} d\mathbf{T}$. Then, for $\mathbf{T}$ lower triangular,

$$\mathrm{tr}(\mathbf{X}) = \mathrm{tr}(\mathbf{TT}^t) = t_{11}^2 + (t_{21}^2 + t_{22}^2) + \cdots + (t_{p1}^2 + \cdots + t_{pp}^2),$$

and $|\mathbf{X}| = |\mathbf{TT}^t| = \prod_{j=1}^{p} t_{jj}^2$. The integral now splits into p integrals in t_{jj}, and $p(p-1)/2$ integrals in $t_{ij}, i > j$, as follows:

$$\Gamma_p(\alpha) = \int_{\mathbf{X}>0} |\mathbf{X}|^{\alpha - \frac{p+1}{2}} \exp\left[-\mathrm{tr}(\mathbf{X})\right] d\mathbf{X}$$

$$= \left\{ \prod_{j=1}^{p} 2 \int_0^\infty (t_{jj}^2)^{\alpha - (p+1)/2} t_{jj}^{p+1-j} e^{-t_{jj}^2} dt_{jj} \right\}$$

$$\times \left\{ \prod_{i>j} \int_{-\infty}^\infty e^{-t_{ij}^2} dt_{ij} \right\}$$

$$= \left\{ \prod_{j=1}^{p} 2 \int_0^\infty (t_{jj}^2)^{\alpha - j/2} e^{-t_{jj}^2} dt_{jj}^2 \right\} \left\{ \prod_{i>j} \int_{-\infty}^\infty e^{-t_{ij}^2} dt_{ij} \right\}$$

$$= \left\{ \prod_{j=1}^{p} \Gamma\left(\alpha - \frac{j-1}{2}\right) \right\} \left\{ (\pi)^{\frac{1}{2}\frac{p(p-1)}{2}} \right\}.$$

$\square$

Note: For $p = 1$, we have the common function gamma $\Gamma(\alpha) = \int_0^\infty x^{\alpha-1} \exp(-x) dx$. Also, since $\Re(\alpha) > (j-1)/2$ for $j = 1, \ldots, p$, we have $\Re(\alpha) > (p-1)/2$.

11.7   Jacobians Associated with Eigenvalues

Eigenvalues intervene in many matrix operations.

Theorem 11.3. $\mathbf{X}$ *is positive definite, with p^2 independent elements, and distinct eigenvalues $\{\lambda_1, \ldots, \lambda_p\}$. Then*

$$\mathbf{Y} = |\mathbf{X}|\,\mathbf{X}^{-1} \to d\mathbf{Y} = (p-1)|\mathbf{X}|^{p(p-2)} d\mathbf{X} = (p-1) \left| \prod_{j=1}^{p} \lambda_j \right|^{p(p-2)} d\mathbf{X}$$

and

$$\mathbf{V} = |\mathbf{X}|^{-1}\mathbf{X} \to d\mathbf{V} = (p-1)|\mathbf{X}|^{-p^2} d\mathbf{X}.$$

If, moreover $\mathbf{X}$ *is symmetric, then*

$$\mathbf{Y} = |\mathbf{X}|\,\mathbf{X}^{-1} \to d\mathbf{Y} = (p-1)|\mathbf{X}|^{p(p-2)/2}d\mathbf{X}$$

$$= (p-1)\left|\prod_{j=1}^{p} \lambda_j\right|^{p(p-2)/2} d\mathbf{X}$$

and

$$\mathbf{V} = |\mathbf{X}|^{-1}\mathbf{X} \to d\mathbf{V} = (p-1)|\mathbf{X}|^{-p^2/2}d\mathbf{X}.$$

11.8 Integration of a Matrix

Let $f(\mathbf{X})$ be a scalar function of the matrix $\mathbf{X}$. Then $\int_\Omega f(\mathbf{X})d\mathbf{X}$ is the iterated integral of $f(\mathbf{X})$ for each entry of $\mathbf{X}$ separately over the region Ω located within the space defined by the simplex bounding the ranges of the entries of $\mathbf{X}$.

Since it is usually very difficult to carry out direct integration over a complex region Ω, integration for more simple regions are usually done by changes of variables, matrix decompositions, and finally identification with known expressions.

We also have the region $\mathbf{0} < \mathbf{X} < \mathbf{I}_p$ as the set of all square matrices such that $\mathbf{X}$ and $\mathbf{I}_p - \mathbf{X}$ are positive definite, which reduces to the continuous variable x being between 0 and 1 for unidimensional space.

11.8.1 *Use of skew-symmetric matrices* $(\mathbf{X}^t = -\mathbf{X})$

Skew-symmetric matrices are frequently used in the decomposition of matrices and the computation of their values.

Since $\mathbf{X}^t = -\mathbf{X}$, diagonal entries of $\mathbf{X}$ are zeros and no eigenvalue is -1. Then $(\mathbf{X} + \mathbf{I}_p)^{-1}$ exists.

By the Vinograd theorem, we know that the following decomposition of a matrix $\mathbf{Y}$ into an orthogonal matrix $\mathbf{Z}$ and a lower triangular matrix $\mathbf{T}$, exists, i.e. $\mathbf{Y} = \mathbf{TZ}$, with $\mathbf{Z} = 2(\mathbf{I}_p + \mathbf{X})^{-1} - \mathbf{I}_p = (\mathbf{I}_p - \mathbf{X})(\mathbf{I}_p + \mathbf{X})^{-1}$ or

$$\mathbf{X} = 2(\mathbf{I}_p + \mathbf{Z})^{-1} - \mathbf{I}_p \tag{11.4}$$

and $\mathbf{ZZ}^t = \mathbf{I}_p$, and variables are selected in such a way that there is unique choice of $\mathbf{Z}$.

Remark: In relation to skew-symmetric matrices $\mathbf{X}$ s.t. $(\mathbf{I}_p + \mathbf{X})^{-1}$ is uniquely chosen, we have

$$\int\limits_{\mathbf{X}} 2^{p(p-1)/2} |\mathbf{I}_p + \mathbf{X}|^{-(p-1)} d\mathbf{X} = \frac{\pi^{\frac{p^2}{2}}}{\Gamma_p(p/2)},$$

which is the same value as $\int_{\mathbf{U}} \wedge \left[\mathbf{U}^t (d\mathbf{U}) \right] = \int_{\mathbf{U}} \wedge \left[(d\mathbf{U}) \mathbf{U}^t \right]$, with $\mathbf{U}$ orthonormal, and uniquely defined.

11.8.2   *Decomposition with one component skew-symmetric*

Theorem 11.4. *Let* $\mathbf{Y}, \mathbf{X}, \mathbf{T}$ *be* $(p \times p)$ *matrices, with* $\mathbf{X}$ *being skew-symmetric and* $\mathbf{T}$ *being lower triangular. Then, in the unique decomposition above,* $\mathbf{Y} = \mathbf{T}(\mathbf{I}_p - \mathbf{X})(\mathbf{I}_p + \mathbf{X})^{-1}$, *we have*

$$d\mathbf{Y} = 2^{p(p-1)/2} |\mathbf{X} + \mathbf{I}_p|^{-(p-1)} d\mathbf{X} \left\{ \prod_{j=1}^{p} |t_{jj}|^{p-j} \right\} d\mathbf{T}. \qquad (11.5)$$

Proof. See Mathai (1997).   □

Conditions on integration variables: We can prove that there are two separate sets of conditions **C1** and **C2** to ensure that $\mathbf{X}, \mathbf{T}$ are uniquely defined.
Conditions C1:

- Conditions on t_{jj}: $t_{jj} \geq 0$, except for last one t_{pp} varying on $(-\infty, \infty)$.
- Conditions on t_{jk}: off-diagonal, t_{jk} varying on $(-\infty, \infty)$.
- Conditions on x_{jk}: x_{jk} varying on $(-\infty, \infty)$.

Conditions C2:

- $-\infty < t_{jk} < \infty$ and the first row elements of $(\mathbf{I}_p + \mathbf{X})^{-1}$ are negative, except the first element.

Example 11.2. We compute $\theta = \int \exp(-\text{tr}(\mathbf{Y}\mathbf{Y}^t)) d\mathbf{Y}$, where $\mathbf{Y}$ is a (3×3) matrix of independent variables. We note that this integral is the generalization to matrices of the basic integral related to the univariate normal distribution, $\int_{-\infty}^{\infty} \exp(-x^2) dx = \pi^{1/2}$.

This integral in $\mathbf{Y}$ seems quite general and no indication on how its value can be obtained.

(a) Direct method

$$\theta = \int_{\mathbf{Y}} \exp(-\mathrm{tr}(\mathbf{Y}\mathbf{Y}^t))d\mathbf{Y}$$

$$= \int \exp\left(-\sum_{j,k} y_{jk}^2\right)d\mathbf{Y}$$

$$= \prod_{j,k}\left(\int_{-\infty}^{\infty} e^{-y_{jk}^2}dy_{jk}\right) = (\sqrt{\pi})^9 = \pi^{9/2}.$$

(b) Using Theorem 11.4
From Eq. (11.5), we have

$$\theta = e^{-\mathrm{tr}(\mathbf{T}\mathbf{T}^t)}2^{p(p-1)/2}\int_{\mathbf{T}}\left\{\prod_{j=1}^{p}|t_{jj}|^{p-j}\right\}$$

$$\times|\mathbf{X}+\mathbf{I}_P|^{-(p-1)}d\mathbf{X}d\mathbf{T}, \quad p=3, \qquad (11.6)$$

$\mathbf{X}^t = -\mathbf{X} \Rightarrow x_{jj} = 0$, and $|\mathbf{X}+\mathbf{I}_p|^{-(p-1)} = (1 + x_{12}^2 + x_{13}^2 + x_{23}^2)^{-2}$ by direct computation, and here, $2^{p(p-1)/2} = 2^3$.

Also, $\prod_{j=1}^{p}|t_{jj}|^{p-j} = |t_{11}|^2\,|t_{22}|$ and $\mathrm{tr}(\mathbf{T}\mathbf{T}^t) = t_{11}^2 + t_{22}^2 + t_{33}^2 + t_{21}^2 + t_{31}^2 + t_{32}^2$.

We consider conditions C1: $t_{11}, t_{22} > 0$, but $-\infty < t_{33} < \infty$. Similarly, $-\infty < t_{jk} < \infty, j > k$ and $-\infty < x_{jk} < \infty, j < k$. We then have

(i) $2^3\int e^{-\mathrm{tr}(\mathbf{T}\mathbf{T}^t)}\prod_{j=1}^{3}|t_{jj}|^{3-j}d\mathbf{T} = 2\left\{\prod_{j=1}^{2}2\int_{0}^{\infty}t_{jj}^{3-j}e^{-t_{jj}^2}dt_{jj}\right\}$

$$\times \int_{-\infty}^{\infty}e^{-t_{33}^2}dt_{33}\left\{\prod_{j>k=1}^{3}e^{-t_{jk}^2}dt_{jk}\right\}$$

$$= \Gamma(1)\Gamma(1/2)(1/2)^3 2\Gamma(3/2)\Gamma(1/2)^3$$
$$= \pi^{5/2}.$$

(ii) On the other hand,

$$\int_{\mathbf{X}} |\mathbf{X} + \mathbf{I}_P|^{-(p-1)} d\mathbf{X}$$

$$= \int_{-\infty}^{\infty} \int_{-\infty}^{\infty} \int_{-\infty}^{\infty} (1 + x_{12}^2 + x_{13}^2 + x_{23}^2)^{-2} dx_{12} dx_{13} dx_{23},$$

$$(11.7)$$

which is a type 2 Dirichlet integral, setting $x_{12}^2 = u_{12}$,

$$\int_0^{\infty} \int_0^{\infty} \int_0^{\infty} (1 + u_{12} + u_{13} + u_{23})^{-2}$$

$$\times u_{12}^{-1/2} u_{13}^{-1/2} u_{23}^{-1/2} du_{12} du_{13} du_{23}$$

$$= \frac{[\Gamma(1/2)]^4}{\Gamma(2)}$$

$$= \pi^2.$$

$$(11.8)$$

Hence, the value of $\int \exp(-\mathrm{tr}(\mathbf{Y}\mathbf{Y}^t)) d\mathbf{Y}$ is $\theta = \pi^{5/2}.\pi^2 = \pi^{9/2}$, as in (a).

Notes:

(1) Using conditions **C2**, instead of **C1**, we have integral over $\mathbf{T} = \frac{\pi^{5/2}}{2}$ in Eq. (11.5) and

$$\int |\mathbf{I}_p + \mathbf{X}|^{-2} d\mathbf{X} = \int_{\mathbf{X}_1} |\mathbf{I} + \mathbf{X}_1|^{-2} d\mathbf{X}_1$$

$$\int_{u_{12}, u_{13} > 0} (1 + u_{12}^2 + u_{13}^2)^{-2} du_{12} du_{13}$$

$$= \pi \cdot \frac{\pi}{2^2} = \frac{\pi^2}{2^2}.$$

The final answer is then $\theta = 2^3 \cdot \frac{\pi^{5/2}}{2} \cdot \frac{\pi^2}{2^2} = \pi^{9/2}$, as before.

(2) In general, for $\mathbf{Y}(p \times p)$, we can prove that $\theta = \pi^{p^2/2}$, and, for a rectangular matrix $\mathbf{Y}(n \times p)$, we have $\theta = \pi^{np/2}$. Also, $\int \exp(-\mathrm{tr}(\mathbf{YY}^t))d\mathbf{Y} = (2\pi)^{np/2}$ for a matrix $\mathbf{Y}(n \times p)$ with independent entries $Y_{ij} \sim N(0,1)$ (see Eq. (11.11)).

11.8.3   *Spectral decomposition*

Theorem 11.5. *Let $\mathbf{X}$ be a $(p \times p)$ symmetric matrix of independent entries, with distinct non-null eigenvalues. Let $\mathbf{D} = diag(\lambda_1, \ldots, \lambda_p)$, with $\lambda_1 > \cdots > \lambda_p$, and let $\mathbf{V}$ be a $(p \times p)$ orthogonal matrix s.t. $\mathbf{X} = \mathbf{VDV}^t$. Then*

$$d\mathbf{X} = \left\{ \prod_{i=1}^{p-1} \prod_{j=i+1}^{p} |\lambda_i - \lambda_j| \right\} d\mathbf{D}d\mathbf{G}, \qquad (11.9)$$

where $d\mathbf{G} = \mathbf{V}^t d\mathbf{V}$.

This result will be used in part (c) of the application of 11.9.

11.8.4   *Singular value decomposition of a square matrix*

Recall this decomposition earlier that states that any real rectangular matrix $\mathbf{X}$ can be decomposed as $\mathbf{X} = \mathbf{UDV}^t$, with $\mathbf{U}, \mathbf{V}$ orthogonal:

$$\mathbf{U}^t\mathbf{U} = \mathbf{I}_p, \mathbf{V}^t\mathbf{V} = \mathbf{I}_p, \mathbf{D} = diag(\mu_1, \ldots, \mu_p),$$

with $\mu_1, \ldots, \mu_p$ being singular values of $\mathbf{X}$. Since $\mathbf{XX}^t = \mathbf{UDD}^t\mathbf{U}^t = \mathbf{UD}^2\mathbf{U}^t$, and, similarly, $\mathbf{X}^t\mathbf{X} = \mathbf{VD}^2\mathbf{V}^t$, we have $\left\{ \lambda_i = \mu_i^2 \right\}_{i=1}^{p}$ as the eigenvalues of $\mathbf{XX}^t$ and of $\mathbf{X}^t\mathbf{X}$. Suppose these values are distinct.

Theorem 11.6. *Under the above conditions, with p^2 independent real variables and $\mathbf{D} = diag(\mu_1, \ldots, \mu_p)$, with $\mu_1 > \mu_2 > \cdots > \mu_p$. Then with the decomposition $\mathbf{X} = \mathbf{UDV}^t$, we have*

$$d\mathbf{X} = \left\{ \prod_{i<j} |\mu_i^2 - \mu_j^2| \right\} d\mathbf{D}d\mathbf{G}d\mathbf{H}, \quad \text{with } d\mathbf{G} = \mathbf{U}^t d\mathbf{U}, d\mathbf{H} = (d\mathbf{V}^t)\mathbf{V}.$$

$$(11.10)$$

Note: $\mathbf{U}, \mathbf{V}$ are not unique. There are also 2^p choices for which $\mathbf{D}$ remains the same. Then if $\wedge[\mathbf{U}^t d\mathbf{U}]$ or $\wedge[(d\mathbf{V}^t)\mathbf{V}]$ is integrated over the full orthogonal group, then the final result has to be divided by 2^p.

11.9   Computation of a Simple Definite Matrix Integral Under Various Hypotheses

Although definite integrals of matrices can look quite abstract, concrete results can be obtained, as seen in the following cases. We suppose here that specific hypotheses can apply to different cases in order to be able to carry out computations.

Application: Computation of $I = \int_0^{\mathbf{I}_p} |\mathbf{X}|^\alpha d\mathbf{X}$, $\Re(\alpha) \geq 0$.

(a) In one-dimensional space, we have

$$\int_0^1 x^\alpha dx = \frac{x^{\alpha+1}}{\alpha+1} \Big|_0^1 = \frac{1}{\alpha+1}.$$

Hence, $\int_0^1 dx = 1$ for $\alpha = 0$.

(b) In two dimensions, we can see that it is a matrix variate beta of first kind

$$I = \int_0^{\mathbf{I}_2} |\mathbf{X}|^\alpha d\mathbf{X} = \text{Beta}_2^I(\alpha + 3/2, 3/2)$$

$$= \frac{\Gamma_2(\alpha + 3/2)\Gamma_2(3/2)}{\Gamma_2(\alpha + 3)}$$

$$= \frac{\pi}{(\alpha+1)(\alpha+2)(2\alpha+3)}.$$

Then $\int_0^{\mathbf{I}_2} d\mathbf{X}$ corresponds to $\alpha = 0$ and the value of the integral is $\frac{\pi}{3!}$, and hence, $\int_0^{\mathbf{I}_2} d\mathbf{X} = \frac{\pi}{6}$, a result different from (a).

(c) We suppose $\mathbf{X}(2 \times 2)$ real symmetric positive definite, with eigenvalues $0 < \lambda_1 < \lambda_2 < 1$, and use Theorem 11.5 on spectral decomposition:

$$\int_0^{\mathbf{I}_2} |\mathbf{X}|^\alpha d\mathbf{X} = \int_{0<\lambda_1<\lambda_2<1} (\lambda_1\lambda_2)(\lambda_2 - \lambda_1)d\lambda_1 d\lambda_2 \left(2^{-2}\int_V d\mathbf{G}\right),$$

with $(d\mathbf{G}) = \mathbf{V}^t(d\mathbf{V})$. We know that $2^{-2}\int_V d\mathbf{G} = \frac{\pi^2}{\Gamma_2(1)} = \pi$. Now,

$$\int\limits_{0<\lambda_1<\lambda_2<1} (\lambda_1\lambda_2)(\lambda_2 - \lambda_1)d\lambda_1 d\lambda_2$$

$$= \int_0^1 \lambda_1^\alpha \left\{ \int_{\lambda_1}^1 \lambda_2^\alpha(\lambda_2 - \lambda_1)d\lambda_2 \right\} d\lambda_1$$

$$= \frac{1}{(\alpha+1)(\alpha+2)(2\alpha+3)}.$$

So, finally $\int_0^{\mathbf{I}_2} |\mathbf{X}|^\alpha d\mathbf{X}$ is the product of these two integrals, and $\int_0^{\mathbf{I}_2} |\mathbf{X}|^\alpha d\mathbf{X} = \frac{\pi}{(\alpha+1)(\alpha+2)(2\alpha+3)}$, as before.

(d) Using Theorem 11.6 on singular value decomposition, $\mathbf{X}$ is not supposed symmetric, but has distinct *singular values* $0 < \mu_2 < \mu_1 < 1$. We show that

$$\int_0^{\mathbf{I}_2} |\mathbf{X}|^\alpha d\mathbf{X} = \frac{4\pi^2}{(\alpha+1)(\alpha+2)(\alpha+3)},$$

using Eq. (11.10). We have

$$\int_0^{\mathbf{I}_2} |\mathbf{X}|^\alpha d\mathbf{X} = \int_D \phi(\mathbf{D})d\mathbf{D} \int_U d\mathbf{H} \int_V d\mathbf{G},$$

with

$$\int_D \phi(\mathbf{D})d\mathbf{D} = \int\limits_{0<\lambda_2<\lambda_1<1} (\lambda_1\lambda_2)^\alpha(\lambda_1^2 - \lambda_2^2)d\lambda_1 d\lambda_2$$

$$= \int_0^1 \lambda_2^\alpha \left[\int_{\lambda_2}^1 (\lambda_1)^\alpha(\lambda_1^2 - \lambda_2^2)d\lambda_1 \right] d\lambda_2$$

$$= \frac{1}{(\alpha+1)(\alpha+2)(\alpha+3)}.$$

Since

$$
\int_U d\mathbf{H} \int_V d\mathbf{G} = 2^{-p} \left[\frac{2^p \pi^{p^2/2}}{\Gamma_p(p/2)} \right]^2 = 4(\pi)\pi = 4\pi^2,
$$

$d\mathbf{H} = \mathbf{U}^t(d\mathbf{U})$, $d\mathbf{G} = (d\mathbf{V}^t)\mathbf{V}$, we have $\int_0^{\mathbf{I}_2} |\mathbf{X}|^\alpha d\mathbf{X} = \frac{4\pi^2}{(\alpha+1)(\alpha+2)(\alpha+3)}$, as required.

11.10   Stiefel Manifold and the Orthogonal Group

When decomposing a random matrix, we can use an orthogonal component, or semi-orthogonal component, in the decomposition. This is the reason why we have to consider the Stiefel manifold and the orthogonal group since we have to integrate over that group in order to get to the general expression.

Definition 11.1. Let $\mathbf{A}$ be a $(q \times p), q \geq p$ matrix of real numbers, with $\mathbf{A}^t\mathbf{A} = \mathbf{I}_p$ (orthonormal column vectors). These matrices form the Stiefel manifold, denoted by $\mathbf{V}_{p,q}$. The number of free elements in $\mathbf{A}$ is then $pq - \frac{p(p+1)}{2}$. When $p = q$, $\mathbf{A}$ is an orthogonal matrix and we have the orthogonal group of square $(p \times p)$ matrices, denoted by $O_{(p)}$, where the number of free elements is now $p(p-1)/2$.

We also have the surface area of the full Stiefel manifold as

$$
\mathrm{Vol}(\mathbf{V}_{p,q}) = \int_{V_{p,q}} \mathbf{H}^t d\mathbf{H} = \int_{V_{p,q}} \bigwedge_{j=1}^p \bigwedge_{k=j+1}^q u_k{}^t(du_j) = \frac{2^p}{\Gamma_p\left(\frac{q}{2}\right)} \pi^{pq/2},
$$

$$\tag{11.11}$$

which becomes $\frac{2^p}{\Gamma_p(\frac{p}{2})}\pi^{p^2/2}$ for the orthogonal group $O_{(p)}$. As we have seen, there are several instances in decompositions, where we have to divide the above result by 2^p, and we have $\frac{\pi^{pq/2}}{\Gamma_p(\frac{q}{2})}$ as the related value.

Proof of relation (11.11): Let $\mathbf{Z}$ be a $(p \times q)$ random matrix, with entries all independent $N_1(0,1)$ rv's. This type of matrix is frequently used in random matrix theory (RMT) in physics (see Chapter 13).

The joint density of $\mathbf{Z}$, i.e. of the pq entries, is

$$(2\pi)^{-pq/2} \exp\left(-\frac{1}{2}\sum_{i=1}^{p}\sum_{j=1}^{q} z_{ij}^2\right) = (2\pi)^{-pq/2}\text{etr}\left(-\frac{1}{2}\mathbf{Z}^t\mathbf{Z}\right),$$

using matrix notation. Hence, let

$$I = \int\limits_{-\infty<z_{ij}<\infty} \text{etr}\left(-\frac{1}{2}\mathbf{Z}^t\mathbf{Z}\right)(d\mathbf{Z}) = (2\pi)^{pq/2}.$$

In the decomposition $\mathbf{Z} = \mathbf{H}_1\mathbf{T}$, where $\mathbf{T}$ is upper triangular, with positive diagonal elements, and $\mathbf{H}_1 \in V_{pq}$, we have $\text{tr}(\mathbf{Z}^t\mathbf{Z}) = \text{tr}(\mathbf{T}^t\mathbf{T}) = \sum_{i\leq j} t_{ij}^2$ (sum of t_{ij}^2 on and off diagonal), and

$$(d\mathbf{Z}) = \prod_{i=1}^{q} t_{ii}^{p-i}(d\mathbf{T}).(\mathbf{H}_1^t d\mathbf{H}_1).$$

Hence,

$$\int\limits_{-\infty<z_{ij}<\infty} \text{etr}\left(-\frac{1}{2}\mathbf{Z}^t\mathbf{Z}\right)(d\mathbf{Z}) = (2\pi)^{pq/2}$$

and I becomes

$$\int\cdots\int \exp\left(-\frac{1}{2}\sum_{i\leq j}^{q} t_{ij}^2\right) \prod_{i=1}^{q} t_{ii}^{p-i} \bigwedge_{i\leq j}^{q} dt_{ij} \left(\int_{V_{p,q}} \mathbf{H}_1^t d\mathbf{H}_1\right).$$

Let the integral concerning t_{ij} be A and $\int_{V_{p,q}} \mathbf{H}_1^t d\mathbf{H}_1 = B$. We have $A.B = I = (2\pi)^{pq/2}$. But A can be seen to be

$$A = \prod_{i<j}^{p}\left[\int_{-\infty}^{\infty} \exp\left(-\frac{t_{ij}^2}{2}\right) dt_{ij}\right].\prod_{i=1}^{p}\left[\int_0^{\infty} \exp\left(-\frac{t_{ii}^2}{2}\right) t_{ii}^{q-i} dt_{ii}\right]$$

$$= \prod_{i<j}^{p}\left[\sqrt{2\pi}\right].\prod_{i=1}^{p}\left[2^{\frac{q-i-1}{2}}\Gamma\left(\frac{q-i+1}{2}\right)\right]$$

$$= (\pi)^{\frac{p(p-1)}{4}} \cdot \prod_{i=1}^{p} 2^{\frac{pq}{2-p}} \cdot \Gamma\left(\frac{q-i+1}{2}\right)$$

$$= \Gamma_p\left(\frac{q}{2}\right) 2^{\frac{pq}{2}-p}.$$

Hence, $\int_{V_{p,q}} \mathbf{H}_1^t d\mathbf{H}_1 = B = I/A = \frac{2^p \pi^{\frac{pq}{2}}}{\Gamma_p\left(\frac{q}{2}\right)}.$

11.11   Conclusion

This chapter presents some basic notions on Jacobians and transformations between matrices in simple cases. It also gives some concrete examples of integral computations of simple matrix functions, using these Jacobians, when some simple matrix decompositions are used. These transformations are applied in Chapter 12. Other applications in multivariate analysis are numerous, but, unfortunately, they are out of the scope of this book.

Bibliography

Farrell, R. (1985). *Multivariate Calculation: Use of the Continuous Groups* (Springer-Verlag, New York).

Mathai, A.M. (1997). *Jacobians of Matrix Transformations and Functions of Matrix Argument* (World Scientific, Singapore).

Muirhead, R. (1983). *Aspects of Multivariate Statistical Theory* (Wiley, New York).

Rudin, W. (1976). *Principles of Mathematical Analysis*, 3rd ed. (Wiley, New York).

Pham-Gia, T. and Dinh, N.T. (2016). Hypergeometric functions, from one scalar variable to several matrix arguments, in statistics and beyond, *Open Journal of Statistics*, **6**(5), 951–994.

Pham-Gia, T., Duong, T.P. and Dinh, N.T. (2019). Distribution of powers of the central matrix beta variate, *Statistical Methods and Applications*, https://doi.org/10.1007/s10260-019-00497-3.

Seber, G. (2007). *A Matrix Handbook for Statisticians* (Wiley-Interscience, New York).

Strang, G. (2006). *Linear Algebra and its Applications*, 4th ed. (Thomson, Brooks/Cole, Belmont, Ca 94002-3098, USA).

Zhang, Z. (2007). Pseudo-inverse multivariate/matrix-variate distributions, *Journal of Multivariate Analysis*, **98**, 1684–1692.

Chapter 12

Matrix Variate Distributions

12.1 Introduction

This chapter uses the information provided by the last chapter to present distributions of variables under their matrix forms, and the normal, gamma and beta, among others, will be studied. We will consider for each of them, when possible, three types of distributions: the entries distribution, the determinant distribution and the eigenvalue distribution. Besides, there are also the central and non-central forms, with the latter treated in Chapter 13.

Theoretically, to each univariate distribution, there is a multivariate distribution of the same type, for a random p-vector, and a matrix variate distribution, also of the same type, for a rectangular random $(n \times p)$ matrix, or a square $(p \times p)$ matrix (see Table 12.1 for the gamma family). Most of them are presented by Gupta and Nagar (2000). But there are distributions defined on matrices that do not have corresponding scalar versions.

The matrix beta distribution is treated here not only because of its relation to the Gaussian matrix but also because of its potential importance at present and in the years ahead. Present importance is because of its close relation with the Wilks's statistic, the latter being widely used in multivariate analysis of variance (MANOVA). Potential importance is in view of the large number of questions/inquiries that our research group AMSARG receives on our papers on operations on the univariate beta. This fact leads us to think that in the years ahead, the matrix beta could also have numerous applications.

Table 12.1   Gamma distributions, with scalar and matrix variates.

Scalar random variable X	Random matrix Y
(a) Central chi-square distribution, $$X \sim \chi_\nu^2.$$ $f(\nu; x) = \frac{x^{\nu/2-1}e^{-x/2}}{2^{\nu/2}\Gamma(\nu/2)}$, $\nu > 0$, $x \geq 0$, or as a Bessel function.	(a) Central Wishart, $\mathbf{Y} \sim W_m(n, \mathbf{\Sigma})$ $$f(n, \mathbf{\Sigma}; \mathbf{Y}) = \frac{|\mathbf{Y}|^{\frac{n-m-1}{2}} \operatorname{etr}(-\frac{\mathbf{\Sigma}^{-1}\mathbf{Y}}{2})}{2^{np/2}\Gamma_m(\frac{n}{2})|\mathbf{\Sigma}|^{n/2}},$$ $\mathbf{Y} > 0$.
(b) Inverse chi-square distribution $Y \sim \chi_\nu^2 \Rightarrow X = 1/Y \sim Inv(\chi_\nu^2)$. Density: $$f(x) = \frac{1}{2^{\nu/2}\Gamma(\nu/2)} x^{\frac{\nu}{2}-1} e^{-\frac{1}{2x}},$$ where $x \in (0, \infty)$. cdf: $$F(x) = \frac{\Gamma(\frac{\nu}{2}, \frac{1}{2x})}{\Gamma(\frac{\nu}{2})}.$$	(b) Inverse Wishart distribution (see Eq. (13.16)) $\mathbf{X} \sim IW_m(\nu, \mathbf{\Omega})$ if $\mathbf{Y} = \mathbf{X}^{-1} \sim W_m(\nu, \mathbf{\Omega}^{-1})$. Density: $$f(\mathbf{X}) = \frac{|\mathbf{\Omega}|^{\nu/2}}{2^{\nu m/2}\Gamma_m(\nu/2)}$$ $$\times |\mathbf{X}|^{-\frac{\nu+p+1}{2}} \operatorname{etr}(-\tfrac{1}{2}\mathbf{\Omega}\mathbf{X}^{-1}),$$ where $\mathbf{X} > 0$.
(c) Non-central chi-square, $$X \sim \chi_p^2(\tau^2), \ p, \tau^2 > 0$$ $f(p, \tau^2; x)$ $$= \frac{e^{-(\frac{\tau^2+x}{2})} x^{\frac{p}{2}-1}}{2^{p/2}} \sum_{k=0}^{\infty} \left(\frac{\tau^2}{4}\right)^k \frac{x^k}{k!\Gamma(\frac{p}{2}+k)},$$ $x \geq 0$. τ^2: Non-centrality parameter or as a Bessel function. It has MGF $$M(t) = \frac{\exp(\frac{\tau^2 t}{1-2t})}{(1-2t)^{p/2}}.$$	(c) Non-central Wishart, $$\mathbf{Y} \sim W_m(n, \mathbf{\Sigma}, \mathbf{\Theta}),$$ $$f(n, \mathbf{\Sigma}, \mathbf{\Theta}; \mathbf{Y}) = \frac{|\mathbf{Y}|^{\frac{n-m-1}{2}}}{2^{np/2}\Gamma_m(\frac{n}{2})|\mathbf{\Sigma}|^{n/2}}$$ $$\times \operatorname{etr}(-\frac{\mathbf{\Theta}}{2} - \frac{\mathbf{\Sigma}^{-1}\mathbf{Y}}{2})$$ $$\times {}_0F_1(\tfrac{n}{2}; \tfrac{1}{4}\mathbf{\Theta}\mathbf{\Sigma}^{-1}\mathbf{Y}),$$ where $\mathbf{Y} > 0$. $\mathbf{\Theta}$: Non-centrality parameter matrix.
(d) Central gamma, $$X \sim \Gamma(a, \lambda),$$ $$f(a, \lambda; x) = \frac{x^{a-1}e^{-x/\lambda}}{\lambda^a \Gamma(a)},$$ $a, \lambda > 0$, $x \geq 0$.	(d) Central matrix gamma, $$\mathbf{Y} \sim \Gamma_m(a, \mathbf{\Lambda}),$$ $$f(a, \mathbf{\Lambda}; \mathbf{Y}) = \frac{\operatorname{etr}(-\mathbf{\Lambda}^{-1}\mathbf{Y})|\mathbf{Y}|^{a-\frac{m+1}{2}}}{\Gamma_m(a)|\mathbf{\Lambda}|^a},$$ $\mathbf{Y} > 0, a > \frac{m+1}{2}$.

(*Continued*)

Table 12.1 (*Continued*)

Scalar random variable X	Random matrix $\mathbf{Y}$
(e) Inverse gamma distribution $X \sim Ivg(\alpha, \beta)$ if $X = 1/Y$ and $Y \sim \text{Gam}(\alpha, \beta)$, with α: shape par.; β: scale par. Density: $f(x) = \frac{\beta^\alpha}{\Gamma(\alpha)} x^{-(\alpha+1)} \exp(-\beta/x)$, where $x \in (0, \infty)$. cdf: $$F(x) = \frac{\Gamma(\alpha, \beta/x)}{\Gamma(\alpha)}.$$	(e) Inverse matrix gamma distribution $\mathbf{X} \sim IMG(\alpha, \beta, \mathbf{\Omega})$, shape par. $\alpha > \frac{p-1}{2}$; $\beta > 0$ scale parameter; $\mathbf{X}, (p \times p)$ real, pos. def. Density: $f(\mathbf{X}) = \frac{|\mathbf{\Omega}|^\alpha}{\beta^{p\alpha}\Gamma_p(\alpha)} |\mathbf{X}|^{-(\alpha + \frac{p+1}{2})}$ $\times \exp(-\frac{1}{\beta}\text{tr}(\mathbf{\Omega}\mathbf{X}^{-1}))$. For $\beta = 2, \alpha = \nu/2$, we have the inverse Wishart distribution.
(f) Non-Central Gamma, $X \sim \Gamma(a, \lambda, \gamma)$ γ: Non-centrality parameter. $f(a, \lambda, \gamma; x) =$ $e^{-\gamma/\lambda} \sum_{m=0}^{\infty} \frac{(\gamma/\lambda)^m}{m!} \frac{e^{-x/\lambda} x^{a+m-1}}{\lambda^{a+m}\Gamma(a+m)}.$	(f) Non-Central Matrix Gamma, $\mathbf{Y} \sim \Gamma_m(a, \mathbf{\Lambda}, \mathbf{K})$ $\mathbf{K}$: Non-centrality parameter matrix. $f(a, \mathbf{\Lambda}, \mathbf{K}; \mathbf{Y}) =$ $\frac{\text{etr}(-\mathbf{K} - \mathbf{\Lambda}^{-1}\mathbf{Y})|\mathbf{Y}|^{a - \frac{m+1}{2}}}{\Gamma_m(a)|\mathbf{\Lambda}|^a}$ $\times {}_0F_1(a, \mathbf{K}\mathbf{\Lambda}^{-1}\mathbf{Y}), \mathbf{Y} > 0.$
(g) Randomized gamma, $X \sim RG(\rho, \mu)$ $f(\rho, \mu; t) = e^{-(\mu+x)} \sqrt{(\frac{x}{\mu})^p} I_\rho(2\sqrt{\mu x})$, $x \geq 0$, where $I_\nu(z) = \sum_{m=0}^{\infty} \frac{1}{m!\Gamma(m+\nu+1)} (\frac{z}{2})^{2m+\nu}.$	

In this chapter, we first talk about random matrices in general. Then the normal matrix variate is studied. The gamma family is then presented and the Wishart distribution has some of its properties reviewed. The generalized variance and its interval estimation using the Meijer functions are considered. The matrix variate gamma is briefly studied. Finally, the central matrix variate betas of both types are introduced under their three forms of distributions. Non-central distributions of these matrix variates will be studied in Chapter 13.

12.2   Square Random Matrices

For a random $(p \times p)$ symmetric matrix, $p \geq 1$, there are usually three associated distributions (see Pham-Gia and Turkkan (2011)). We essentially distinguish between the following:

(a) The distribution of its elements (i.e. its $p(p+1)/2$ independent elements e_{ij}, called here as its *elements distribution* or *entries distribution* (with matrix input and matrix output) for convenience. This is a mathematical expression relating the components of the matrix, but usually it is too complex to be expressed as an equation (or several equations) on the elements e_{ij} themselves, and hence, most often it is expressed as an expression of its determinant.

(b) The univariate distribution of its determinant (denoted here by $|\mathbf{X}|$ instead of $\det(\mathbf{X})$) (with matrix input and real output) is called *determinant distribution. Being univariate, this distribution is more convenient to study.* Also, frequently, this distribution can be expressed as a G-function and can be numerically computed.

(c) The distribution of its latent roots, or eigenvalues (with matrix input and p-vector output), called *latent roots distribution*, together with its sum and product, together with their product. This distribution is acquiring increasing importance with the development of random matrix theory in theoretical physics (see Chapter 13).

These three distributions evidently fuse into a single one when the matrix variate becomes a scalar univariate variable, then called its density. The literature is mostly concerned with the first distribution than with the other two. It is noted that since for a square symmetric matrix, the product of its eigenvalues is its determinant, while the sum of its eigenvalues is its trace, we are in fact studying the product of the latent roots besides studying their distribution.

In general, the density of the determinant of a random matrix $|\mathbf{A}|$ can be obtained from its entries distribution $f(.)$, in some simple cases, by using following relation for differentials: $d|\mathbf{A}| = |\mathbf{A}|\text{tr}[\mathbf{A}^{-1}(d\mathbf{A})]$ (see Mathai (1997, p. 150)). In practice, frequently, we have to manipulate $|\mathbf{A}|$ directly, often using an orthogonal transformation, to arrive at a product of independent diagonal and off-diagonal elements.

Distributions of ratios of latent roots, and the distributions of some associated statistics, remain very complicated, and in this chapter, we just study some of their basic properties. For the density of the latent roots $\{\ell_1, \ldots, \ell_p\}$, we have, using the entries density $f(.)$,

$$h(\ell_1, \ldots, \ell_p) = \frac{\pi^{\frac{p^2}{2}}}{\Gamma_p(p/2)} \prod_{i<j}^{p} (\ell_i - \ell_j) \int_{O(p)} f(\mathbf{HLH}^t)(d\mathbf{H}),$$

$$(\ell_1 > \cdots > \ell_p), \tag{12.1}$$

where $O(p)$ is the orthogonal group, $\mathbf{H}$ is an orthogonal $(p \times p)$ matrix, $(d\mathbf{H})$ is the Haar invariant measure on $O(p)$, and $\mathbf{L} = diag(\ell_1, \ldots, \ell_p)$ (see Muirhead (1983, p. 105)). But the integral within this expression is often uncomputable.

12.3 Rectangular Random Matrices

A few random matrices of this type can be studied, but we cannot consider their determinants, nor their eigenvalues, except in the singular value decomposition (see Chapter 11). The matrix variate normal distribution is such a case, and the same remark applies to the matrix t-distribution.

12.4 Normal Matrix Variate

In Chapter 9, we have studied the p-dimension normal vector, denoted by $\mathbf{X} \sim N_p(\boldsymbol{\mu}, \boldsymbol{\Sigma})$. Here, we take another step forward and consider a $(p \times n)$ random matrix $\mathbf{X}$. But the density of a normal random matrix is defined via its vector form, as shown in the following.

Definition 12.1. The rectangular random matrix $\mathbf{X}(p \times n)$ is called normal, or Gaussian, with mean matrix $\mathbf{M}(p \times n)$, and covariance matrix $\boldsymbol{\Sigma} \otimes \boldsymbol{\Psi}$, with $\boldsymbol{\Sigma}$ being a positive definite $(p \times p)$ matrix and $\boldsymbol{\Psi}(n \times n)$ being also positive definite, if

$$\text{Vec}(\mathbf{X}^t) \sim N_{pn}(\text{Vec}(\mathbf{M}^t), \boldsymbol{\Sigma} \otimes \boldsymbol{\Psi}), \tag{12.2}$$

where the operator $\text{Vec}(\mathbf{Y})$ transforms a matrix $\mathbf{Y}$ into a vector by piling the columns of $\mathbf{Y}$ to make a vector (see Chapter 11). So, we have two distinct notations: $\mathbf{X} \sim N_{[p,n]}(\mathbf{M}; \boldsymbol{\Sigma}, \boldsymbol{\Psi})$ for its matrix form and $\mathbf{X} \sim N_{p.n}(\mathbf{M}, \boldsymbol{\Sigma} \otimes \boldsymbol{\Psi})$ for its vector form.

Let $\mathbf{X} \sim N_{[p,n]}(\mathbf{M}; \boldsymbol{\Sigma}, \boldsymbol{\Psi})$. Then its density is

$$\int(\mathbf{X}) = \frac{1}{(2\pi)^{np/2}|\boldsymbol{\Sigma}|^{n/2}|\boldsymbol{\Psi}|^{p/2}}$$

$$\times etr\left\{-\frac{1}{2}\boldsymbol{\Sigma}^{-1}(\mathbf{X} - \mathbf{M})\boldsymbol{\Psi}^{-1}(\mathbf{X} - \mathbf{M})^t\right\}, \quad (12.3)$$

with $\mathbf{X} \in \mathbb{R}^{p \times n}$. Also, since the density of the normal p-vector $\mathbf{X}$, considered in Chapter 9, can also be written, using its $(p, 1)$ matrix form, as

$$f(\mathbf{x}) = \frac{1}{(2\pi)^{p/2}|\boldsymbol{\Sigma}|^{1/2}}etr\left\{-\frac{1}{2}\boldsymbol{\Sigma}^{-1}(\mathbf{x} - \boldsymbol{\mu})(\mathbf{x} - \boldsymbol{\mu})^t\right\},$$

$$\mathbf{x}, \boldsymbol{\mu} \in \mathbb{R}^p, \boldsymbol{\Sigma} > \mathbf{0}.$$

We can show that (12.3) can be written as

$$f(\mathbf{X}) = \frac{1}{(2\pi)^{np/2}|\boldsymbol{\Sigma}|^{n/2}|\boldsymbol{\Psi}|^{p/2}} \exp\left\{-\frac{1}{2}[\text{Vec}(\mathbf{X}) - \text{Vec}(\mathbf{M})]^t\right.$$

$$\left. \times (\boldsymbol{\Sigma} \otimes \boldsymbol{\Psi})^{-1}[\text{Vec}(\mathbf{X}) - \text{Vec}(\mathbf{M})]\right\}.$$

12.4.1 *Properties*

When sampling from a multivariate normal population, we obtain a matrix variate normal distribution if observation vectors are displayed as column vectors alongside, i.e.

$$\mathbf{X} = [\mathbf{X}_1 \mathbf{X}_2 \dots \mathbf{X}_N] = \begin{bmatrix} x_{11} & x_{12} & \dots & x_{1N} \\ x_{21} & x_{22} & \dots & x_{2N} \\ \dots & \dots & \dots & \dots \\ x_{p1} & x_{p2} & \dots & x_{pN} \end{bmatrix}.$$

We have the following properties:

(a) If $\mathbf{X} \sim N_{[p,n]}(\mathbf{M}; \mathbf{\Sigma}, \mathbf{\Psi})$, then for its transpose, we have $\mathbf{X}^t \sim N_{[n,p]}(\mathbf{M}^t; \mathbf{\Psi}, \mathbf{\Sigma})$.

(b) If $\mathbf{X} \sim N_{[p,n]}(\mathbf{M}; \mathbf{\Sigma}, \mathbf{\Psi})$, then, partitioning $\mathbf{X}, \mathbf{M}, \mathbf{\Sigma}$ and $\mathbf{\Psi}$ as

$$\mathbf{X} = \begin{bmatrix} \mathbf{X}_{11} & \mathbf{X}_{12} \\ \mathbf{X}_{21} & \mathbf{X}_{22} \end{bmatrix} \begin{matrix} m, \\ p-m, \end{matrix} \qquad \mathbf{M} = \begin{bmatrix} \mathbf{M}_{11} & \mathbf{M}_{12} \\ \mathbf{M}_{21} & \mathbf{M}_{22} \end{bmatrix} \begin{matrix} m, \\ p-m, \end{matrix}$$
$$\phantom{\mathbf{X} = }\begin{matrix} t & n-t \end{matrix} \qquad\qquad\qquad \begin{matrix} t & n-t \end{matrix}$$

$$\mathbf{\Sigma} = \begin{bmatrix} \mathbf{\Sigma}_{11} & \mathbf{\Sigma}_{12} \\ \mathbf{\Sigma}_{21} & \mathbf{\Sigma}_{22} \end{bmatrix} \begin{matrix} m, \\ p-m, \end{matrix} \qquad \mathbf{\Psi} = \begin{bmatrix} \mathbf{\Psi}_{11} & \mathbf{\Psi}_{12} \\ \mathbf{\Psi}_{21} & \mathbf{\Psi}_{22} \end{bmatrix} \begin{matrix} t, \\ n-t, \end{matrix}$$
$$\phantom{\mathbf{\Sigma} = }\begin{matrix} m & p-m \end{matrix} \qquad\qquad\qquad \begin{matrix} t & n-t \end{matrix}$$

we have $\mathbf{X}_{11} \sim N_{[m,t]}(\mathbf{M}_{11}; \mathbf{\Sigma}_{11}, \mathbf{\Psi}_{11})$.

(c) If $\mathbf{X} \sim N_{[p,n]}(\mathbf{M}; \mathbf{\Sigma}, \mathbf{\Psi})$, then we have

$$\mathrm{tr}\left\{ \mathbf{\Sigma}^{-1}(\mathbf{X} - \mathbf{M})\mathbf{\Psi}^{-1}(\mathbf{X} - \mathbf{M})^t \right\} \sim \chi^2_{np}, \qquad (12.4)$$

where tr is the trace.

12.4.2 *Derived matrix distributions*

(1) There are several special cases of the matrix variate normal distribution that are more frequently encountered and used than the general normal matrix. They are as follows: the symmetric matrix variate normal, the restricted matrix variate normal, the matrix variate-generalized normal, etc. They can be seen in Gupta and Nagar (2000).

(2) We have seen the Wishart distribution derived from the multivariate normal. It can also be derived from the matrix variate normal, as seen in the following: Let $\mathbf{X} \sim N_{[p,n]}(\mathbf{0}; \mathbf{\Sigma}, \mathbf{I}_n)$ and $\mathbf{S} = \mathbf{X}\mathbf{X}^t, n \geq p$. Then $\mathbf{S} \sim W_p(n, \mathbf{\Sigma})$.

12.5 From the Normal to the Gamma Family of Distributions

As seen in the diagram relating various distributions in Chapter 2, only the normal is one of the basic ones, i.e. we have to define it.

Others are derived from the normal, or from other basic distributions, by unary or binary operations. The gamma distribution could be another basic one, although one of its particular cases, the χ_1^2, with 1 dof, is the square of the standard univariate normal. It generalizes to the central and non-central chi-square with p dof and to the central and non-central univariate gamma, and to p-dimensions, then to the matrix Wishart and non-central Wishart variables. The latter generalizes to the matrix variate gamma and to the non-central matrix variate gamma (see Table 12.1 which shows, for gamma distributions, the correspondence between scalar and random matrix variables).

For scalar distributions, we have the following:

(i)   Univariate gamma family (see Chapter 2);
(ii) Bivariate gamma distributions (see Chapter 13);
(iii) Multivariate gamma distributions (see Chapter 13).

12.6   The Wishart Distribution

12.6.1   *Its entries distribution*

Let $\mathbf{A}$ be a $(p \times p)$ symmetric matrix. We say that $\mathbf{A} \sim W_p(n, \boldsymbol{\Sigma})$, $n \geq p$, i.e. $\mathbf{A}$ follows a central Wishart distribution with matrix $\boldsymbol{\Sigma}$ and n dof, if the density of $\mathbf{A}$ (or its entries distribution) is

$$f(\mathbf{A}) = K.\text{etr}\left(-\frac{1}{2}\boldsymbol{\Sigma}^{-1}\mathbf{A}\right)|\mathbf{A}|^{\frac{n-p-1}{2}}, \qquad (12.5)$$

with $\mathbf{A} > 0, K = \frac{1}{2^{np/2}\Gamma_p(n/2)|\boldsymbol{\Sigma}|^{n/2}}$, and

$$\Gamma_p(\alpha) = \pi^{\frac{p(p-1)}{4}}\Gamma(\alpha)\Gamma\left(\alpha - \frac{1}{2}\right)\cdots\Gamma\left(\alpha - \frac{p-1}{2}\right), \alpha > \frac{p-1}{2}$$

(see Chapter 11).

The Wishart distribution is the generalization of the univariate chi-square distribution to matrix variate distributions.

Theorem 12.1. *Let* $\mathbf{X} \sim N_p(\boldsymbol{\mu}, \boldsymbol{\Sigma})$ *be a normal vector and* $\{\mathbf{X}_1, \ldots, \mathbf{X}_N\}$ *be a sample taken from this population. Classical* *results show that the sample mean vector*

$$\overline{\mathbf{X}} = \frac{1}{N} \sum_{i=1}^{N} \mathbf{X}_i$$

is independent of the sample covariance matrix $\mathbf{S}$, *where*

$$n\mathbf{S} = \mathbf{V} = \sum_{j=1}^{N} (\mathbf{X}_j - \overline{\mathbf{X}})(\mathbf{X}_j - \overline{\mathbf{X}})^t, \qquad (12.6)$$

with $n = N - 1 \geq p$, *and they are independent, and distributed, respectively, as a normal vector* $\overline{\mathbf{X}} \sim N_p(\boldsymbol{\mu}, \boldsymbol{\Sigma}/N)$ *and a central Wishart matrix* $\mathbf{S} \sim W_p(n, \boldsymbol{\Sigma}/n)$, *but*

$$\mathbf{V} \sim W_p(n, \boldsymbol{\Sigma}). \qquad (12.7)$$

Note: The maximum likelihood estimates of $\boldsymbol{\mu}$ and $\boldsymbol{\Sigma}$ are $\overline{\mathbf{X}}$ and $\frac{n}{N}\mathbf{S}$, respectively. If $n < p$, the distribution is singular and no density exists. The case of pseudo-Wishart matrices will not be considered here.

Proof. The proof of this theorem can be done using the classical methods based on Jacobians (Anderson, 1984) or using exterior algebras and a special integral on Stiefel manifold (Muirhead, 1983). In both cases, advanced mathematics are required.

Remark: For $p = 1$, we have $\boldsymbol{\Sigma} = \sigma^2, \mathbf{S} = s^2$ and the Wishart density becomes

$$\left(\frac{n}{2\sigma^2}\right)^{n/2} \frac{1}{\Gamma(n/2)} \exp\left(-\frac{ns^2/\sigma^2}{2}\right) (s^2)^{n/2-1} = \frac{e^{-v/2}v^{n/2-1}}{2^{n/2}\Gamma(n/2)},$$

when setting $v = ns^2/\sigma^2$, or a χ_n^2 density with n dof, as expected.

12.6.2   *Characteristic function*

For the univariate case, as seen above, if $A \sim W_1(n, \sigma^2)$, then $A/\sigma^2 \sim \chi_n^2$ and chf of A is

$$\varphi_A(t) = \frac{1}{(1 - 2it\sigma^2)^{n/2}}.$$

Generalizing this result, we have, for $\mathbf{A} \sim W_p(n, \mathbf{\Sigma})$, the chf of $\mathbf{A}$, i.e. of the $p(p+1)/2$ elements,

$$\varphi_{\mathbf{A}}(\mathbf{\Theta}) = \det (\mathbf{I}_p - i\mathbf{\Gamma}\mathbf{\Sigma})^{-n/2},$$

with $\mathbf{\Gamma} = [\gamma_{ij}]$, where $\gamma_{ij} = [(1 + \delta_{ij})\theta_{ij}]$, $i, j = 1, \ldots, p$, where $\theta_{ij} = \theta_{ji}$, and δ_{ij} is the Kronecker symbol.

The proof of this result is elaborate.

12.6.3   *Additivity*

We consider s random square $(p \times p)$ matrices which are independent Wishart with same covariance matrix $\mathbf{\Sigma}$. Then their sum is also Wishart, i.e. $\sum_{j=1}^{s} \mathbf{A}_j \sim W_p(\sum n_j, \mathbf{\Sigma})$. The proof uses the expression of the chf obtained above, which can be written as a product of p individual chf's.

12.6.4   *The diagonal submatrices of a Wishart matrix*

Theorem 12.2. *Let* $\mathbf{A} \sim W_p(n, \mathbf{\Sigma})$, *and let* $\mathbf{A} = \begin{bmatrix} \mathbf{A}_{11} & \mathbf{A}_{12} \\ \mathbf{A}_{21} & \mathbf{A}_{22} \end{bmatrix}$, $\mathbf{\Sigma} = \begin{bmatrix} \mathbf{\Sigma}_{11} & \mathbf{\Sigma}_{12} \\ \mathbf{\Sigma}_{21} & \mathbf{\Sigma}_{22} \end{bmatrix}$, *where* $\mathbf{A}_{11}$ *and* $\mathbf{\Sigma}_{11}$ *are* $(r \times r)$, *with* $r + s = p$. *Then* $\mathbf{A}_{11} \sim W_r(n, \mathbf{\Sigma}_{11})$. *Furthermore,* $\mathbf{A}_{11}, \mathbf{A}_{22}$ *are independent if* $\mathbf{\Sigma}_{12} = \mathbf{0}$.

Proof. The proof is based on the chf obtained earlier.   □

Corollary 12.1. *The diagonal elements of* $\mathbf{A}$ *arc all independent, with* $a_{ii}/\sigma_{ii} \sim \chi_n^2$.

We can see that their sum, or $\mathrm{tr}(\mathbf{A}) = \sum_{i=1}^{p} \frac{a_{ii}}{\sigma_{ii}}$, has a χ_{np}^2 distribution.

12.6.5 *Distributions of some ratios*

Let $\mathbf{A} \sim W_p(n, \boldsymbol{\Sigma})$ and $\mathbf{Y}$ be a $(p \times 1)$ random vector independent of $\mathbf{A}$. Then we have the following:

(a) The ratio $\frac{\mathbf{Y}^t \mathbf{A} \mathbf{Y}}{\mathbf{Y}^t \boldsymbol{\Sigma} \mathbf{Y}} \sim \chi_n^2$ is independent of $\mathbf{Y}$.

(b) $\frac{\mathbf{Y}^t \boldsymbol{\Sigma}^{-1} \mathbf{Y}}{\mathbf{Y}^t \mathbf{A}^{-1} \mathbf{Y}} \sim \chi_{n-p+1}^2$ is independent of $\mathbf{Y}$.

(c) The ratio $\frac{\overline{\mathbf{X}}^t \mathbf{V} \overline{\mathbf{X}}}{\overline{\mathbf{X}}^t \boldsymbol{\Sigma} \overline{\mathbf{X}}} \sim \chi_n^2$ is independent of $\overline{\mathbf{X}}$, where $\mathbf{V}$ is given by expression (12.6).

The proofs of these related results are more elaborate and we will not give them here.

12.6.6 *Bartlett decomposition*

Bartlett's theorem here informs us on the entries of a Wishart matrix.

Theorem 12.3. *Let $\mathbf{A} \sim W_p(n, \mathbf{I}_p)$, $n \geq p$ and let $\mathbf{A} = \mathbf{T}^t \mathbf{T}$ be the Cholesky decomposition of $\mathbf{A}$, with $\mathbf{T}$ being upper triangular. Then all elements of $\mathbf{T}$ are independent, each element off-diagonal is $N(0,1)$, and on the diagonal, it is $t_{ii}^2 \sim \chi_{n-i+1}^2$.*

Proof. See Gupta and Nagar (2000, p. 91). $\square$

We also have $\mathrm{tr}(\mathbf{A}) = \sum_{i \leq j}^p t_{ij}^2 \sim \chi_{np}^2$ (see Pham-Gia *et al.* (2015)), while $|\mathbf{A}| = |\boldsymbol{\Sigma}| \prod_{i=1}^p t_{ii}^2 \sim |\boldsymbol{\Sigma}| \prod_{i=1}^p \chi_{n-i+1}^2$. The above results can then be directly derived.

12.6.7 *Determinant distribution*

Since the generalized sample variance is the determinant of the sample covariance matrix, we refer to Section 12.7.

12.6.8 *Latent roots of a Wishart matrix*

Latent roots, or eigenvalues, of a random matrix play an important role in multivariate statistical analysis and in random matrix theory (RMT) in theoretical physics.

We establish this density for the case $\mathbf{A} \sim W_p(n, \alpha \mathbf{I}_p)$. There is a theorem that gives the expression of $f(\lambda_1, \ldots, \lambda_p)$ in function of the entries density of the square matrix $\mathbf{X}$ as follows:

$$f(\lambda_1, \ldots, \lambda_p) = \frac{\pi^{p^2/2}}{\Gamma_p(p/2)} \prod_{i<j} (\lambda_i - \lambda_j) \int_{O(p)} f(\mathbf{HLH}^t)(d\mathbf{H}),$$

where $\mathbf{L} = diag(\lambda_1, \ldots, \lambda_p), \lambda_1 > \cdots > \lambda_p$.

Proof. For $\boldsymbol{\Sigma} = \alpha \mathbf{I}_p$, we can then prove that

$$f(\lambda_1, \ldots, \lambda_p) = \frac{\pi^{p^2/2}}{(2\lambda)^{np/2} \Gamma_p(p/2) \Gamma_p(n/2)} \exp\left(-\sum_{i=1}^{p} \lambda_i/2\alpha\right)$$
$$\times \prod_{i<j} (\lambda_i - \lambda_j) \prod \lambda_i^{(n-p-1)/2}. \tag{12.8}$$

For $\boldsymbol{\Sigma} \neq \alpha \mathbf{I}_p$, we do not have a closed form, but the density can be expressed in terms of zonal polynomials. $\qquad \square$

Note: Zonal polynomials were developed by Hua (1963) and James (1961) and, basically, served to the development of $(\mathrm{tr}\mathbf{S})^k$, where tr is the trace as follows:

Let $\mathbf{S}(p \times p)$ be a symmetric matrix and let $\mathbf{V}_k$ be the vector space of homogeneous polynomials $\varphi(\mathbf{S})$ of degree k in $p(p+1)/2$ entries of $\mathbf{S}$. We know that $\mathbf{V}_k$ can be decomposed into a direct sum of irreducible invariant subspaces $\mathbf{V}_k$, with $\kappa = (k_1, \ldots, k_p)$, $k_1 \geq \cdots \geq k_p, \sum_{j=1}^{p} k_j = k$.

We have $(\mathrm{tr}\mathbf{S})^k = \sum_\kappa C_\kappa(\mathbf{S})$, where the zonal polynomial $C_\kappa(\mathbf{S})$ is the component of $(\mathrm{tr}\mathbf{S})^k$ in the subspace $\mathbf{V}_\kappa$.

Zonal polynomials come into play in matrix variate distributions and provide a good theoretical tool to handle non-central matrix variate distributions. However, they are very difficult to compute numerically, and hence, their usefulness is, for the time being, limited.
Note: For the non-central Wishart, see Chapter 13.

12.7 The Generalized Variance

12.7.1 *Expression of the generalized variance*

In multivariate statistics, the generalized variance is just the determinant of the covariance matrix, either for the population, $|\boldsymbol{\Sigma}|$ or for the sample, $|\mathbf{S}|$. It is the equivalent of σ^2 and s^2 in univariate statistics, and its geometric interpretation also follows loosely those of these two measures. Introduced by Wilks (1932), it provides a single number summary of the whole matrix $\mathbf{S}$, and although convenient and useful, it is undoubtedly subject to several weaknesses. Setting $\mathbf{A} = n\mathbf{S}$, we have

$$\mathbf{A} \sim W_p(n, \mathbf{\Sigma}), \ n \geq p \quad \text{and} \quad n^p \, |\mathbf{S}| \, / \, |\mathbf{\Sigma}| = |\mathbf{A}| \, / \, |\mathbf{\Sigma}| \sim \prod_{i=1}^{p} \chi^2_{n-i+1}.$$

(12.9)

The latest product can be expressed as a G-function (Pham-Gia, 2008).

For a normal population, distributional properties of $|\mathbf{S}|$ can be established (see, e.g. Anderson (1984, Section 7.5)), but until recently, there were limitations on the computation of its density, resulting in an absence of numerical applied methods and results related to this multivariate dispersion measure. This situation, which lasted for a long period of time, was very likely due to the fact that the distribution of $|\mathbf{S}|$, given under the form of a Meijer G-function, could not be easily computed, since the values of this function rely on the theory of residues in complex analysis. Some approximate distributions have been suggested, for example, by Hoel (1937), who proposed using a two-parameter gamma as its approximate density.

Fortunately, the latest advances in mathematical computing have allowed the computation of several special functions that can be handled only theoretically in the past. Here, we first derive the density of $|\mathbf{S}|$ in terms of Meijer G-functions and use the software Maple (or Mathematica) to compute and graph it.

12.7.2 *Expression of the density of the sample generalized variance*

Theorem 12.4. *The ratio* $Y = \frac{|\mathbf{A}|}{|\mathbf{\Sigma}|} = n^p \frac{|\mathbf{S}|}{|\mathbf{\Sigma}|}$, *related to a random sample of size* $N = n + 1$, *has its density given by, for* $y > 0$,

$$h(y) = \frac{1}{2^p} \left[\prod_{j=0}^{p-1} \frac{1}{\Gamma(\frac{n-j}{2})} \right] G_{0 \ p}^{p \ 0} \left[\frac{y}{2^p} \, \middle| \, \frac{(n-2)}{2}, \right.$$

$$\left. \frac{(n-3)}{2}, \dots, \frac{(n-(p+1))}{2} \right]. \quad (12.10)$$

Proof. See Pham-Gia (2010).   □

12.7.3   *Estimation of $|\Sigma|$*

A simple and direct approach is suggested by Anderson (1984, p. 266), based on the result that, asymptotically, $\sqrt{n}\left(\frac{|\mathbf{S}|}{|\Sigma|} - 1\right) \sim N(0, 2p)$. This would lead to the approximate confidence interval:

$$\left(\frac{|\widehat{\mathbf{S}}|}{1 - z_{\alpha/2}\sqrt{\frac{2p}{n}}}, \frac{|\widehat{\mathbf{S}}|}{1 + z_{\alpha/2}\sqrt{\frac{2p}{n}}}\right),$$

where $\mathbb{P}(Z \leq z_{\alpha/2}) = \alpha/2$, and $\mathbf{S}$ is previously defined, with $\widehat{\mathbf{S}}$ denoting its estimated value.

This formula requires, however, that n be sufficiently large, i.e. $n > 2p(z_{\alpha/2})^2$, in order for the upper bound to be positive. With the results in the previous sections, however, we will be able to improve on this confidence interval and deal with any sample size.

If we use the sample covariance matrix $\mathbf{S}$, the density of $n^p \frac{|\mathbf{S}|}{|\Sigma|}$ is given by (12.10). Let $h_{\alpha/2}$ and $h_{1-(\alpha/2)}$ be the lower and upper $100(\alpha/2)$th percentiles of this density, which can be numerically determined with precision. We then have

$$h_{\alpha/2} \leq \frac{n^p |\mathbf{S}|}{|\Sigma|} \leq h_{1-\alpha/2}.$$

Hence, the confidence interval for $|\Sigma|$ is

$$\left(\frac{n^p |\widehat{\mathbf{S}}|}{h_{1-\alpha/2}}, \frac{n^p |\widehat{\mathbf{S}}|}{h_{\alpha/2}}\right). \tag{12.11}$$

12.7.4   *Numerical illustration*

Let's consider in $\mathbb{R}^4$ the distribution $N_4(\boldsymbol{\mu}_1, \boldsymbol{\Sigma}_1)$ with known parameters,

$$\boldsymbol{\mu}_1{}^t = (6.35, 6.20, 5.55, 5.23)$$

and

$$\Sigma_1 = \begin{pmatrix} 1.80 & 0.140 & 0.023 & 0.010 \\ 0.140 & 0.900 & 0.070 & 0.030 \\ 0.023 & 0.070 & 0.700 & 0.030 \\ 0.010 & 0.030 & 0.030 & 0.600 \end{pmatrix}.$$

This distribution arises in the analysis of Fisher's iris data (Pham-Gia *et al.*, 2008). Supposing $|\Sigma_1|$ unknown, we will estimate it by a confidence interval, using simulation. We generate $N = 9$ observations from this distribution, using the classical approach, as given by Tanizaki (2004), and compute the 95% confidence interval for $|\Sigma_1|$, based on the observed sample covariance matrix $\widehat{\mathbf{S}}_1$ obtained from that sample.

Results:

(a) The nine generated observations lead to the estimated sample mean vector

$$\widehat{\boldsymbol{\mu}}_1^t = \{6.485, 6.479, 5.102, 5.014\}$$

and the estimated sample covariance matrix

$$\widehat{\mathbf{S}}_1 = \begin{pmatrix} 2.465 & 0.694 & -0.162 & 0.017 \\ 0.694 & 0.578 & -0.097 & -0.004 \\ -0.162 & -0.097 & 0.857 & 0.506 \\ 0.017 & -0.004 & 0.506 & 0.521 \end{pmatrix},$$

with determinant value, $|\widehat{\mathbf{S}}_1| = 0.1708$.

(b) Density $h(x)$ is given by Eq. (12.10), where $n = 8$, $p = 4$.
The exact 95% confidence interval $\Rightarrow \alpha = 0.05 \Rightarrow$
$\begin{cases} \alpha/2 = 0.025, \\ 1 - \alpha/2 = 0.975, \end{cases}$
$h_{\alpha/2}$: lower $100(\alpha/2)$th percentile of $h(x) \Rightarrow h_{\alpha/2} = 65.3448$ (using Maple).
$h_{1-\alpha/2}$: upper $100(\alpha/2)$th percentile of $h(x) \Rightarrow h_{1-\alpha/2} = 7{,}926.0490$ (using Maple).

Also,

$$n^p|\widehat{\mathbf{S}}| = 0.1708 \times 8^4 = 699.5968.$$

The confidence interval for $|\boldsymbol{\Sigma}|$ (using Eq. (12.11)

$$\left(\frac{n^p\left|\widehat{\mathbf{S}}\right|}{h_{1-\alpha/2}}, \frac{n^p\left|\widehat{\mathbf{S}}\right|}{h_{\alpha/2}}\right) = \left(\frac{699.5968}{7,926.0490}, \frac{699.5968}{65.3448}\right)$$

$$= (0.0883, 10.7188),$$

which contains the true value of $|\boldsymbol{\Sigma}_1| = 0.6645$, as expected (Fig. 12.1).

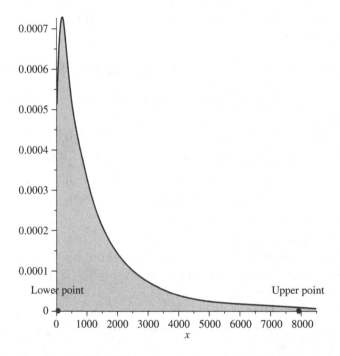

Fig. 12.1   Density of $n^p|\mathbf{S}_1|/|\boldsymbol{\Sigma}_1|$, where lower point is 65.3448 and upper point is 7,926.0490.

12.8   The Matrix Variate Gamma Distribution

In the preceding sections, we started with the Wishart distribution. However, it is more general to consider the matrix variate gamma distribution, $\mathbf{W} \sim \text{Gam}_p(a, \mathbf{C})$ defined as follows:

Let $\mathbf{W}$ be a positive definite $(p \times p)$ matrix. We say that $\mathbf{W} \sim \text{Gam}_p(a, \mathbf{C})$ if its density is

$$f(\mathbf{W}) = \frac{1}{|\mathbf{C}|^{-a}\Gamma_p(a)}\text{etr}(-C\mathbf{W})|\mathbf{W}|^{a-(p+1)/2}, \quad \mathbf{W} > 0, \quad (12.12)$$

where $\mathbf{C}(p \times p) > 0$, $a > (p-1)/2$. So, a Wishart matrix can be expressed as a gamma matrix:

$$\mathbf{X} \sim W_p(n, \boldsymbol{\Sigma}) \to \mathbf{X} \sim \text{Gam}_p(n/2, \boldsymbol{\Sigma}^{-1}/2).$$

12.9   The Central Matrix Variate Beta Distributions

Recall the following results in univariate distributions: Let X_1, X_2 be two independent $\chi^2_{k_1}$ and $\chi^2_{k_2}$ variables in the univariate case. Then $V_1 = X_1 + X_2$, $V_2 = X_1/V_1$, $V_3 = X_1/X_2$ are three independent variables having, respectively, the $\chi^2_{k_1+k_2}$, $\beta^I_1(\alpha, \beta)$, and $\beta^{II}_1(\alpha, \beta)$ distributions.

We have similar results for matrix variate distributions, which have an important role in theoretical multivariate analysis. For applications, it would not be surprising to see it widely applied in the future, just like its counterpart in one dimension, as mentioned in Chapter 4.

12.9.1   *Entries distributions*

Definition 12.2.

(a) The random symmetric positive definite matrix $\mathbf{U}$ is said to be a *central beta matrix variate of the first kind, with parameters* $a, b > \frac{p-1}{2}$, denoted $\mathbf{U} \sim \beta^I_p(a, b)$, if its density (i.e. the relationships between its matrix elements) is given by

$$f(\mathbf{U}) = \frac{|\mathbf{U}|^{a-\frac{p+1}{2}}|\mathbf{I}_p - \mathbf{U}|^{b-\frac{p+1}{2}}}{\text{Beta}_p(a, b)}, \quad 0 < \mathbf{U} < \mathbf{I}_p, \quad (12.13)$$

where $\text{Beta}_p(a, b)$ is the beta function in $\mathbb{R}^p$, i.e.

$$\text{Beta}_p(a, b) = \frac{\Gamma_p(a)\Gamma_p(b)}{\Gamma_p(a + b)},$$

with

$$\Gamma_p(a) = \pi^{\frac{p(p-1)}{4}} \prod_{j=1}^{p} \Gamma\left(a - \frac{j-1}{2}\right).$$

(b) The random symmetric positive definite matrix $\mathbf{V}$ is said to be a *central beta matrix variate of the second kind, with parameters* $a, b > \frac{p-1}{2}$, denoted $\mathbf{V} \sim \beta_p^{II}(a, b)$, if its density is given by

$$g(\mathbf{V}) = \frac{|\mathbf{V}|^{a-\frac{p+1}{2}}|\mathbf{I}_p + \mathbf{V}|^{-(a+b)}}{\text{Beta}_p(a, b)}, \quad \mathbf{V} > \mathbf{0}. \tag{12.14}$$

Chapter 5 of Gupta and Nagar (2000) gives the basic properties of these two matrix variates. We have simple transformations relating them, as in the univariate case:

- If $\mathbf{U} \sim \beta_p^{I}(a, b)$, then $\mathbf{V} = (\mathbf{I}_p - \mathbf{U})^{-\frac{1}{2}}\mathbf{U}(\mathbf{I}_p - \mathbf{U})^{-\frac{1}{2}} \sim \beta_p^{II}(a, b)$.
- If $\mathbf{V} \sim \beta_p^{II}(a, b)$, then $\mathbf{U} = (\mathbf{I}_p + \mathbf{V})^{-\frac{1}{2}}\mathbf{V}(\mathbf{I}_p + \mathbf{V})^{-\frac{1}{2}} \sim \beta_p^{I}(a, b)$.

These distributions can be obtained from two samples of the multivariate normal, which give rise to two independent Wishart matrices $\mathbf{X}_1$ and $\mathbf{X}_2$. We have

$$\mathbf{Y}_1 = (\mathbf{X}_1 + \mathbf{X}_2)^{-\frac{1}{2}}\mathbf{X}_1(\mathbf{X}_1 + \mathbf{X}_2)^{-\frac{1}{2}} \sim \beta_p^{I}\left(\frac{k_1}{2}, \frac{k_2}{2}\right)$$

and

$$\mathbf{Y}_2 = (\mathbf{X}_2)^{-\frac{1}{2}}\mathbf{X}_1(\mathbf{X}_2)^{-\frac{1}{2}} \sim \beta_p^{II}\left(\frac{k_1}{2}, \frac{k_2}{2}\right).$$

The density of $\mathbf{Y}_1$ is given by (12.13), with $a = k_1/2, b = k_2/2$, and similarly, the density of $\mathbf{Y}_2$ is given by (12.14), with $a = k_1/2, b = k_2/2$.

12.9.2 *Determinant distribution*

Let $\mathbf{U} \sim \beta_p^I(a, b)$ and $\mathbf{V} \sim \beta_p^{II}(a, b)$. To find the densities of determinants $|\mathbf{U}|$ and $|\mathbf{V}|$, we can use Section 12.9.5 on Wilk's statistics.

12.9.3 *Latent roots distributions*

The latent roots of matrix variates $\mathbf{U}$ and $\mathbf{V}$ are used in several statistical tests. Their distributions have been found almost simultaneously by five well-known statisticians (Fisher, Hsu, Mood, Roy, and Girshick) between 1939 and 1951. They have been used in hypothesis testing, in relation to the equality of two covariance matrices, in multivariate analysis of variance, in canonical correlation, among other topics.

The density of the latent roots $\{f_1, \ldots, f_p\}$ of $\mathbf{U} \sim \beta_p^I(a, b)$ is given by

$$\frac{\pi^{p^2/2}}{\Gamma_p(p/2) B_p(a, b)} \left[\prod_{j=1}^{p} f_j^{a - \frac{p+1}{2}} (1 - f_j)^{b - \frac{p+1}{2}} \right] \prod_{i<j} (f_i - f_j), \quad (12.15)$$

which is defined in sector $1 > f_1 > f_2 > \cdots > f_p > 0$.

The density of the latent roots $\{\ell_1, \ldots, \ell_p\}$ of $\mathbf{V} \sim \beta_p^{II}(a, b)$ is given by

$$\frac{\pi^{p^2/2}}{\Gamma_p(p/2) B_p(a, b)} \left[\prod_{j=1}^{p} \ell_j^{a - \frac{p+1}{2}} (1 + \ell_j)^{-(a+b)} \right] \prod_{i<j} (\ell_i - \ell_j), \quad (12.16)$$

where $\ell_1 > \ell_2 > \cdots > \ell_p > 0$.

We also have $\ell_j = f_j / (1 - f_j)$.

Note: Latent roots distributions of powers of beta matrix variates, and their estimation, have been investigated by Pham-Gia *et al.* (2019).

12.9.4 *Decomposition of a beta matrix variate*

The following decomposition of the matrix beta distribution is similar to the one for the Wishart, as given by Theorem 12.3.

Let $\mathbf{U} \sim \text{beta}_p^I(a, b)$, we have (Gupta and Nagar, 2000) $\mathbf{U} = \mathbf{T}\mathbf{T}^t$, where $\mathbf{T} = (t_{ij})$ is an upper triangular matrix with diagonal elements

$t_{11}^2, \ldots, t_{pp}^2$, independently distributed, and $t_{ii}^2 \sim \beta_1^I(a - \frac{p-i}{2}, b)$ for $i = 1, \ldots, p$. Therefore, its determinant $|\mathbf{U}|$ can be expressed as a product of independent univariate betas of the first kind, $\prod_{i=1}^p t_{ii}^2$. Its density is given by Pham-Gia (2008) as follows, for $0 < u < 1$,

$$f_{|\mathbf{U}|}(u) = \prod_{j=1}^p \frac{\Gamma\left(a + b - \frac{j-1}{2}\right)}{\Gamma\left(a - \frac{j-1}{2}\right)} G_{p\,p}^{p\,0}$$

$$\times \left[u \left| \begin{matrix} a + b - 1, a + b - \frac{3}{2}, \ldots, a + b - \frac{p+1}{2} \\ a - 1, a - \frac{3}{2}, \ldots, a - \frac{p+1}{2} \end{matrix} \right. \right], \quad (12.17)$$

where $G_{r\,q}^{m\,n}(.)$ is the Meijer G-function.

For two independent matrix gammas, $\mathbf{W}_1, \mathbf{W}_2$, the entries distribution of $\mathbf{F} = \mathbf{W}_2^{-\frac{1}{2}} \mathbf{W}_1 \mathbf{W}_2^{-\frac{1}{2}}$ is discussed by Perlman (1977) to obtain $\mathbf{F} \sim F_p(\alpha_1, \alpha_2)$. For the ratio of the two determinants, there is no difficulty, and

$$|\mathbf{F}| = \frac{|\mathbf{W}_1|}{|\mathbf{W}_2|} = \frac{\prod_{i=1}^p X_i}{\prod_{i=1}^p Y_i} \sim \prod_{i=1}^p T_i, \quad (12.18)$$

with independent $T_i \sim \beta_1^{II}(\alpha_1 - \frac{i-1}{2}, \alpha_2 - \frac{i-1}{2})$.

Hence, the distribution of $|\mathbf{F}|$, with $\mathbf{F} \sim F_p(\alpha_1, \alpha_2)$ has density

$$\prod_{i=1}^p \beta_1^{II}\left(\alpha_1 - \frac{i-1}{2}, \alpha_2 - \frac{i-1}{2}\right). \quad (12.19)$$

Notes:

(1) For non-central matrix beta variate distributions, see Chapter 13.

(2) We see that, like the Wishart matrix which factorizes into normal matrices, the beta type I and type II factorize into inverted-t and t matrices.

(3) There is also a factorization result of the matrix beta type I, $\beta_p^I(a, b)$, into a product of p independent univariate betas type I, $\prod_{j=0}^{p-1} \beta_1^I(a - \frac{j}{2}; b)$ and $(p-1)$ independent Dirichlet variables, and a similar result holds for $\beta_p^{II}(a, b)$ (see Tan (1960)).

12.9.5 *Wilks's distributions*

Wilks's statistic associated with MANOVA is precisely the determinant of the central beta variable of the first kind. But it corresponds to the beta variable in univariate ANOVA instead of the Fisher–Snedecor F_{ν_1,ν_2} used there. The matrix variate $\mathbf{F}$ has been considered in several recent publications, and it is proven that its determinant can serve as well as Wilks' statistic (see Duong *et al.* (2019a, 2019b)).

The determinant of $\mathbf{U}$ has the generalized Wilks's distribution of the first type, denoted $\Lambda^I(2(a+b), p, 2b)$.

Similarly for $\mathbf{V} \sim \beta_p^{II}(a, b)$, its determinant $|\mathbf{V}|$ has the generalized Wilks's distribution of the second type, denoted $\Lambda^{II}(2(a+b), p, 2b)$, and expressed as a product of p independent univariate beta primes. Its density is given by Pham-Gia and Turkkan (2011) as

$$
f_{|\mathbf{V}|}(v) = \frac{1}{\prod_{j=1}^{p} \Gamma\left(a - \frac{j-1}{2}\right) \Gamma\left(b - \frac{j-1}{2}\right)}
$$
$$
\times G_{p\ p}^{p\ p}\left[v \left|
\begin{array}{c}
-b, -\left(b - \frac{1}{2}\right), \ldots, -\left(b - \frac{p-1}{2}\right) \\
a - 1, a - \frac{3}{2}, \ldots, a - \frac{p+1}{2}
\end{array}
\right.\right] \tag{12.20}
$$

for $v > 0$.

12.10 Conclusion

We have presented the main matrix variate distributions, which are related to the matrix normal distribution. These distributions have been of interest for many years, starting with Wishart's publication on the normal covariance matrix in 1928. The normal matrix and its distribution occupy a central role in the theory of matrix distributions, but its applications seem only to be present lately.

Bibliography

Anderson, T.W. (1984). *Introduction to Multivariate Analysis*, 2nd ed. (Wiley, New York).

Duong, T.P., Pham-Gia, T. and Dinh, N.T. (2019a). Exact distributions of two non-central generalized Wilks's statistics, *Journal of Statistical Computation and Simulation*, **89**(10), 1798–1818.

Duong, T.P., Pham-Gia, T. and Dinh, N.T. (2019b). Exact distribution of the non-central Wilks's statistic of the second kind, *Statistics & Probability Letters*, **153C**, 80–89.

Gupta, A.K. and Nagar, D.K. (2000). *Matrix Variate Distributions* (Chapman and Hall/CRC, Boca Raton).

Hoel, P.G. (1937). A significance test for components analysis, *Annals of Mathematical Analysis*, **8**, 149–158.

Hua, L.K. (1963). *Harmonic Analysis of Functions of Several Complex Variables in Classical Domains* (Moscow (in Russian) English translation by American Mathematical Society Translation of Math Monographs).

James, A.T. (1961). Zonal polynomials of the real positive definite symmetric matrices, *Annals of Mathematics*, **74**, 456–469.

Mathai, A.M. (1997). *Jacobians of Matrix Transformations and Functions of Matrix Argument* (World Scientific, Singapore).

Muirhead, R. (1983). *Aspects of Multivariate Statistical Theory* (Wiley, New York).

Perlman, M.D. (1977). A note on the matrix-variate F-Distribution, *Sankhya: The Indian Journal of Statistics*, Series A, **39**, 290–298.

Pham-Gia, T. (2008). Exact expression of Wilks's statistic and applications, *Journal of Multivariate Analysis*, **1999**, 1698–1716.

Pham-Gia, T. and Turkkan, N. (2010). Exact expression of the density of the generalized sample variance and applications, *Statistical Papers*, **51**(4), 931–945; Erratum (2011), **52**(3), 749.

Pham-Gia, T. and Turkkan, N. (2011). Distributions of the ratio: From Random Variables to Random Matrices, *Open Journal of Statistics*, **1**, 93–100.

Pham-Gia, T., Dinh, N.T. and Duong, T.P. (2015). Trace of the Wishart matrix and applications, *Open Journal of Statistics*, **5**, 173–190.

Pham-Gia, T., Duong, T.P. and Dinh, N.T. (2019). Distribution of powers of the central matrix beta variate, *Statistical Methods and Applications*, https://doi.org/10.1007/s10260-019-00497-3.

Pham-Gia, T., Turkkan, N. and Vovan, T. (2008). Statistical discrimination analysis using the maximum function, *Communications in Statistics — Simulation and Computation*, **37**(2), 320–336.

Tan, W.I. (1960). Note on the multivariate and generalized multivariate beta distribution, *Journal of American Statistical Association*, **64**, 230–241.

Tanizaki, H. (2004). *Computational Methods in Statistics and Econometrics* (Marcel Dekker, New York).

Chapter 13

Complements and Supplementary Information

13.1 Introduction

This last chapter presents information and knowledge non-essential to the comprehension of the materiel presented in the previous chapters, which it wishes to complement. Readers with a modest mathematics/statistics background can and will, probably, ignore this chapter. But other readers can refer to this chapter to review the basic approaches to some integral transforms (Laplace, Fourier, and Mellin), to the residue theorem and the complex inversion formula using a Bromwich path, to several distributions mentioned in the book, and finally also to several forms of non-central distributions mentioned. Scalar, multivariate, and matrix forms are presented, sometimes related to the latest results published. As supplementary information, we present the normal distribution in the complex case, and its various generalizations, as well as basic notions on random matrix theory (RMT) presently in use in theoretical physics.

13.2 Basic Integral Transforms

13.2.1 *The Laplace, Fourier, and Mellin transforms*

These three transforms play key roles in this book. These involve three steps:

- **Step 1:** Defining the integral transform in the real domain.
- **Step 2:** Performing all the operations/modifications, including integrating in the complex domain, along a chosen path and applying the residue theorem, if needed.
- **Step 3:** Finally coming back to the real domain with the inverse integral transform.

13.2.1.1   *The Laplace transform*

For a function $f(x)$ such that $\int_0^\infty e^{-kx}|f(x)|dx < \infty$ for some real value k, the Laplace transform of $f(x)$, $x \geq 0$, is

$$L_t\{f(x)\} = L_f(t) = \int\limits_0^\infty e^{-rx} f(x)dx, \qquad (13.1)$$

where r is a complex variable. This is a function defined on $\mathbb{C}$, with complex values. Conversely, if $L_t\{f(x)\}$ is analytic, of order $O(r^{-k})$ in some half-plane $\Re(r) > c$, with c, k real and $k > 1$, then its inverse is $f(x)$ and is uniquely determined by

$$f(x) = \frac{1}{2i\pi} \lim_{\beta \to \infty} \int\limits_{w-i\beta}^{w+i\beta} e^{tx} L_f(t)dt, \quad t \in \mathbb{C}, \qquad (13.2)$$

evaluated over any line $\Re(r) = w > c$ in the complex plane. This is a real-valued function defined on $\mathbb{R}$. Two functions with the same Laplace transform are identical (see Section 13.2.2 on complex inversion formula).

Note: If $f(x)$ is the density of X, $M_t\{f(x)\} = \int_0^\infty e^{tx} f(x)dx$ is the *moment-generating function* of X. It has the advantage of being completely in the real domain.

13.2.1.2   *The Fourier transform*

Let $f(x)$, $-\infty < x < \infty$, be s.t. $\int_{-\infty}^\infty |f(x)|e^{ikx}dx < \infty$ for some real k. The Fourier transform of $f(x)$ is defined by

$$F_t\{f(x)\} = F_f(t) = \int\limits_{-\infty}^\infty e^{-itx} f(x)dx.$$

This is a complex-valued function defined on $\mathbb{R}$, and its inverse is real-valued:

$$f(x) = \frac{1}{2\pi} \int_{-\infty}^{\infty} e^{itx} F_f(t) dt.$$

The Fourier transform is well utilized in probability, under the form of characteristic function of the random variable, denoted chf, which permits one to derive many results in an elegant way. In signal processing and imaging, it provides another approach to treat several problems, commonly called the *frequency domain*, as opposed to the *real time domain*.

13.2.1.3 *The Mellin transform*

The Mellin transform of $f(x), x \geq 0$, where $\int_0^\infty x^{k-1}|f(x)|dx < \infty$ for some real k, is defined by

$$\mathbf{M}_s\{f(x)\} = \int_0^{\infty} x^{s-1} f(x) dx. \tag{13.3}$$

This is a real function and its inverse Mellin transform, also a real function, is independent of the real value w used in the following relation:

$$f(x) = M_x^{-1}[\mathbf{M}_s(f(x))] = \frac{1}{2i\pi} \int_{w-i\infty}^{w+i\infty} x^{-s} \mathbf{M}_s(f(x)) ds, \quad s \in \mathbb{C}. \tag{13.4}$$

But Eq. (13.4) is valid under the condition that (13.3) exists as an analytic function of the complex variable s for $c_1 \leq \Re(s) = w \leq c_2$.

There are well-known relations between these three integral transformations. Also, they have been extended to fractional calculus and to matrices.

Since the Mellin transform generates the moments of a function, it is often more convenient to use it than the moment-generating function (mgf) just considered or other transforms.

13.2.2 *The complex inversion formula*

When integration is made in the complex plane, along an axis like in (13.2), the use of the residue theorem provides an elegant approach. In the following, we compute the inverse Laplace transform.

If $f(s) = L\{F(t)\}$, then $L^{-1}\{f(s)\}$ is given by

$$F(t) = \frac{1}{2\pi i} \int_{\gamma-i\infty}^{\gamma+i\infty} e^{st} f(s) ds, \quad t > 0, \tag{13.5}$$

and $F(t) = 0$ for $t < 0$. This method is called the *complex inversion integral*, using a *Bromwich's contour*. The result provides a direct means for obtaining the inverse Laplace transform of a given function $f(s)$.

The integration in (13.5) is to be performed along a line $s = \gamma$ in the complex plane, where $s = x + iy$. The real number γ is chosen so that $s = \gamma$ lies to the right of all the singularities (poles, branch points, or essential singularities).

There are several results and questions related to the complex inversion integral. Due to lack of space, we refer to Springer (1979) for integration along a Bromwich path, and the application of the residue theorem in order to use Eq. (13.5), and Spiegel (1974) can be consulted for the case where the Bromwich path has to be modified because of the presence of branch points.

13.3 Meijer *G*-Functions and Fox *H*-Functions

This section just gives a short introduction to Meijer *G*-functions and Fox *H*-functions (see also Section 4.5 of Chapter 4).

G- and *H*-functions are considered to be among the most general functions defined in mathematical analysis. They provide powerful tools in theoretical and applied mathematics, as well as in theoretical physics and several branches of applied sciences. Here, they are used in the expressions of the densities of several positive random variables and in the distributions of determinants of random matrices, as shown by Pham-Gia (2008). For example, when considering a random beta matrix variate, its determinant has its density expressed as a *G*-function, since it is a product of independent univariate betas. Products and ratios of independent random matrices can hence have their determinant densities expressed as G-functions as well, and so do several test statistics in multivariate analysis (Pham-Gia, 2014). The three types of G-functions mostly encountered here are $G^{m\ 0}_{m\ m}$, $G^{m\ 0}_{0\ m}$, $G^{m\ n}_{p\ q}$. But, as Mathai and Saxena (1973, Section 5.6), have remarked, we often have here the cases where the parameters differ by integers, and computations of residues have to be adjusted accordingly.

13.3.1   Definitions of Meijer G-functions and Fox H-functions

These functions are defined as integrals, in the complex plane, from a generalization of the Mellin–Barnes integral. There are three possible paths of integration and the theorem of residues can be applied to obtain their numerical values. We define the H-function, using the Mellin–Barnes formula, and consider the ratio of two products of gamma functions as integrand. Fox's H-function is hence defined as the integral along the complex contour L, of the expression $\varphi(\cdot)x^{-s}$, i.e.

$$
H^{m\ \ r}_{p\ \ q}\left[x\left|\begin{array}{l}(a_1,\alpha_1),\ldots,(a_p,\alpha_p)\\(b_1,\beta_1),\ldots,(b_q,\beta_q)\end{array}\right.\right]
$$

$$
= \frac{1}{2\pi i}\int_L \varphi\left(\begin{array}{l}(a_1,\alpha_1),\ldots,(a_p,\alpha_p)\\(b_1,\beta_1),\ldots,(b_q,\beta_q)\end{array}\right)x^{-s}ds, \qquad (13.6)
$$

where

$$
\varphi\left(\begin{array}{l}(a_1,\alpha_1),\ldots,(a_p,\alpha_p)\\(b_1,\beta_1),\ldots,(b_q,\beta_q)\end{array}\right)
$$

$$
= \frac{\displaystyle\prod_{j=1}^{m}\Gamma(b_j+\beta_j s)\prod_{j=1}^{r}\Gamma(1-a_j-\alpha_j s)}{\displaystyle\prod_{j=m+1}^{q}\Gamma(1-b_j-\beta_j s)\prod_{j=r+1}^{p}\Gamma(a_j+\alpha_j s)}. \qquad (13.7)
$$

Some readers might wonder how this expression of $\varphi\left(\begin{smallmatrix}(a_1,\alpha_1),\ldots,(a_p,\alpha_p)\\(b_1,\beta_1),\ldots,(b_q,\beta_q)\end{smallmatrix}\right)$ comes into play. As explained by Pham-Gia and Dinh (2016), these functions are generalizations of Gauss hypergeometric function $_pF_q(.)$, but the convergent series, originally defining $_pF_q(.)$, has been replaced by (13.6), according to the Mellin–Barnes integral, which has a similar expression, so that the theorem of residues can be applied. The condition to be satisfied is that no pole of $\Gamma(b_j+\beta_j s)$, $j=1,\ldots,m$ coincides with any pole of $\Gamma(1-a_j-\alpha_j s)$, $j=1,\ldots,r$. Furthermore, we suppose that

$$\alpha_j(\nu + b_h) \neq \beta_h(a_j - 1 - \lambda),$$

$$\lambda, \nu = 0, 1, 2, \ldots, \quad j = 1, \ldots, r, \quad h = 1, \ldots, m.$$

In particular, we have

$$\exp(x) = G_{0\ \ 1}^{1\ \ 0}(-x|0),$$

and hence,

$$\exp(-x) = G_{0\ \ 1}^{1\ \ 0}(x|0),$$

equalities that will be used for the normal distribution.

13.3.2 *Convergence*

The G-function converges when L is taken as one of the two paths $L_\infty, L_{-\infty}$ encircling the right poles (related to a_j), or the left poles (related to b_j), defining $G(z)$ for $|z| < 1$ and $|z| \geq 1$, respectively, depending on the values of p and q, or a third path L^* can be taken as the vertical axis, separating them, for $p + q < 2(m + n)$ and $|\arg(z)| < \delta\pi$, with $\delta = m+n-(p+q)/2$, so that Jordan's lemma is satisfied. For discussions on the G-function, see Mathai and Saxena (1973), and for discussions on the H-function, see Springer (1979), which also treats some uses of these functions in statistics, as well as some computational issues. We wish to mention the following points:

- The three paths of integration are similar to those of $_pF_q(\cdot)$, and the convergence of H and G now depends on $\beta = \sum_{j=1}^{p} a_j - \sum_{j=1}^{q} b_j$ and also on $\lambda = \prod_{j=1}^{p} \alpha_j^{\alpha_j} / \prod_{j=1}^{q} \beta_j^{\beta_j}$. Mathai *et al.* (2010) consider H and G as functions of $z \in \mathbb{C}$, and hence, the path L^* is included.
- There are numerous properties of the Meijer G-functions: contiguity, relations with themselves, derivatives, integral transforms, etc., which we cannot list here due to space limitations. They can

be seen in Mathai and Saxena (1973).

$$G_{p\ \ q}^{m\ \ n}\left[z\left|\begin{matrix}a_1,\ldots,a_p\\b_1,\ldots,b_q\end{matrix}\right.\right] = H_{p\ \ q}^{m\ \ n}\left[z\left|\begin{matrix}(a_1,1),\ldots,(a_p,1)\\(b_1,1),\ldots,(b_q,1)\end{matrix}\right.\right]$$

$$= \frac{1}{2\pi i}\int_L \frac{\prod\limits_{j=1}^{m}\Gamma(b_j+s)\prod\limits_{j=1}^{n}\Gamma(1-a_j-s)}{\prod\limits_{j=m+1}^{q}\Gamma(1-b_j-s)\prod\limits_{j=n+1}^{p}\Gamma(a_j+s)}z^{-s}ds.$$

We note that (13.7) is just one among several ways to generalize the integrand. But if no two b_j differ by an integer, all related poles are of first order and for $|z| < 1$, $G_{p\ q}^{m\ n}(.)$ can be expressed as a combination of $_pF_{q-1}(.)$. Similar conclusions can be drawn for a_k, $k = 1,\ldots,n$ and the domain $|z| > 1$.

13.3.3 *Computation*

The H-function can be brought to the G-function for computation, when all a_i, b_j are positive rational numbers, by a simple change of variable, and using the multiplication formula for gamma functions.

From (13.6), we can see that G- and H-functions are inverse Mellin transforms of $\varphi(.)$ and that the Meijer function $G(x)$ is a special case, when $\alpha_i = \beta_j = 1$, $\forall i, j$, of $H(x)$.

Under some mild conditions on a_i, b_j, the $G_{p\ q+1}^{1\ \ p}(.)$ function can be expressed as a $_pF_q(.)$ function and conversely:

$$G_{p\ \ q+1}^{1\ \ \ p}\left[-z\left|\begin{matrix}(1-a_1,\ldots,1-a_p)\\0,1-b_1,\ldots,1-b_q\end{matrix}\right.\right]$$

$$= \frac{\prod\limits_{j=1}^{p}\Gamma(a_j)}{\prod\limits_{j=1}^{q}\Gamma(b_j)}\cdot {_pF_q}\left[\begin{matrix}a_1,\ldots,a_p\\b_1,\ldots,b_q\end{matrix};z\right].$$

13.4 Some Properties of the *H*-Function

$$H_{p\ \ q}^{m\ \ n}\left[z\left|\begin{matrix}(a_1,\alpha_1),\ldots,(a_p,\alpha_p)\\(b_1,\beta_1),\ldots,(b_q,\beta_q)\end{matrix}\right.\right]$$

$$= H_{q\ \ p}^{n\ \ m}\left[\frac{1}{z}\left|\begin{matrix}(1-b_1,\beta_1),\ldots,(1-b_q,\beta_q)\\(1-a_1,\alpha_1),\ldots,(1-a_p,\alpha_p)\end{matrix}\right.\right]$$

for $z \neq 0$,

$$
H^{m \ n}_{p \ q} \left[z \left| \begin{array}{c} (a_1, \alpha_1), \ldots, (a_p, \alpha_p) \\ (b_1, \beta_1), \ldots, (b_q, \beta_q) \end{array} \right. \right]
$$

$$
= k H^{m \ n}_{p \ q} \left[z^k \left| \begin{array}{c} (a_1, k\alpha_1), \ldots, (a_p, k\alpha_p) \\ (b_1, k\beta_1), \ldots, (b_q, k\beta_q) \end{array} \right. \right]
$$

for $k > 0$, and

$$
z^\sigma H^{m \ n}_{p \ q} \left[z \left| \begin{array}{c} (a_1, \alpha_1), \ldots, (a_p, \alpha_p) \\ (b_1, \beta_1), \ldots, (b_q, \beta_q) \end{array} \right. \right]
$$

$$
= H^{m \ n}_{p \ q} \left[z \left| \begin{array}{c} (a_1 + \sigma\alpha_1, \alpha_1), \ldots, (a_p + \sigma\alpha_p, \alpha_p) \\ (b_1 + \sigma\alpha_1, \beta_1), \ldots, (b_q + \sigma\alpha_p, \beta_q) \end{array} \right. \right].
$$

Remarks:

(1) The characteristic function is also a H-function with a complex argument, and the cdf too, since

$$
F(y) = 1 - \frac{1}{2i\pi} \int_{c-i\infty}^{c+i\infty} y^{-s} s^{-1} M_{s+1}(h(y)) ds.
$$

(2) Maple and Mathematica now compute numerically the G-function.

(3) Mathai *et al.* (2010, p. 119) give the generalized gamma density, with expression

$$
f(x) = \frac{\beta a^{\frac{\alpha}{\beta}}}{\Gamma\left(\frac{\alpha}{\beta}\right)} x^{\alpha-1} \exp(-ax^\beta), \quad x > 0, \quad a, \alpha, \beta > 0,
$$

which reduces to the folded normal for $\alpha = 1, \beta = 2$, and other well-known densities for different values of α, β. These densities can then be expressed as G-functions. Furthermore, for the product $P = X_1, \ldots, X_k$, where each independent X_i has a H-function density, the density of P can also be obtained as a G- or H-function.

Calculus of residues is an important chapter in complex analysis and its application in computing values of integrals is of much interest.

13.5 Example of the Use of Calculus of Residues

We have

$$G^{2\ 1}_{1\ 2}\left(z\left|\begin{matrix}\frac{1}{2}\\3,-3\end{matrix}\right.\right) = \frac{1}{2\pi i}\int_L \Gamma(3+s)\Gamma(-3+s)$$

$$\times\Gamma\left(1-\frac{1}{2}-s\right)x^{-s}ds.$$

With Mathematica: The function is infinite at poles of integrand (Fig. 13.2).

(L), the integration curve, separates poles of $\Gamma(1-1/2-s)$ from those of $\Gamma(3+s)$, $\Gamma(-3+s)$ (Figs. 13.1 and 13.2).

Since poles can come in infinite numbers, one question we might ask is: How many poles can one choose? The answer depends on

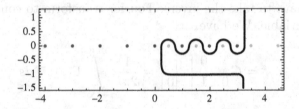

Fig. 13.1 The integration curve L on the complex plane.

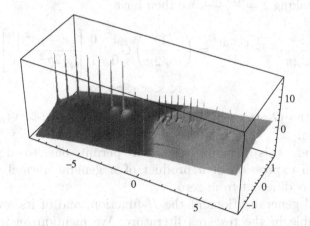

Fig. 13.2 The integration curve L in R^3.

our desired accuracy, since more poles would increase the numerical result accuracy.

13.6   The Normal Distribution and the G-Function

G- and H-functions, as defined above by Mathai *et al.* (2010), are functions of a complex variable, and results obtained are valid in a part, or in the whole, of the complex plane.

In probability and statistics however, we deal mostly with positive functions of real variables, which can take positive or negative values. Depending on the sign of the variable, appropriate adjustments have to be made, as seen in Chapter 4, unlike the case of the complex variable. Hence, the use of H- and G-functions in probability and statistics has to be adapted from the results obtained in the complex plane, when the real variable becomes a complex variable, which happens when we take the Fourier or Laplace transform, for example. We then have to take the inverse Fourier transform to come back to the real variable. We have

$$e^{-z} = G\begin{matrix} 1 & 0 \\ 0 & 1 \end{matrix}\left[z\,\middle|\,0\right] = \frac{1}{2\pi i}\int\limits_{\gamma-i\infty}^{\gamma+i\infty}\Gamma(s)z^{-s}ds$$

for any complex z.

In $\mathbb{R}$, taking $z = \frac{(x-\mu)^2}{2\sigma^2}$, we then have

$$\frac{1}{\sigma\sqrt{2\pi}}e^{-\frac{1}{2\sigma^2}(x-\mu)^2} = \left(\frac{1}{\sigma\sqrt{2\pi}}\right)G\begin{matrix} 1 & 0 \\ 0 & 1 \end{matrix}\left[\frac{(x-\mu)^2}{2\sigma^2}\,\middle|\,0\right],$$

$$-\infty < x < \infty, \tag{13.8}$$

which is the G-representation of the density of $X \sim N(\mu, \sigma^2)$ for μ, σ^2 given.

However, this formula does not permit one to derive the G-function expression of a product of n general normal variables with means different from zero.

Several generalizations of the H-function, and of its evaluation, are available in the research literature. We mention an important one by Cook and Barnes (1981), where linear combinations of products, quotients, and rational powers of independent rv's are also considered, and some improvements are made in the computation of residues. In Sections 13.7 and 13.8, we give the expressions of several distributions encountered earlier.

In what follows, we give a list of distributions we have mentioned in the book, without going into details. But here too, we can only give the expressions of their densities, and how they are derived, since space does not allow us to present more elaborate arguments. They are listed as follows:

(A) **Central distributions**

   (A.1) **Univariate distributions:** Gamma, inverse gamma, inverse chi-square, chi and chi-square, generalized Student t, Cauchy, generalized normal 1, and generalized normal 2.

   (A.2) **Multivariate distributions:** Multivariate gamma, multivariate beta, and multivariate F.

   (A.3) **Matrix variate distributions:** Wishart, inverse Wishart, and matrix variate t.

(B) **Non-central distributions**

   (B.1) **Univariate distributions:** Non-central chi-square, non-central gamma, non-central F, doubly non-central F, non-central t, and non-central beta.

   (B.2) **Matrix variate distributions:** Non-central Wishart (entries distribution, determinant distribution, trace distribution), non-central matrix gamma, and non-central matrix betas (entries distribution, determinant distribution).

13.7   Some Central Distributions

There are distributions that we have used, or mentioned, in the previous chapters without going into details. Following sections discuss some of these distributions.

13.7.1   *Univariate distributions*

13.7.1.1   *Gamma distribution*

The gamma distribution (see Chapter 12) is a flexible distribution for positive real-valued rv's, $X > 0$. It is defined in terms of two parameters. There are two common parameterizations:

$$\text{Gam}(x|\text{shape} = a, \text{ rate} = b) = \frac{b^a}{\Gamma(a)} x^{a-1} e^{-bx}, \quad x, a, b > 0.$$

The second parameterization is

$$\text{Gam}(x|\text{shape} = \alpha, \text{ scale} = \beta) = \frac{1}{\beta^\alpha \Gamma(\alpha)} x^{\alpha-1} e^{-x/\beta}.$$

13.7.1.2 *Inverse Gamma distribution*

Let $X \sim \text{Gam}(\text{shape} = a, \text{rate} = b)$ and $Y = 1/X$. Then it is easy to show that $Y \sim \text{IG}(\text{shape} = a, \text{scale} = b)$, where the density of the inverse gamma distribution is given by

$$f(y) = \frac{b^a}{\Gamma(a)} y^{-(a+1)} e^{-b/x}, \quad y, a, b > 0.$$

The distribution has the following properties:

- Mean: $\frac{b}{a-1}$, $a > 1$.
- Mode: $\frac{b}{a+1}$.
- Variance: $\frac{b^2}{(a-1)^2(a-2)}$, $a > 2$.

13.7.1.3 *Inverse chi-square distribution*

The inverse chi-square distribution, written as χ_ν^{-2}, corresponds to $\text{IG}(a = \nu/2, b = \text{scale} = 1/2)$. If $X \sim \chi_\nu^2$, then $1/X = Y \sim \chi_\nu^{-2}$, with density

$$f(y) = \frac{2^{-\nu/2}}{\Gamma(\nu/2)} y^{\nu/2-1} e^{-1/(2y)}, \quad 0 < y,$$

and cdf

$$\frac{\Gamma(\nu/2, 1/(2y))}{\Gamma(\nu/2)}.$$

13.7.1.4 *Chi and chi-square distributions*

The chi distribution with ν dof, $\chi_\nu = \sqrt{\chi_\nu^2}$, has density

$$f(x) = \frac{x^{\nu-1}\exp(-x^2/2)}{2^{(\nu-2)/2}\Gamma(\nu/2)}$$

and moments

$$\mu_r = 2^{r/2}\frac{\Gamma((\nu+r)/2)}{\Gamma(\nu/2)}.$$

For $r = 2$, it is called the Rayleigh distribution.

A characteristic of the χ_2^2 distribution is that its cdf, for $\nu = 2$, is the exponential function, i.e. in the special case of $\nu = 2$: $\mathbb{P}(X \le x) = 1 - e^{-x/2}$, while for the chi distribution with 2 dof, $Y \sim \chi_2 = \sqrt{\chi_2^2}$, we have $\mathbb{P}(Y \le y) = 1 - e^{-y^2/2}$.

13.7.1.5 *Generalized Student t-distribution*

Let

$$T = \frac{U}{\sqrt{\frac{V}{\nu}}},$$

where $U \sim N(0,1)$ is independent of $V \sim \chi_\nu^2$ with ν dof. Its density is (see Chapter 4)

$$f(t) = \frac{\Gamma\left(\frac{\nu+1}{2}\right)}{\sqrt{\nu\pi}\Gamma\left(\frac{\nu}{2}\right)\left(1 + \frac{t^2}{\nu}\right)^{\frac{\nu+1}{2}}}, \quad -\infty < t < \infty.$$

The generalized t-distribution has density

$$f(x|\mu,\sigma^2) = c\left[1 + \frac{1}{\nu}\left(\frac{x-\mu}{\sigma}\right)^2\right]^{-\frac{\nu+1}{2}}, \tag{13.9}$$

where $c = \frac{\Gamma(\nu/2+1/2)}{\Gamma(\nu/2)}\frac{1}{\sqrt{\nu\pi}\sigma}$, where c is the normalized constant, μ is the mean, ν is the number of dof, and σ is the scale. We have the following:

- Mean: μ for $\mu, \nu > 0$.
- Mode: μ.
- Variance: $\frac{\nu\sigma^2}{\nu-2}, \nu > 2$.

Note: If $X \sim t_\nu(\mu, \sigma^2)$, then $\frac{x-\mu}{\sigma} \sim t_\nu$, which corresponds to a standard t-distribution, with $\mu = 0$, $\sigma^2 = 1$:

As $\nu \to \infty$, t approaches a Gaussian distribution.

t-distributions are like Gaussian distributions, but with heavy tails. Hence, they are more robust to outliers.

It can be shown that the t-distribution can be expressed as an infinite sum of Gaussians, where each Gaussian has a different precision:

$$p(x|\mu, a, b) = \int \mathcal{N}(x|\mu, \tau^{-1}) Ga(\tau|a, \text{rate} = b) d\tau = t_{2a}(x|\mu, b/a).$$

13.7.1.6 *Cauchy distribution*

If $\nu = 1$ in Eq. (13.9), this is called a Cauchy distribution. This is an interesting distribution since if $X \sim$ CAU with density

$$f(x) = \frac{1}{\pi(1+x^2)}, \quad -\infty < x < \infty,$$

then $\mathbb{E}[X]$ does not exist, since the corresponding integral diverges. Essentially, this is because the tails are so heavy that samples from the distribution can get very far from the center μ. The sum of k independent Cauchy has density (see Chapter 4)

$$f(x) = \frac{k}{\pi(k^2 + x^2)}, \quad -\infty < x < \infty.$$

13.7.1.7 *Generalized normal distribution in $\mathbb{R}^1$*

There have been several efforts to generalize the general univariate normal distribution in $\mathbb{R}^1$. We mention the following ones.

In the statistical literature, there are now two different approaches to generalizing the univariate standard normal on the real line. Both approaches add a shape parameter to it.

Approach 1 (all new densities are symmetric): The family has density

$$f(x; \mu, \alpha, \beta) = \frac{\beta}{2\alpha \Gamma(1/\beta)} \exp\left\{ -\left(\frac{|x-\mu|}{\alpha}\right)^\beta \right\}, \quad -\infty < x < \infty,$$

$$(13.10)$$

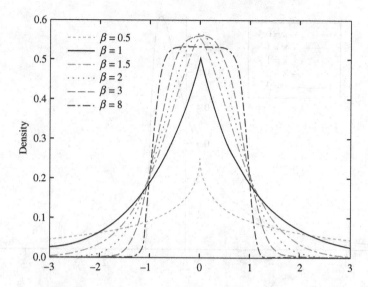

Fig. 13.3 Generalized normal distributions, family 1.

with Mean = Median = Mode = μ and Variance = $\frac{\alpha^2\Gamma(3/\beta)}{\Gamma(1/\beta)}$. Densities are always symmetrical wrt the vertical axis. The family contains the normal distribution when $\beta = 2$, with mean μ and variance $\alpha^2/2$, and the Laplace distribution when $\beta = 1$ (see Figs. 13.3 and 13.4).

Approach 2: Densities can have an asymmetry wrt the vertical axis. The right skewness or left skewness wrt the normal is due to the shape parameter $\lambda.\eta$ is the location parameter and α is the scale parameter. It has density

$$f(x; \alpha, \lambda, \eta) = \frac{\phi(h(x))}{\alpha - \lambda(x - \eta)}, \tag{13.11}$$

where ϕ is the standard normal pdf, and

$$h(x) = \begin{cases} -\dfrac{1}{\lambda}\log\left\{1 - \dfrac{\lambda(x - \eta)}{\alpha}\right\}, & \lambda \neq 0, \\ \dfrac{x - \eta}{\alpha}, & \lambda = 0. \end{cases}$$

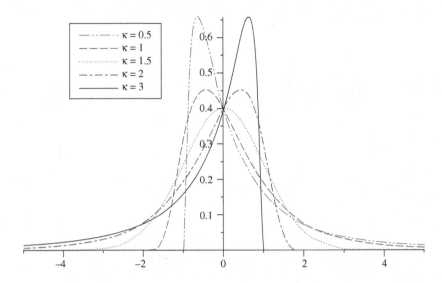

Fig. 13.4   Generalized normal distributions, family 2.

Its definition domain is

$$
\begin{cases}
\left(-\infty, \eta + \dfrac{\alpha}{\lambda}\right) & \text{if } \lambda > 0, \\[2mm]
(-\infty, \infty) & \text{if } \lambda = 0, \\[2mm]
\left(\eta + \dfrac{\alpha}{\lambda}, \infty\right) & \text{if } \lambda < 0.
\end{cases}
$$

Its mean is $\eta - \frac{\alpha}{\lambda}(e^{\lambda^2/2} - 1)$ and variance is $(\frac{\alpha}{\lambda})^2 e^{\lambda^2}(e^{\lambda^2} - 1)$. Both families of densities have found applications in statistics.

13.7.2   *Multivariate distributions*

These distributions concern a random vector. They are usually defined in such a way that each component has a distribution of the same type, but there are, in general, several ways that a univarite distribution can be generalized to a multivariate one.

13.7.2.1   *Multivariate gamma distributions*

There are several of them (see also Chapter 12), and some are given as follows:

(1) The correlated gamma variables, with density

$$f(x_1, \ldots, x_p) = \frac{\Gamma(\alpha)}{\Gamma(\alpha + \alpha_0)} \left(\sum_{i=1}^{p} x_i \right)^{\alpha_0} \prod_{i=1}^{p} \frac{x_i^{\alpha_i - 1}}{\Gamma(\alpha_i)} e^{-x_i}, \quad x_i > 0,$$

(13.12)

where $\alpha_0, \alpha_1, \ldots, \alpha_p > 0$.

(2) The multivariate chi-square, obtained from a sample of size n taken from the normal $N_p(\boldsymbol{\mu}, \boldsymbol{\Sigma})$, provides another multigamma distribution: Each statistic $S_j = \sum_{j=1}^{n} (X_{ij} - \overline{X}_{\bullet j})^2$ has a χ^2_{n-1} distribution and their joint distribution is a multivariate chi-square.

(3) The most important is the p-variate standard gamma, in the sense of Krishnamoorthy and Parthasarathy, which can be defined by its characteristic function

$$\psi(t_1, \ldots, t_p; \alpha, \mathbf{R}) = |\mathbf{I}_p - i\mathbf{R}\mathbf{T}|^{-\alpha},$$

with $\mathbf{I}_p$ being the identity matrix and $\mathbf{R}$ being the $(p \times p)$ correlation matrix and $\mathbf{T} = diag(t_1, \ldots, t_p)$.

13.7.2.2 The multivariate beta

(1) **The Dirichlet distribution:** We proceed similar to the above case. Let $\{X_j\}$, $j = 0, 1, 2, \ldots, m, m \geq 1$, be independent $\chi^2_{\nu_j}$ variables. We set

$$\left\{ Y_j = X_j \Big/ \sum_{i=0}^{m} X_i \right\}, \quad j = 1, \ldots, m, Y_0 = \sum_{i=0}^{m} X_i$$

and consider the joint density of $\{Y_j\}$, $j = 1, \ldots, m$, which is

$$\frac{\prod_{j=1}^{m} y_j^{\frac{\nu_j}{2} - 1} \left(1 - \sum_{j=1}^{m} y_j \right)^{\frac{\nu_0}{2} - 1}}{K}, \quad y_j \geq 0,$$

$$\sum_{j=1}^{m} y_j \leq 1, \quad \nu_0, \nu_1, \ldots, \nu_m > 0,$$

where $K = \frac{\prod_{j=0}^{m} \Gamma(\nu_j/2)}{\Gamma(\sum_{j=0}^{m} \nu_j/2)}$. This is the Dirichlet distribution, considered as a generalization of the univariate beta and where the parameter $\nu_j/2$ can be replaced by θ_j, to obtain the more general expression:

$$f(y_1,\ldots,y_m) = \frac{\Gamma\left(\sum_{j=0}^{m}\theta_j\right)}{\prod_{j=0}^{m}\Gamma(\theta_j)} \prod_{j=1}^{m} y_j^{\theta_j-1} \left(1 - \sum_{j=1}^{m} y_j\right)^{\theta_0-1},$$

$$(13.13)$$

where $y_j \geq 0, \sum_{j=1}^{m} y_j \leq 1$.

Note: Y_j has a standard beta of the first kind, $\beta_1^I(\theta_j, \sum_{k=0}^{m}\theta_k - \theta_j)$.

(2) **Multivariate F:** A generalization of the univariate Fisher–Snedecor distribution $F_{\nu_1,\nu_2} = \frac{X_1/\nu_1}{X_2/\nu_2}$ is provided by the multivariate F distribution, with density

$$h(f_1,\ldots,f_m) = \frac{\Gamma(\nu/2) \prod_{j=0}^{m} \nu_j^{\nu_j/2}}{\prod_{j=0}^{m}\Gamma(\nu_j/2)} \frac{\prod_{j=1}^{m} f_j^{\nu_j/2-1}}{(\nu_0 + \sum_{j=1}^{m}\nu_j f_j)^{\nu/2}}, \quad f_j > 0,$$

$$\nu = \sum_{j=0}^{m}\nu_j. \qquad (13.14)$$

(3) **Multivariate t:** A vector $\mathbf{X}$ has a multivariate t-distribution, with parameters n and Σ, denoted $\mathbf{X} \sim \mathbf{t}_p(n,\boldsymbol{\mu},\Sigma)$, if its density is

$$f(\mathbf{t}) = \frac{\Gamma\left(\frac{n+p}{2}\right)}{(n\pi)^{p/2}\Gamma\left(\frac{n}{2}\right)}|\Sigma|^{-1/2}\left(1 + \frac{\mathbf{t}^t\Sigma^{-1}\mathbf{t}}{n}\right)^{-\frac{n+p}{2}}, \quad \mathbf{t} \in \mathbb{R}^p.$$

$$(13.15)$$

There are, in fact, several other definitions for the multivariate t. Any subset of $\mathbf{X}$ has a multivariate t-distribution.

$\mathbf{Y} \sim \mathbf{t}_p(1,\mathbf{0},\mathbf{I}_p)$ is the multivariate Cauchy distribution.

13.7.3 *Matrix variate distributions*

We study only the *entries distribution* of the following matrix variate distributions.

13.7.3.1 *Wishart distribution*

Let $\mathbf{X}$ be a $(p \times p)$ symmetric positive definite matrix. The Wishart is the matrix generalization of the univariate gamma and of the chi-square. It was studied in Chapter 12.

13.7.3.2 *Inverted Wishart*

This is the matrix variate generalization of the univariate inverse gamma. Consider a $(p \times p)$ positive definite (covariance) matrix $\mathbf{X}$, a number of dof ν, and a $(p \times p)$ matrix $\mathbf{\Psi}$.

We define $\mathbf{X} \sim IW_p(\nu, \mathbf{\Psi})$ if its density is

$$f(\mathbf{X}) = \frac{|\mathbf{\Psi}|^{\frac{\nu-p-1}{2}}}{2^{\frac{\nu-p-1}{2}} \Gamma_p\left(\frac{\nu-p-1}{2}\right) |\mathbf{X}|^{\nu/2}} \mathrm{etr}\left(-\frac{\mathbf{X}^{-1}\mathbf{\Psi}}{2}\right), \quad \mathbf{X} > 0,$$

(13.16)

where $\nu > 2p$, $\mathbf{\Psi} > 0$.

We then have the correspondence between the inverted Wishart and the Wishart as follows:

$$\mathbf{X} \sim IW_p(\nu, \mathbf{\Psi}) \Rightarrow \mathbf{X}^{-1} \sim W_p(\nu - p - 1, \mathbf{\Psi}^{-1}),$$

where $W_p(.)$ is the Wishart matrix of dim $(p \times p)$.

13.7.3.3 *Matrix variate t*

Let the rectangular $(p \times m)$ matrix $\mathbf{X}$ have density

$$f(\mathbf{X}) = K|\mathbf{\Sigma}|^{-m/2}|\mathbf{\Omega}|^{-p/2}|\mathbf{I}_p$$
$$+ \mathbf{\Sigma}^{-1}(\mathbf{X} - \mathbf{M})\mathbf{\Omega}^{-1}(\mathbf{X} - \mathbf{M})^t|^{-(n+m+p-1)/2},$$

where

$$K = \frac{\Gamma_p((n + m + p - 1)/2)}{\pi^{mp/2}\Gamma_p((n + p - 1)/2)},$$

with $\mathbf{X}, \mathbf{\Omega} \in \mathbb{R}^{p \times m}$, $\mathbf{\Omega}(m \times m)$, $\mathbf{\Sigma}(p \times p)$, $n > 0$. Then $\mathbf{X}$ has a matrix variate t-distribution, with parameters $\mathbf{M}, \mathbf{\Sigma}, \mathbf{\Omega}, n$, denoted by $\mathbf{X} \sim \mathbf{T}(n, \mathbf{M}, \mathbf{\Sigma}, \mathbf{\Omega})$. We have $\mathbb{E}(\mathbf{X}) = \mathbf{M}$, $\mathrm{Cov}(\mathrm{vec}(\mathbf{X}^t)) = \mathbf{\Sigma} \otimes \mathbf{\Omega}/(n-2)$.

13.8　Some Non-central Distributions

Non-central distributions occur when we deal with non-null hypotheses in hypothesis testing, and its power, and sample size determination. These are often quite complex questions, especially when dealing with matrix variates, where hypergeometric functions of matrix arguments are still numerically difficult to compute.

Graphs of the non-central densities are often deformations of those of the central ones.

13.8.1　*Univariate distributions*

13.8.1.1　*Non-central chi-square variate*

If $Z_1, Z_2, \ldots, Z_\nu$ are independently normally distributed random variables, each having zero mean and unit standard deviation, and if $a_1, a_2, \ldots, a_\nu$ are constants with $\lambda = \sum_{i=1}^{\nu} a_i^2$, then $Y = \sum_{i=1}^{\nu} (Z_i + a_i)^2 \Rightarrow Y \sim \chi^2_{(nc),\nu,\lambda}$ is a *non-central chi-square variate*, with ν dof and non-centrality parameter λ (subscript (nc) serves to denote a non-central distribution). The probability density function of Y has been obtained in a number of different ways:

(a)
$$f_{(nc)}(y) = e^{-\lambda/2} \sum_{r=0}^{\infty} \frac{\left(\frac{1}{2}\lambda\right)^r e^{-y/2} (y)^{\nu/2+r-1}}{2^{\nu/2+r} r! \Gamma\left(\frac{1}{2}\nu + r\right)},$$

$0 < y < \infty$. We can see that this density is distributed as a mixture of central $\chi^2_{\nu+2j}$ distributions $j = 0, 1, \ldots$, with Poisson weights $e^{-\lambda/2}(\frac{1}{2}\lambda)^j / j!$.

(Hence, $X \sim \chi^2_{(nc),m,\delta}$ has as density

$$f(x) = \sum_{k=0}^{\infty} \mathbb{P}(K = k) h_{m+2k}(x),$$

where $K \sim Poi(\delta/2)$, so that

$$\mathbb{P}(K = k) = \exp(-\delta/2) \frac{(\delta/2)^k}{k!}, \quad k = 0, 1, \ldots,$$

and $h_r(x) = \exp(-x/2) x^{r/2-1} / 2^{r/2} \Gamma(r/2))$.

(b)

$$f_{\text{nc}}(w) = e^{-\lambda/2} \left(\frac{1}{2^{\nu/2}\Gamma(\nu/2)} e^{-w/2} w^{\frac{\nu}{2}-1} \right) \cdot {}_0F_1 \left(\frac{\nu}{2}; \frac{\lambda w}{4} \right)$$

$$= e^{-\lambda/2} \cdot {}_0F_1 \left(\frac{\nu}{2}; \frac{\lambda w}{4} \right) \cdot f_c(w),$$

where the hypergeometric function ${}_0F_1(.)$ is defined by an infinite series and $f_c(w)$ denotes the density of the central $W \sim \chi_\nu^2$.

A nice property of hypergeometric functions, especially ${}_1F_1$ and ${}_1F_2$, is that they could provide, by mere multiplication with the central density, the expression of the non-central density.

(c) If $\mathbf{X} \sim N_p(\boldsymbol{\mu}, \mathbf{I}_p)$, then the random variable $W = \mathbf{X}^t\mathbf{X}$ has a non-central chi-square distribution $\chi_{p,\lambda}^2$, with p dof and non-centrality parameter $\lambda = \boldsymbol{\mu}^t\boldsymbol{\mu}$.

Remark: MATLAB computes the cdf of the non-central chi-square using the subroutine **ncx2cdf(x,v,delta)**.

13.8.1.2 *Non-central gamma*

For $\{X_1, \ldots, X_n\}$, with $X_i \sim N(\mu_i, \sigma^2)$, we have $W = \sum_{j=1}^n X_i^2 \sim$ $\text{Gam}(n/2, 2\sigma^2, \lambda)$ non-central, with non-centrality parameter $\lambda = \sum_{j=1}^n \mu_j^2$. Table 12.1 of Chapter 12 summarizes different information on the gamma.

Let $X \sim \text{Gam}_{(\text{nc})} a, \lambda, \gamma(x)$, where γ is the non-centrality parameter. Its density is

$$f_{(\text{nc})}(a, \lambda, \gamma; x) = e^{-\gamma/\lambda} \sum_{m=0}^\infty \frac{(\gamma/\lambda)^m}{m!} \frac{e^{-x/\lambda} x^{a+m-1}}{\lambda^{a+m} \Gamma(a+m)}.$$

13.8.1.3 *Non-central F*

We make the following considerations:

(a) The density of the central F, $X \sim F_{\nu_1,\nu_2}$, defined here as the ratio of two independent central chi-square, $F_{\nu_1,\nu_2} = \frac{\chi_{\nu_1}^2}{\chi_{\nu_2}^2}$, is

$$h_c(\nu_1, \nu_2; x) = \frac{1}{B(\nu_1/2, \nu_2/2)} \frac{x^{\frac{\nu_1}{2}-1}}{(1 + \frac{\nu_1}{\nu_2}x)^{\frac{\nu_1+\nu_2}{2}}}. \tag{13.17}$$

(b) If $Z_1, Z_2, \ldots, Z_{v_1+v_2} \overset{\text{iid}}{\sim} N_1(0,1)$ and if $a_1, a_2, \ldots, a_{v_1}$ are v_1 constants with $\lambda = \sum_{i=1}^{v_1} a_i^2$, then the sum $W = \sum_{i=1}^{v_1} (Z_1 + a_i)^2$ has the non-central chi-square distribution with v_1 dof and non-centrality parameter λ, i.e. $W \sim \chi^2_{(\text{nc}),v_1,\lambda}$.

The non-central $F_{(\text{nc})}$ ratio is defined, for $0 < F_{(\text{nc})} < \infty$, as

$$F_{(\text{nc})} = \frac{v_2 \sum_{i=1}^{v_1} (Z_i + a_1)^2}{v_1 \sum_{i=v_1+1}^{v_1+v_2} Z_i^2} = \frac{\chi^2_{(\text{nc}),v_1,\lambda}/v_1}{\chi^2_2/v_2}.$$

This is the ratio of a non-central chi-square variate to an independent central chi-square variate.

(c) The probability density function $f_{\text{nc}}(w)$, or more precisely, $f_{(\text{nc}),\nu_1,\nu_2;\lambda}(w)$ of the non-central F-distribution, with dof v_1 and v_2, and non-centrality parameter λ, denoted by $F_{\nu_1,\nu_2;\lambda}$, is given by

$$f_{\nu_1,\nu_2;\lambda}(w) = e^{-\lambda/2} \sum_{r=0}^{\infty} \frac{\frac{\left(\frac{1}{2}\lambda\right)^r}{r!} \left(\frac{v_1}{v_2}\right)^{v_1/2+r_1}}{B\left(\frac{1}{2}v_1, \frac{1}{2}v_2\right)} \times \frac{(w)^{v_1/2+r-1}}{\left(1 + \frac{v_1}{v_2}w\right)^{(v_1+v_2)/2+r}}$$

$$= \frac{e^{-\lambda/2} \left(\frac{\nu_1}{\nu_2}\right)^{\nu_1/2} (w)^{v_1/2-1}}{B\left(\frac{1}{2}v_1, \frac{1}{2}v_2\right) \left(1 + \frac{v_1}{v_2}w\right)^{(v_1+v_2)/2}}$$

$$\times {}_1F_1\left(\frac{v_1 + v_2}{2}; v_1/2; \frac{\frac{v_1\lambda w}{2v_2}}{1 + \frac{v_1}{v_2}w}\right)$$

$$= e^{-\lambda/2} f_{c,v_1,v_2}(w) \times {}_1F_1\left(\frac{v_1 + v_2}{2}; v_1/2; \frac{\frac{v_1\lambda w}{2v_2}}{1 + \frac{v_1}{v_2}w}\right), \qquad (13.18)$$

where $f_{c,\nu_1,\nu_2}(w)$ is given by Eq. (13.17) and the confluent hypergeometric function ${}_1F_1$ is defined as

$$ {}_1F_1(y; \gamma, \delta) = \sum_{k=0}^{\infty} \frac{(\gamma, k)y^k}{(\delta, k)k!}.$$

Note: The non-central $F_{\nu_1,\nu_2;\lambda}$ generalizes into the *doubly non-central* $F_{\nu_1,\nu_2;\lambda_1,\lambda_2}$, and the latter is defined as

$$F_{\nu_1,\nu_2;\lambda_1,\lambda_2} = \left(\frac{\chi_{nc,1}^2}{\nu_1}\right)\Big/\left(\frac{\chi_{nc,2}^2}{\nu_2}\right),$$

where $\chi_{nc,2}^2$ is also a non-central chi-square, with ν_2 dof and λ_2 non-centrality parameter.

13.8.1.4 *The non-central t*

In the ratio $T = \frac{U}{\sqrt{V/\nu}}$, let $U \sim N_1(\delta, 1)$ and $V \sim \chi_\nu^2$. T now has the non-central univariate t-distribution, with 1 dof, and non-centrality parameter δ. It can be shown that the density of T is

$$f_{(nc)}(t) = \exp(-\delta^2/2)\psi(t) \sum_{j=0}^{\infty} \frac{\Gamma\left(\frac{m+j+1}{2}\right)}{\Gamma\left(\frac{m+1}{2}\right)} \frac{\left(\frac{t\delta\sqrt{2}}{\sqrt{m+t}}\right)^j}{j!}, \qquad (13.19)$$

where $\psi(t)$ is the density of the central t.

13.8.1.5 *Non-central univariate beta distributions*

(a) Let χ_{ν_1,λ_1}^2 be a non-central chi-square variable, with non-central parameter λ_1, independent of $\chi_{\nu_2}^2$, a central chi-square variable. Then $Z = \frac{\chi_{\nu_1,\lambda_1}^2}{\chi_{\nu_1,\lambda_1}^2 + \chi_{\nu_2}^2}$ has a non-central beta distribution of the first kind, with $\alpha = \nu_1/2$, $\beta = \nu_2/2$, denoted $Z \sim \beta_{nc}^I(\alpha, \beta; \lambda_1)$, with density

$$f_{(nc)}(z) = \exp(-\lambda_1) \sum_{i=0}^{\infty} \lambda_1^i \frac{\text{beta}(z; \alpha + i, \beta)}{i!}$$

$$= \exp(-\lambda_1) \frac{z^{\alpha-1}(1-z)^{\beta-1}}{B(\alpha, \beta)} \sum_{i=0}^{\infty} \frac{(\alpha + \beta, i)(\lambda_1 z)^i}{(\alpha, i)i!},$$

$$(13.20)$$

$0 \le z \le 1$.

Hence, we can write

$$f_{\mathrm{nc}}(z; \alpha, \beta; \lambda_1) = \exp(-\lambda_1) f_{cb}(z; \alpha, \beta) \cdot {}_1F_1(\lambda_1 z; \alpha + \beta, \alpha),$$

where the confluent hypergeometric function ${}_1F_1(.)$ has been defined previously, and $f_{cb}(z; \alpha, \beta)$ refers to the central beta of the first kind $\beta_1^I(\alpha, \beta)$.

$1 - Z = \dfrac{\chi^2_{\nu_2}}{(\chi^2_{\nu_1, \lambda_1} + \chi^2_{\nu_2})}$ has a complementary non-central beta distribution, with properties similar to those of $\beta_{\mathrm{nc}}^I(\alpha, \beta; \lambda_1)$.

Finally, the ratio $V = \dfrac{\chi^2_{\mathrm{nc}, \nu_1, \lambda_1}}{\chi^2_{\nu_2}}$ is a non-central beta of the second kind, denoted $\beta_{\mathrm{nc}}^{II}(\alpha, \beta; \lambda_1)$.

(b) If $\chi^2_{\nu_2}$ too is non-central in the above ratio, i.e. $\chi^2_{\nu_2} = \chi^2_{\nu_2, \lambda_2}$, then Z has a doubly non-central distribution, with density

$$f_{(\mathrm{dnc})}(z; \alpha_1, \alpha_2; \lambda_1, \lambda_2)$$

$$= \exp(-(\lambda_1 + \lambda_2)) \times \sum_{j_2=0}^{\infty} \sum_{j_1=0}^{\infty} \frac{\lambda_1^{j_1}}{j_1!} \frac{\lambda_2^{j_2}}{j_2!} \mathrm{Beta}(\alpha_1 + j_1, \alpha_2 + j_2).$$

We then have

$$f_{(\mathrm{dnc})}(z; \alpha_1, \alpha_2; \lambda_1, \lambda_2) = \exp(-(\lambda_1 + \lambda_2)) f_{cb}(z; \alpha_1, \alpha_2)$$

$$\times \Theta(\lambda_1 z, \lambda_2(1 - z); \alpha_1 + \alpha_2; \alpha_1, \alpha_2), \tag{13.21}$$

where the second Humbert function (or second confluent hypergeometric function in two variables) is defined by

$$\Theta(x, y; \gamma; \delta, \varepsilon) = \sum_{k=0}^{\infty} \sum_{j=0}^{\infty} \frac{(\gamma, j+k)}{(\delta, j)(\varepsilon, k)} \frac{x^j y^k}{j! k!}.$$

13.8.2   *Non-central matrix variate distributions*

We consider first the non-central Wishart.

13.8.2.1   *Non-central Wishart*

Expressions of the density of the non-central Wishart: For $\mathbf{X} \sim N_p(\boldsymbol{\mu}, \boldsymbol{\Sigma})$, if we take $\mathbf{A} = \sum_{i=1}^{N} \mathbf{X}_i \mathbf{X}_i^t$, $\Omega = \sum_{i=1}^{N} \boldsymbol{\mu}_i \boldsymbol{\mu}_i^t$, then

A has a non-central Wishart distribution with non-central matrix parameter $\boldsymbol{\Omega}$, i.e. $\mathbf{A} \sim W_{p,(\text{nc})}(n, \boldsymbol{\Sigma}, \boldsymbol{\Omega})$, and its two distributions are as follows:

(a) **Entries distribution:**

$$f_{(\text{nc})}(\mathbf{w}) = K|\boldsymbol{\Sigma}|^{-n/2}|\mathbf{w}|^{\frac{n-p-1}{2}} \exp\left[-\frac{\text{tr}\left\{\boldsymbol{\Sigma}^{-1}(\mathbf{w} + \boldsymbol{\Omega})\right\}}{2}\right]$$

$$\times {}_0F_1\left(\frac{n}{2}; \frac{1}{4}\boldsymbol{\Omega}\boldsymbol{\Sigma}^{-1}\mathbf{w}\right), \tag{13.22}$$

(see Mathai (1980)), where ${}_0F_1(.)$ is a hypergeometric function of a matrix argument, and

$$K = \frac{\pi^{\frac{p(p-1)}{4}}}{2^{\frac{np}{2}}} \prod_{i=1}^{p} \Gamma\left(\frac{n+1-i}{2}\right).$$

When $\boldsymbol{\Omega} = \mathbf{0}$, it reduces to the density of the central Wishart distribution.

(b) **Its determinant distribution** can be shown to be

$$f_{(\text{nc})}(x) = \frac{\pi^{p(p-1)/4}}{\Gamma_p(n/2)} \exp(-\text{tr}\boldsymbol{\Omega}) \sum_{k}\sum_{\kappa}\left\{\frac{C_\kappa(\boldsymbol{\Omega})}{k!\left(\frac{n}{2}\right)_\kappa}\right\}\frac{1}{x}$$

$$\times G_{0\ \ p}^{p\ \ 0}\left[x\Big|\frac{n}{2}+k_1, \frac{n-1}{2}+k_2,\ldots,\right.$$

$$\left.\frac{n-p+1}{2}+k_p\right], \tag{13.23}$$

where $x > 0$. Here, $C_\kappa(\mathbf{Z})$ is the zonal polynomial of degree k and the Meijer's function G is of a matrix argument. Details can be obtained from Mathai and Saxena (1973). We can see that setting $\boldsymbol{\Omega} = \mathbf{0}$, this expression reduces to the central case (see Chapter 12).

(c) **Trace of the non-central Wishart distribution:** Its trace T can be shown to be a non-central chi-square in some cases.

First, a simple case is the linear non-central case, where the non-centrality parameter is concentrated at one component. It can be treated as the central case. For a normal vector $\mathbf{X}$, this will happen when only the first component of $\boldsymbol{\mu}$ is different from 0 and for a normal $(p \times m)$ matrix $\mathbf{X}$, when only the first line of $\mathbf{M}$ is different from 0.

More precisely, let $\mathbf{X} \sim W_p(n, \mathbf{I}_p, \boldsymbol{\Theta})$, where $\boldsymbol{\Theta} = \mathrm{diag}(\theta, 0, \ldots, 0)$. Then there is a decomposition of $\mathbf{X} = \mathbf{TT}^t$, where $\mathbf{T}$ is lower triangular with independent elements such that only the first element is a non-central chi-square, i.e.

$$t_{11}^2 \sim \chi_n^2(\theta), t_{ii}^2 \sim \chi_{n-i+1}^2, \quad i = 2, \ldots, p$$

and

$$t_{ij} \sim N(0, 1), \quad 1 \le j < i \le p.$$

Hence, we have

$$T \sim \chi_{np}^2(\theta)$$

(see Pham-Gia *et al.* (2015)).

13.8.2.2   *Non-central matrix variate gamma*

For the non-central matrix gamma, its density is

$$f_{(\mathrm{nc})}(\mathbf{X}) = \frac{1}{|\mathbf{C}|^{-a}\Gamma_p(a)} \mathrm{etr}(-\boldsymbol{\Omega} - \mathbf{CW})|\mathbf{W}|^{a-(p+1)/2}$$
$$\times {}_0F_1(\mathbf{a}, \boldsymbol{\Omega}\mathbf{CW}), \quad \mathbf{W} > \mathbf{0}, \tag{13.24}$$

which means that, for a non-central Wishart matrix, we have the relation

$$\mathbf{X} \sim W_p(n, \boldsymbol{\Sigma}, \boldsymbol{\Omega}) \to \mathbf{X} \sim Ga_p(n/2, \boldsymbol{\Sigma}^{-1}/2, \boldsymbol{\Omega}/2)$$

(see Table 12.1).

13.8.2.3   *Non-central matrix variate betas*

There are four types of non-central beta distributions, classified as follows (we classify them here slightly differently from Gupta and Nagar (2000)).

Type I

- **Type I(A): $\mathbf{U} \sim \beta_p^{I(A)}(a, b; \boldsymbol{\Theta})$** has density

$$f_{(\mathrm{nc})}(\mathbf{U}) = \frac{|\mathbf{U}|^{a-\frac{p+1}{2}}|\mathbf{I_p}-\mathbf{U}|^{b-\frac{p+1}{2}}}{\mathrm{Beta}_p(a, b)} \exp(-\boldsymbol{\Theta})$$
$$\times {}_1F_1(a + b; b; \boldsymbol{\Theta}(\mathbf{I}_p - \mathbf{U})), \quad 0 < \mathbf{U} < \mathbf{I}_p,$$

where $a, b > (p - 1)/2$,

$$\mathrm{Beta}_p(a, b) = \frac{\Gamma_p(a)\Gamma_p(b)}{\Gamma_p(a + b)},$$

and

$$_1F_1(a + b; b; \boldsymbol{\Theta}(\mathbf{I}_p - \mathbf{U}))$$

is the confluent hypergeometric function.

- **Type I(B): $\mathbf{W} \sim \beta_p^{I(B)}(a, b; \boldsymbol{\Theta})$** has density

$$f_{(\mathrm{nc})}(\mathbf{W}) = \frac{\mathbf{W}^{a-\frac{p+1}{2}}|\mathbf{I_p} - \mathbf{W}|^{b-\frac{p+1}{2}}}{\mathrm{Beta}_p(a, b)} \exp(-\boldsymbol{\Theta})$$
$$\times {}_1F_1(a + b; a; \boldsymbol{\Theta}\mathbf{W}), \quad 0 < \mathbf{W} < \mathbf{I}_p.$$

where $a, b > (p - 1)/2$.

Type II

- **Type II(A): $\mathbf{T} \sim \beta_p^{II(A)}(a, b; \boldsymbol{\Theta})$** has density

$$f_{(\mathrm{nc})}(\mathbf{T}) = \frac{\mathrm{etr}(-\boldsymbol{\Theta})|\mathbf{T}|^{a-\frac{p+1}{2}}}{\mathrm{Beta}_p(a, b)|\mathbf{I}_p + \mathbf{T}|^{a+b}}$$
$$\times {}_1F_1(a + b; b; \boldsymbol{\Theta}(\mathbf{I}_p + \mathbf{T})^{-1}), \quad \mathbf{V} > 0,$$

where $a, b > (p - 1)/2$.

- **Type II(B): $\mathbf{V} \sim \beta_p^{II(B)}(a, b; \boldsymbol{\Theta})$** has density

$$
f_{(nc)}(\mathbf{V}) = \frac{|\mathbf{V}|^{a - \frac{p+1}{2}} \operatorname{etr}(-\boldsymbol{\Theta})}{\operatorname{Beta}_p(a, b)|\mathbf{I}_p + \mathbf{V}|^{a+b}}
$$
$$
\times \, {}_1F_1(a + b; a; \boldsymbol{\Theta}\mathbf{V}(\mathbf{I}_p + \mathbf{V})^{-1}), \quad \mathbf{V} > 0,
$$

where $a, b > (p - 1)/2$.

Properties of these non-central betas can be obtained from Gupta and Nagar (2000, Section 5.5, Chapter 5).

Note: Some recent results obtained by AMRSAG, concerning the non-central matrix Wilks's distributions, which are distributions of the determinants of the above matrices, have been published in the *Journal of Statistical Computation and Simulation* (see Duong *et al.* (2019a)) and *Statistics & Probability Letters* (see Duong *et al.* (2019b)).

13.9   Complex Case for the Normal Distribution

We have considered only real random scalar variables, real random vector, and real random matrices. But these variables can also be complex, and most commonly, can be expressed as a sum $X + iY$, with the two variables X and Y real, and $i = \sqrt{-1}$.

13.9.1   *One-dimensional case*

We first consider the univariate complex normal variable $\tilde{z} = x + iy$. It can be shown that its density is

$$
f(\tilde{z}) = \frac{\exp\left(-\frac{1}{\sigma^2}\left\{(\tilde{z} - \mathbb{E}(\tilde{z}))^2\right\}\right)}{\pi\sigma^2}
$$
$$
= \frac{\exp\left(-\frac{1}{\sigma^2}\left\{(x - \mu_x)^2 + (y - \mu_y)^2\right\}\right)}{\pi\sigma^2}, \quad -\infty < x, y < \infty.
$$

On the other hand, it can also be considered as a bivariate real normal, with density

$$f(x, y) = \frac{\exp\left(-\frac{1}{2}[(x - \mu_x)(y - \mu_y)] \begin{bmatrix} \sigma^2/2 & 0 \\ 0 & \sigma^2/2 \end{bmatrix} \begin{bmatrix} x - \mu_x \\ y - \mu_y \end{bmatrix}\right)}{\pi \sigma^2}$$

$$= \frac{\exp\left(-\frac{1}{\sigma^2}\{(x - \mu_x)^2 + (y - \mu_y)^2\}\right)}{\pi \sigma^2}.$$

13.9.2 *Vector case*

A complex random vector is normal, $\tilde{\mathbf{Z}} \sim N_p(\tilde{\boldsymbol{\mu}}, \tilde{\boldsymbol{\Sigma}})$, if it has density

$$f(\tilde{\mathbf{z}}) = \frac{1}{(\pi)^p |\tilde{\boldsymbol{\Sigma}}|} \exp\left[-(\tilde{z} - \tilde{\boldsymbol{\mu}})^* \tilde{\boldsymbol{\Sigma}}^{-1}(\tilde{z} - \tilde{\boldsymbol{\mu}})\right],$$

where $\tilde{\boldsymbol{\Sigma}} = \boldsymbol{\Sigma}_1 + i\boldsymbol{\Sigma}_2$, where $\boldsymbol{\Sigma}_1, \boldsymbol{\Sigma}_2$ are real matrices, with $\boldsymbol{\Sigma}_1$ symmetric positive definite, while $\boldsymbol{\Sigma}_2$ is skew-symmetric, and $*$ denotes the complex conjugate.

It can be shown that if $\tilde{\mathbf{Z}} = \mathbf{X} + i\mathbf{Y}$ and the real $(p \times 1)$ vectors $\mathbf{X}$ and $\mathbf{Y}$ have a joint normal density, then this density is the same as $f(\tilde{\mathbf{z}})$ above.

13.9.3 *Matrix case*

The complex random matrix $\tilde{\mathbf{X}}$ is said to have a complex matrix variate non-singular normal density if

$$f(\tilde{\mathbf{X}}) = \frac{1}{\pi^{pq} |\mathbf{V}|^p |\mathbf{W}|^q} \exp\left\{-\mathrm{tr}\left[\mathbf{V}^{-1}(\tilde{\mathbf{X}} - \mathbf{M})^* \mathbf{W}^{-1}(\tilde{\mathbf{X}} - \mathbf{M})\right]\right\},$$

where $\mathbf{V}$ and $\mathbf{W}$ are $(q \times q)$ and $(p \times p)$ Hermitian positive definite matrices, $\mathbf{M}$ is a $(p \times q)$ matrix of constants, and $\tilde{\mathbf{X}}$ is a $(p \times q)$ matrix of pq independent rv's (see Johnson and Kotz (1972)).

13.10 Elliptical Contour Distribution

A generalization of the multivariate normal, called *elliptically contoured distribution*, is obtained by replacing $\exp[-\frac{1}{2}(\mathbf{x} - \boldsymbol{\mu})^t \boldsymbol{\Sigma}^{-1}(\mathbf{x} - \boldsymbol{\mu})]$ by $g((\mathbf{x} - \boldsymbol{\mu})^t \boldsymbol{\Sigma}^{-1}(\mathbf{x} - \boldsymbol{\mu})), \mathbf{x} \in \mathbb{R}^p$, in the expression of the density of the normal vector, where $g(y)$, like $\exp(-y/2)$, is a non-increasing function of y, here of the square Mahalanobis distance. The multivariate *t*-distribution can be obtained this way.

Elliptical contour distributions have many properties shared with the multivariate normal, which is one of its particular cases. They can be obtained from Anderson and Kai-Tai Fang (1990), and we refer the reader to this reference.

The last step in generalization is to consider the class of *matrix variate elliptically* contour distributions defined as follows.

Let $\mathbf{X}(p \times n)$ belong to $RS_{p,n}(\phi)$ (RS = *right spherical* with characteristic function ϕ) and $\mathbf{M}(p \times m), \mathbf{B}(n \times m)$ be constant. Then the random matrix $\mathbf{W}(p \times m)$, defined by $\mathbf{W} = \mathbf{M} + \mathbf{X}\mathbf{B}$, is said to have a *matrix variate elliptically contoured distribution*, where $\mathbf{\Sigma} = \mathbf{B}^t\mathbf{B}$, if the density of $\mathbf{W}(p \times m)$ is

$$g(\mathbf{W}) = \frac{1}{|\mathbf{\Sigma}|^{p/2}} f((\mathbf{W} - \mathbf{M})^t \mathbf{\Sigma}^{-1}(\mathbf{W} - \mathbf{M})),$$

where f is similar to g of the preceding section.

This book will not be complete if we do not mention some new developments in the probability theory and the normal distribution.

13.11   Random Matrix Theory

It can be stated that *random matrix theory* (RMT) started with the work of Wishart in 1928. However, in the 1950s, Wigner gave it a completely different direction in his study of Hamiltonians $\mathbf{H}$ living in an infinite-dimensional Hilbert space governed by physical laws. Shrodinger equation is $\mathbf{H}v = \lambda v$, where $\{\lambda, v\}$ is the eigenvalue–eigenvector pair associated with $\mathbf{H}$, which is, in most cases, unknown, with the above equation unsolvable. Wigner succeeded in computing the spectral characteristics of random symmetric real matrices with iid entries. The underlying idea is that the characteristic energies of chaotic systems could behave locally like eigenvalues of a very large matrix, with randomly distributed elements.

He also proved the limit semi-circular law and provided insights into the separations between neighboring eigenvalues of such matrices.

In the 1980s and 1990s, results in nuclear physics and quantum chaos have helped the advancement of RMT. Previously, in mathematics, there were results by Marcenko and Pastur (1967) on the spectrum of a large covariance matrix and by Arnold (1967) on eigenvalue distributions in ensembles of RMT with independent entries. Serious studies revealed a similarity between eigenvalues of GUE/CUE matrices with those of zeros of the Riemann zeta function. Mehta (2004) can be consulted for basic results on RMT.

In the 1990s with the development of free probability by Voiculescu, the situation changed rapidly. This new field studies non-commutative random variables and its relation with RMT is a key reason for the wide use of free probability in other subjects. Several directions for research were initiated and the statistics of the largest eigenvalue in Gaussian ensembles attracted considerable attention. The development of RMT is now made on several fronts, particularly in *theoretical physics, mathematical analysis* and *probability/statistics*.

Wigner and Dyson were interested in approximating $\mathbf{H}$ by an ensemble of finite large $(p \times p)$ Hermitian matrices $\mathbf{H}_p$, with a probability density of the form $p(\mathbf{H}_p) = A \exp(-\beta \mathrm{tr}\,[V(\mathbf{H}_p)])$, with V being a function of $\mathbf{H}_p$. A possible choice is

$$V(\mathbf{H}_p) = a\mathbf{H}_p{}^2 + b\mathbf{H}_p + c, \quad a > 0.$$

Example: Let $V(\mathbf{H}_p) = \delta \mathbf{H}_p{}^2$. Let the diagonal entries $\{H_{ii}\}$, $i = 1, \ldots, \infty$ be iid $N_1(0, 2)$ while the off-diagonal $\{H_{ii}\}$, $i = 1 \leq i < j \leq \infty$ be iid $N_1(0, 1)$ and assume they are all independent. If n is a positive integer, then the random symmetric matrix

$$\begin{pmatrix} H_{11} & H_{12} & \cdots & H_{1p} \\ H_{12} & H_{22} & \cdots & H_{2p} \\ \cdots & \cdots & \ddots & \cdots \\ H_{1p} & H_{2p} & \cdots & H_{pp} \end{pmatrix}$$

is called a *Gaussian orthonormal ensemble* (GOE(p)). Using the equality $\mathrm{tr}(\mathbf{H}_P{}^2) = \sum_i \sum_j H_{ij}^2$, the joint distribution of $\{H_{ij}\}$ can be shown to be

$$\frac{1}{\sqrt{2}(2\pi)^{\frac{n(n+1)}{4}}} \exp\left(-\frac{1}{4}\mathrm{tr}(\mathbf{H}_p{}^2)\right).$$

We define a "time-reversal" transformation as $\mathbf{H}_p \rightarrow \mathbf{U}\mathbf{H}_p\mathbf{U}^{-1}$, where $\mathbf{U}$ is orthogonal. This brings us to the three Gaussian ensembles introduced by Dyson.

13.11.1 *Gaussian ensembles*

These are models for random symmetric matrices whose entries are independent normal rv's.

13.11.1.1 *Gaussian orthogonal ensemble*

Consider $\mathbf{A}$ a $(p \times p)$ matrix with entries iid $N(0,1)$. Let $\mathbf{H}_p = (\mathbf{A}+\mathbf{A}^t)/2$. Hence, the diagonal entries of $\mathbf{H}_p$ are iid $N(0,1)$, while the off-diagonal are iid $N(0,1/2)$. $\mathbf{H}_p$ is also called the Wigner matrix. This model suits Hamiltonians with time-reversal symmetry.

13.11.1.2 *Gaussian unitary ensemble*

$\mathbf{H}_p = (\mathbf{A}+\mathbf{A}^*)/2$, where $\mathbf{A}^*$ is the Hermitian transpose of the complex matrix $\mathbf{A}$. Here, the diagonal entries of $\mathbf{H}_p$ are iid $N(0,1)$, while the off-diagonal are iid $N(0,1/2)$. We can write $\mathbf{H}_p = (H_{lm})$, where $H_{lm} = U_{lm} + iV_{lm}$, with $U_{lm}, V_{lm} \sim N(0,1/2)$, $1 \leq l < m \leq n$ and $H_{mm} \sim N(0,1)$, $1 \leq m \leq n$. This model suits Hamiltonians lacking time-reversal symmetry. Its density is $K \exp\left(-\frac{1}{2}\mathrm{tr}(\mathbf{H}_p{}^2)\right)$.

13.11.1.3 *Gaussian simplectic ensemble*

$\mathbf{H}_p = (\mathbf{A}+\mathbf{A}^D)/2$, where $\mathbf{A}^D$ is the dual transpose of the quaternion matrix $\mathbf{A}$. Here, the diagonal entries of $\mathbf{H}_p$ are iid $N(0,1)$, while the off-diagonal are iid $N_1(0,1/2)$. Hence, we can write $\mathbf{H}_p = (H_{lm})$, where

$$H_{lm} = U_{lm} + iV_{lm} + jW_{lm} + kZ_{lm},$$

with

$$U_{lm}, V_{lm}, W_{lm}, Z_{lm} \sim N(0,1/2), \quad 1 \leq l < m \leq n$$

and $H_{mm} \sim N(0,1), 1 \leq m \leq n$. This model suits Hamiltonians with time-reversal symmetry but no rotational symmetry. Its density is $K \exp(-\mathrm{tr}(\mathbf{H}_p{}^2))$.

Note: There are also circular ensembles, introduced by Dyson, similar to the above ensembles, called circular orthogonal ensemble (COE), circular unitary ensemble (CUE), and circular sympletic ensemble (CSE), with eigenvalues confined to the unit circle.

13.11.2 Bulk and Edges

Spectrum of random matrices: As stated before, eigenvalues of random matrices have a particular place in theoretical energy physics. We distinguish between the bulk, which consists of most of the eigenvalues, and the edges (or extremes), which contain the largest and the smallest.

13.11.2.1 Bulk

The Wigner matrix $\mathbf{H}_p = (\mathbf{A} + \mathbf{A}^t)/2$ is a member of the Gaussian orthogonal ensemble. For finite p, the exact distribution of the eigenvalues $\lambda_1 > \cdots > \lambda_p$ is

$$p(\lambda_1, \ldots, \lambda_p) = c_p \prod \sqrt{e^{-\lambda_j^2}} \prod_{1 \leq j < k \leq p} |\lambda_j - \lambda_k|.$$

For general β-Gaussian ensembles, where $\beta = 1$ (for GOE), $\beta = 2$ (for GUE), $\beta = 4$ (for GSE), the joint density of the eigenvalues of $\mathbf{H}_p$ is

$$p_\beta(\lambda_1, \ldots, \lambda_p) = K_p \prod_{j=1}^{p} \sqrt{w_{n,p,\beta}(\lambda_j)} \prod_{1 \leq j < k \leq p} |\lambda_j - \lambda_k|^\beta, \qquad (13.25)$$

with

$$w_{n,p,\chi,\beta}(\lambda) = \lambda^{\beta(n-p+1)-2} e^{-\beta\lambda}$$

and

$$K_p = (2\pi)^{-\beta n p/2} \prod_{i=1}^{n} \frac{\Gamma\left(1 + \frac{\beta}{2}\right)}{\Gamma\left(1 + \frac{\beta j}{2}\right)\Gamma\left(\frac{\beta}{2}(p - n + i)\right)}, \qquad \lambda \geq 0.$$

13.11.2.2   *Wigner semi-circle law*

For limit distribution of the eigenvalues of a Wigner matrix $\mathbf{H}_p$ as $p \to \infty$, we have

$$\frac{1}{p}\# \{i : \lambda_i \leq \lambda\} \overset{a.s.}{\to} G(\lambda), \quad p \to \infty,$$

with density $g(x) = \frac{\sqrt{4-\lambda^2}}{2\pi}, |\lambda| \leq 2$. This is a semi-circle (see Fig. 13.5).

13.11.2.3   *Marcenko–Pastur law (also called quarter circle law)*

Let $p \to \infty$, $n \to \infty$, such that $p/n \to \gamma \in [0, \infty)$. Then the empirical distribution of the eigenvalues follows the Marcenko–Pastur law, i.e.

$$\frac{1}{p}\# \left\{i : \widehat{\lambda}_i \leq x\right\} \overset{a.s.}{\to} G(x),$$

whose density is

$$g(x) = \frac{1}{2\pi\gamma x} \sqrt{(b_+ - x)(x - b_-)} I_{[b_-, b_+]}(x),$$

with $b_\pm = (1 \pm \sqrt{\gamma})^2$.

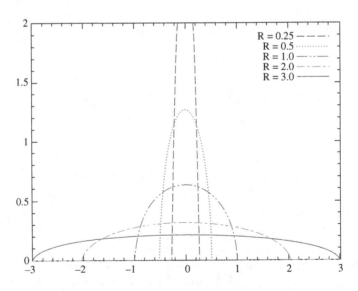

Fig. 13.5   Wigner semi-circle pdf.

Bandeira *et al.* (2017), for example, show that under certain conditions, Kendall's rank correlation matrix converges to the Marcenko–Pastur law.

13.11.2.4 *Edges of spectrum*

Results obtained by Tracy and Widom (1994) appeared to be highly universal and have several applications.

For the GUE, Tracy–Widom showed that the largest eigenvalue has the limiting distribution:

$$F_2(t) = \lim_{p \to \infty} \mathbb{P} \left\{ \frac{\widehat{\lambda}_1 - 2\sqrt{p}}{p^{1/6}} \leq t \right\},$$

where $F_2(.)$ is the Tracy–Widom law of order 2,

$$F_2(t) = \exp\left(- \int_t^\infty (x - t)(q(x))^2 dx \right),$$

where $q(x)$ is the solution of Painlevé II differential equation:

$$q''(x) = xq(x) + 2[q(x)]^3,$$

with $q(x) \sim \text{Ai}(x), x \to \infty$. The Airy function is the solution of

$$(\text{Ai}(x))'' = x.\text{Ai}(x),$$

with boundary condition:

$$\text{Ai}(x) \sim \frac{\exp\left(-\frac{2x^{3/2}}{3}\right)}{2\sqrt{\pi}x^{1/4}}, \quad x \to \infty$$

(see Fujikoshi *et al.* (2010)).

13.11.3 *Convergence*

Tracy–Widom law of order 1 (see also Section 3.11 of Chapter 3): Let Ω be a $(r \times p)$ matrix with iid entries $N(0,1)$. Let $\mathbf{S} = \Omega\Omega^t$ which has $W_p(p, \mathbf{I}_r)$, i.e. the Wishart distribution, with non-negative eigenvalues. Let $\widehat{\lambda}_1$ be the largest. Let the centering and scaling factors be

$$\mu_{pr} = (\sqrt{p-1} + \sqrt{r})^2, \quad \sigma_{pr} = (\sqrt{p-1} + \sqrt{r})\left(\frac{1}{\sqrt{p-1}} + \frac{1}{\sqrt{r}}\right)^{1/3}.$$

Then, for $r/p \to \gamma \in (0, \infty)$,

$$\left(\frac{\lambda_1 - \mu_{pr}}{\sigma_{pr}}\right) \xrightarrow{d} X_1 \sim F_1,$$

where F_1 is the Tracy–Widom law of order 1, with distribution

$$F_1(t) = \sqrt{[F_2(t)]} \exp\left(-\int_t^\infty q(x)dx/2\right).$$

There are the real and complex cases to consider for these laws, which have found applications in random permutations, in sorting cards in the game of Solitaire, in determining spacings of consecutive zeros of the Riemann zeta function, etc. The reader can consult Izenman (2008) for some of these applications. These limit laws in RMT have been compared to the Gaussian limit results in the classical statistical theory.

13.12   Conclusion

This chapter presents some information on distributions that we had not had the chance to discuss in detail in the previous chapters. They have some importance in various sub-domains of probability, statistics, and theoretical physics.

Other information presented here concern topics related to the normal or belonging to a neighboring domain. They are included for the sake of completeness.

Bibliography

Anderson, T.W. and Kai-Tai Fang (1990). Theory and applications of elliptically contoured and related distributions, Technical Report No. 24 September 1990, U.S. Army Research Office Contract DAAL03-89-K-0033, Department of Statistics, Stanford University, Stanford, California.

Arnold, L. (1967). On asymptotic distributions of eigenvalues of random matrices, *Journal of Mathematical Analysis and Applications*, **20**(2), 262–265.

Bandeira, A.S., Lodhia, A. and Rigollet, P. (2017). Marcenko–Pastur law for Kendall's Tau, *Electronic Communications in Probability*, **22**(32), 1–7.

Cook, I.D. and Barnes, J.W. (1981). Evaluation of the *H*-Function integral and of probability distributions of functions of independent random variables, *American Journal of Mathematical and Management Sciences*, **1**(4), 293–339.

Duong, T.P., Pham-Gia, T. and Dinh, N.T. (2019a). Exact distributions of two non-central generalized Wilks's statistics, *Journal of Statistical Computation and Simulation*, **89**(10), 1798–1818.

Duong, T.P., Pham-Gia, T. and Dinh, N.T. (2019b). Exact distribution of the non-central Wilks's statistic of the second kind, *Statistics & Probability Letters*, **153C**, 80–89.

Fujikoshi, Y., Ulyanov, V.V. and Shimizu, R. (2010). *Multivariate Statistics, High-Dimensional and Large-Sample Approximations* (Wiley, New York).

Gupta, A.K. and Nagar, D.K. (2000). *Matrix Variate Distributions* (Chapman and Hall/CRC, Boca Raton).

Izenman, A.J. (2008). Introduction to Random Matrix Theory.

Johnson, N. and Kotz, S. (1972). *Continuous Multivariate Distributions* (Wiley, New York).

Marcenko, V.A. and Pastur, L.A. (1967). Distribution of eigenvalues in certain sets of random matrices, *Matematicheskii Sbornik*, **114**(4), 507–536.

Mathai, A.M. (1980). Moments of the trace of a non-central Wishart matrix, *Communication in Statistics — Theory and Methods*, **A9**(8), 795–801.

Mathai, A.M. and Saxena, R.K. (1973). *Generalized Hypergeometric Functions with Applications in Statistics and Physical Sciences* (Springer-Verlag, New York).

Mathai, A.M., Saxena R.K. and Haubold, H.J. (2010). *The H-Function: Theory and Applications* (Springer, New York).

Mehta, M.L. (2004). *Random Matrices* (Elsevier, New York).

Pham-Gia, T. (2008). Exact expression of Wilks's statistic and applications, *Journal of Multivariate Analysis*, **1999**, 1698–1716.

Pham-Gia, T. and Choulakian, V. (2014). Distribution of the sample correlation matrix and applications, *Open Journal of Statistics*, **4**(5), 330–344.

Pham-Gia, T. and Dinh, N.T. (2016). Hypergeometric functions, from one scalar variable to several matrix arguments, in statistics and beyond, *Open Journal of Statistics*, **6**(5), 951–994.

Pham-Gia, T., Dinh, N.T. and Duong, T.P. (2015). Trace of the Wishart matrix and applications, *Open Journal of Statistics*, **5**, 173–190.

Spiegel, M. (1974). *Laplace Transforms, Schaum Outline Series* (McGraw-Hill, New York).

Springer, M. (1979). *The Algebra of Random Variables* (Wiley, New York).

Tracy, C.A. and Widom, H. (1994). Level-spacing distributions and the Airy kernel, *Communications in Mathematical Physics*, **159**(1), 151–174.

Postface

Hypergeometric Functions: From One Scalar Variable to Several Matrix Arguments, in Statistics and Beyond, by *T.Pham-Gia and D. Ngoc Thanh, Open Journal of Statistics, 2016, 6, 951–994.*

Hypergeometric functions, **G** and **H**, have been increasingly present in several disciplines, including Statistics, but there is much confusion on their proper uses, as well as on their existence and domain, and definition. We try to clarify several points and give a general overview of the topic, going from the univariate case to the matrix case, in one and then in several arguments.

In Multivariate analysis, as reported by Bose (1977), Gauss hypergeometric function was used by Fisher as early as in 1928, in the determination of the density of the sample multiple correlation coefficient R^2.

Leon Ehrenpreis (1991) wrote: *"Hypergeometric functions pervade many branches of mathematics because they are at the confluence of three fundamental viewpoints"*.

The versatility of hypergeometric functions is due to the fact that they can be expressed as an infinite series, or as very different forms of integrals. The three basic formats, Euler, Laplace, and Mellin–Barnes, can then be studied and extended, using mathematical analysis tools.

In statistics, understandably, hypergeometric functions are not often developed, but used, mostly in distribution theory. Let us look at the following topics in the above paper:

1. Integral representation
2. Hypergeometric series and functions in one scalar variable

3. Hypergeometric series and functions in several independent scalar variables

   3.1 Functions in one matrix variate

   3.2 Hypergeometric function in two matrix variates

   We have Gauss hypergeometric series in three and in several parameters.

4. Among Hypergeometric Functions in Matrix Arguments, we have three proposed approaches

   4.1 Laplace Transform Approach

   4.2 Zonal Polynomials Approach

   4.3 Mathai's (1991) Approach

5. Computation of the hypergeometric functions, and old and new relations managed by computer

   Here, Milgram (2006) used computer algorithms to test all closed forms $_3F_2(1)$ identities given in a well-known reference book, and found 89 final identities, with only 23 on the original set.

Finally, we wish to conclude our book here, by citing a special article Yoshida (1997), which clearly shows that hypergeometric functions can create an image deeply affecting a researcher's feelings.

Bibliography

Bose, R. C. (1977). Early history of multivariate statistical analysis. In Krishnaiah, P. R. (ed.), Analysis IV, pp. 3–22 (North-Holland, Amsterdam).

Ehrenpreis, L. (1991). Hypergeometric functions. *ICM-90 Satellite Conference Proceedings-Special Functions*, Masaki Kashiwara, Tetsuji Miwa (ed.), Springer, pp. 78–89.

Mathai, A. M. (1991). Special functions of matrix arguments and statistical distributions. *Indian Journal of Pure and Applied Mathematics*, **22**, 887–903.

Milgram, M. S. (2006). On Hypergeometric $_3F_2(1)$, ArXiv: math/0603096.

Yoshida, M. (1997). Hypergeometric functions, my Love. *Aspects of Mathematics Series,* vol 32 (Springer, New York). http://dx.doi.org/10.1007/978-3-332-90166-8.

Index

439